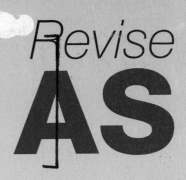

Chemistry

Rob Ritchie

Contents

Specification list   4

AS/A2 Level Chemistry courses   11

Different types of questions in AS examinations   13

Exam technique   15

What grade do you want?   17

Four steps to successful revision   18

Chapter 1 Atomic structure

1.1 Atoms and isotopes   19

1.2 The electronic structure of the atom   21

1.3 Experimental evidence for electronic structures   25

Sample question and model answer   28

Practice examination questions   29

Chapter 2 Atoms, moles and equations

2.1 Atomic mass   30

2.2 The mole   33

2.3 Formulae, equations and reacting quantities   37

Sample question and model answer   41

Practice examination questions   42

Chapter 3 Bonding and structure

3.1 Chemical bonding   44

3.2 Shapes of molecules   50

3.3 Electronegativity, polarity and polarisation   52

3.4 Intermolecular forces   55

3.5 Bonding structure and properties   58

Sample question and model answer   62

Practice examination questions   63

Chapter 4 The Periodic Table

4.1 The modern Periodic Table   64

4.2 Redox reactions   68

4.3 The s-block elements: Group 1 and Group 2   71

4.4	The halogens and their compounds: Group 7	76
4.5	Extraction of metals	81
	Sample question and model answer	85
	Practice examination questions	86

Chapter 5   Chemical energetics

5.1	Enthalpy changes	87
5.2	Determination of enthalpy changes	90
5.3	Bond enthalpy	94
	Sample question and model answer	96
	Practice examination questions	97

Chapter 6   Rates and equilibrium

6.1	Reaction rates	98
6.2	Catalysis	102
6.3	Chemical equilibrium	104
6.4	Acids and bases	109
	Sample question and model answer	111
	Practice examination questions	112

Chapter 7   Organic chemistry

7.1	Basic concepts	113
7.2	Hydrocarbons from oil	118
7.3	Alkanes	120
7.4	Alkenes	122
7.5	Alcohols	127
7.6	Halogenoalkanes	132
	Sample question and model answer	135
	Practice examination questions	136

Practice examination answers		138
Periodic Table		143
Index		144

Specification lists

AQA Chemistry

MODULE	SPECIFICATION TOPIC	CHAPTER REFERENCE	STUDIED IN CLASS	REVISED	PRACTICE QUESTIONS
Module 1 (M1) Atomic Structure, Bonding and Periodicity	Atomic structure	1.1, 1.2, 1.3			
	Amount of substance	2.1, 2.2, 2.3			
	Bonding	3.1, 3.2, 3.3, 3.4, 3.5			
	Periodicity	4.1, 4.3			
Module 2 (M2) Foundation Physical and Inorganic Chemistry	Energetics	5.1, 5.2, 5.3			
	Kinetics	6.1, 6.2			
	Equilibria	6.3			
	Redox reactions	4.2			
	Group VII: the halogens	4.4			
	Extraction of metals	4.5			
Module 3 (M3) Introduction to Organic Chemistry	Nomenclature and isomerism	7.1			
	Petroleum and alkanes	7.2, 7.3			
	Alkenes and epoxyethane	7.4			
	Haloalkanes	7.6			
	Alcohols	7.5			

Examination analysis

The specification comprises three unit tests. All questions are compulsory.

Unit 1	Structured questions: short and extended answers	1 hr test	30%
Unit 2	Structured questions: short and extended answers	1 hr test	30%
Unit 3	comprises two components:		
	3(a) Structured questions: short and extended answers	1 hr test	25%
	3(b) Centre-assessed coursework **or** practical examination	2 hr test	15%

OCR Chemistry

MODULE	SPECIFICATION TOPIC	CHAPTER REFERENCE	STUDIED IN CLASS	REVISED	PRACTICE QUESTIONS
Module 2811 (M1) Foundation Chemistry	Atoms, molecules and stoichiometry	2.1, 2.2, 2.3			
	Atomic structure	1.1, 1.2, 1.3			
	Chemical bonding and structure	3.1, 3.2, 3.3, 3.4, 3.5			
	The Periodic Table: introduction	4.1			
	The Periodic Table: Group 2	4.2, 4.3			
	The Periodic Table: Group 7	4.4			
Module 2812 (M2) Chains and Rings	Basic concepts	7.1			
	Hydrocarbons: Alkanes	7.3			
	Hydrocarbons: Fuels	7.2			
	Hydrocarbons: Alkenes	7.4			
	Alcohols	7.5			
	Halogenoalkanes	7.6			
Module 2813/1 (M3) How Far, How Fast?	Enthalpy changes	5.1, 5.2, 5.3			
	Reaction rates	6.1, 6.2			
	Chemical equilibrium	6.3, 6.4			

Examination analysis

The specification comprises three unit tests. All questions are compulsory.

Unit 2811 Structured questions: short and extended answers	1 hr test	30%
Unit 2812 Structured questions: short and extended answers	1 hr test	30%
Unit 2813 comprises two components:		
• 2813/1: Structured questions: short and extended answers	45 min test	20%
• 2813/2: Centre-assessed coursework **or**		
• 2813/3: Practical examination	1 hr 30 min test	20%

OCR Chemistry, Salters

MODULE	SPECIFICATION TOPIC	CHAPTER REFERENCE	STUDIED IN CLASS	REVISED	PRACTICE QUESTIONS
Module 2850 (M1) Chemistry for life	The elements of life	1.1, 1.2, 1.3, 2.1, 2.2, 2.3, 3.1, 3.2, 4.1, 4.3			
	Developing fuels	2.2, 2.3, 5.1, 5.2, 5.3, 6.2, 7.1, 7.2, 7.3			
Module 2848 (M2) Chemistry of Natural Resources	Minerals to elements	1.2, 2.3, 3.5, 4.2, 4.4, 6.4			
	The atmosphere	6.1, 6.2, 6.3, 7.1, 7.3, 7.6			
	The polymer revolution	3.3, 3.4, 7.1, 7.4, 7.5			

Note that the Salters approach 'drip-feeds' concepts throughout the course. For example, you will study different aspects of Chapter 3: Bonding and structure in different topics. You may need to revisit the chapters and practice questions in this study guide throughout your AS course.

Examination analysis

The specification comprises three unit tests. All questions are compulsory.

Unit 2850 Structured questions: short and extended answers		1 hr 15 min test	30%
Unit 2848 Structured questions: short and extended answers		1 hr 30 min test	40%

Unit 2853 comprises two components:

- Open-book paper   Externally assessed coursework   15%
- Experimental skills   Internally assessed coursework   15%

The open-book paper is taken over a two week period during the third term of the AS course at a date specified by OCR. Marks for this paper take into account quality of written communication.

Edexcel Chemistry

MODULE	SPECIFICATION TOPIC	CHAPTER REFERENCE	STUDIED IN CLASS	REVISED	PRACTICE QUESTIONS
Module 1 (M1) *Structure, bonding and main group chemistry*	Atomic structure	1.1, 1.2, 1.3			
	Formulae, equations and moles	2.1, 2.2, 2.3			
	Structure and Bonding	3.1, 3.2, 3.3, 3.4, 3.5			
	The Periodic Table Introduction	4.1			
	Introduction to reduction and oxidation	4.2			
	Group 1 and Group 2	4.3			
	Group 7	4.4			
Module 2 (M2) *Introductory organic chemistry, energetics, kinetics and equilibrium and applications*	Energetics I	5.1, 5.2, 5.3			
	Organic chemistry (Introduction)	7.1, 7.3, 7.4, 7.5, 7.6			
	Kinetics I	6.1, 6.2			
	Chemical Equilibria I	6.3			
	Industrial inorganic chemistry	4.4, 4.5, 6.3			

Examination analysis

The specification comprises three unit tests. All questions are compulsory.

Unit 1	Structured questions: short answers	*1 hr test*	*30%*
Unit 2	Structured questions: short and extended answers	*1 hr test*	*30%*
Unit 3	Laboratory skills comprises two components:		
3A	Centre-assessed coursework **or** practical examintion	*1 hr 45 min test*	*20%*
3B	Structured questions	*1hr test*	*20%*

Unit 3B assesses a students' ability to interpret information generated and drawn from experimental situations in the laboratory.

Edexcel Chemistry (Nuffield)

MODULE	SPECIFICATION TOPIC	CHAPTER REFERENCE	STUDIED IN CLASS	REVISED	PRACTICE QUESTIONS
Module 1 (M1) Introductory chemistry	Metal compounds	1.1, 2.1, 2.2, 2.3, 5.1			
	Organic chemistry: alcohols	7.1, 7.5			
	The Periodic Table and atomic structure	1.2, 1.3, 3.1, 3.5, 4.1			
	Acid-base reactions and Group 2	2.2, 2.3, 4.1, 4.3, 6.4			
	Energy changes and reactions	5.1, 5.2			
Module 2 (M2) Bonding and reactions	Redox reactions and the halogens	4.2, 4.4			
	Covalency and bond breaking	3.1, 3.2, 3.3, 5.1, 5.2, 5.3, 6.1, 6.2, 6.3			
	Organic chemistry: hydrocarbons	7.1, 7.2, 7.3, 7.4			
	Intermolecular forces	3.4, 3.5			
	Organic chemistry: halogenoalkanes	7.6			

Examination analysis

The specification comprises three unit tests. All questions are compulsory.

Unit 1	Structured questions: short and extended answers	1 hr 15 min test	30%
Unit 2	Structured questions and comprehension exercise	1 hr 30 min test	40%

    Section 2A will contain structured questions, based on topics 6 to 10 with a prerequisite of understanding of topics 1 to 5.

    Section 2B will consist of a summary/comprehension exercise based on a passage. This exercise will include marks for quality of written communication.

Unit 3	Centre-assessed coursework		30%

WJEC Chemistry

MODULE	SPECIFICATION TOPIC	CHAPTER REFERENCE	STUDIED IN CLASS	REVISED	PRACTICE QUESTIONS
Module CH1 (CH1) Physical – Inorganic	Atomic structure	1.1, 1.2, 1.3			
	The mole and stoichiometry	2.1, 2.2, 2.3			
	Structure and bonding	3.1, 3.2, 3.3, 3.4			
	Gases, liquids and solids	3.5			
	The Periodic Table	4.1, 4.2, 4.3, 4.4			
Module CH2 (CH2) Physical – Organic	Principles of energetics	5.1, 5.2, 5.3			
	Organic compounds	7.1, 7.5, 7.6			
	Hydrocarbons and petroleum	7.2, 7.3, 7.4			
	Chemical and acid-base equilibria	2.3, 6.3, 6.4			
	Chemical kinetics	6.1, 6.2			
	Industrial and environmental aspects	6.2, 6.3			

Examination analysis

The specification comprises three unit tests. All questions are compulsory.

Unit CH1	Structured questions and objective questions		1 hr 30 min test	35%
Unit CH2	Structured questions and objective questions		1 hr 30 min test	35%
Unit CH3	Experimental skills comprising two components:			
	CH3a	Structured questions: short answers	1 hour test	10%
	CH3b	Centre-assessed coursework		20%
	Unit CH3a assesses the link between theory and experiment, using the content of teaching modules CH1 and CH2 and other related experiments.			

NICCEA Chemistry

MODULE	SPECIFICATION TOPIC	CHAPTER REFERENCE	STUDIED IN CLASS	REVISED	PRACTICE QUESTIONS
Module M1 (M1) General Chemistry	Atomic structure	1.1, 1.2, 1.3, 2.1			
	Bonding	3.1, 3.3, 3.5			
	Shapes of molecules	3.2			
	Intermolecular forces	3.4, 3.5			
	Energetics	5.1, 5.2, 5.3			
	Redox	4.2			
	Periodic Table	4.1			
	Group 7	4.4			
	Titrations	2.3, 4.4			
Module M2 (M2) Organic, Physical and Inorganic Chemistry	Isomerism	7.1			
	Hydrocarbons	7.2			
	Alkanes	7.3			
	Alkenes	7.4			
	Halogenoalkanes	7.6			
	Alcohols	7.5			
	Equilibrium	6.3			
	Kinetics	6.1, 6.2			
	Group 2	4.3			

Note that the use of balanced equations and calculations involving the mole concept
(Topic refs: 2.1, 2.2 and 2.3) are general aspects that apply to all the molecules, particularly Module M1.

Examination analysis

The specification comprises three unit tests. All questions are compulsory.

Unit 1	Multiple-choice questions and structured questions	1 hr 30 min test	35%
Unit 2	Multiple-choice questions and structured questions	1 hr 30 min test	35%
Unit 3	Practical examination	2 hr 30 min test	30%

AS/A2 Level Chemistry courses

AS and A2

All Chemistry A Level courses being studied from September 2000 are in two parts, with three separate modules in each part. Students first study the AS (Advanced Subsidiary) course. Some will then go on to study the second part of the A Level course, called A2. Advanced Subsidiary is assessed at the standard expected halfway through an A Level course: i.e., between GCSE and Advanced GCE. This means that new AS and A2 courses are designed so that difficulty steadily increases:

- AS Chemistry builds from GCSE Science
- A2 Chemistry builds from AS Chemistry.

How will you be tested?

Assessment units

For AS Chemistry, you will be tested by three assessment units. For the full A Level in Chemistry, you will take a further three units. AS Chemistry forms 50% of the assessment weighting for the full A Level.

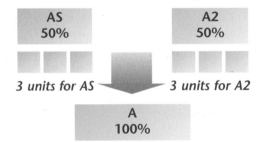

Each unit can normally be taken in either January or June. Alternatively, you can study the whole course before taking any of the unit tests. There is a lot of flexibility about when exams can be taken and the diagram below shows just some of the ways that the assessment units may be taken for AS and A Level Chemistry.

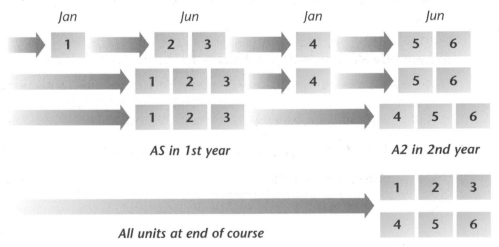

If you are disappointed with a module result, you can resit each module once. You will need to be very careful about when you take up a resit opportunity because you will have only one chance to improve your mark. The higher mark counts.

A2 and Synoptic assessment

After having studied AS Chemistry, you may wish to continue studying Chemistry to A Level. For this you will need to take three further units of Chemistry at A2. Similar assessment arrangements apply except some units, those that draw together different parts of the course in a synoptic assessment, have to be assessed at the end of the course.

Coursework

Coursework may form part of your A Level Chemistry course, depending on which specification you study. Where students have to undertake coursework, it is usually for the assessment of practical skills but this is not always the case. See pages 4–10.

Key Skills

To gain the key skills qualification, which is equivalent to an AS Level, you will need to collect evidence together in a 'portfolio' to show that you have attained a satisfactory level in Communication, Application of number and Information technology. You will also need to take a formal testing in each key skill. You will have many opportunities during AS Chemistry to develop your key skills.

What skills will I need?

The assessment objectives for AS Chemistry are shown below.

Knowledge with understanding

- recall of facts, terminology and relationships
- understanding of principles and concepts
- drawing on existing knowledge to show understanding of the responsible use of chemistry in society
- selecting, organising and presenting information clearly and logically

Application of knowledge and understanding, analysis and evaluation

- explaining and interpreting principles and concepts
- interpreting and translating, from one form into another, data presented as continuous prose or in tables, diagrams and graphs
- carrying out relevant calculations
- applying knowledge and understanding to familiar and unfamiliar situations
- assessing the validity of chemical information, experiments, inferences and statements

Experimental and investigative skills

Chemistry is a practical subject and part of the assessment of AS Chemistry will test your practical skills. You will be assessed on four main skills:

- planning
- implementing
- analysing evidence and drawing conclusions
- evaluating evidence and procedures.

The skills may be assessed in the context of separate practical exercises, although more than one skill could be assessed in any one exercise. They may also be assessed all together in the context of a single 'whole investigation'. An investigation may be set by your teacher or you may be able to pursue an investigation of your choice. Alternatively, you may take a practical examination

You will receive guidance about how your practical skills will be assessed from your teacher. This study guide concentrates on preparing you for the written examinations testing the subject content of AS Chemistry.

Different types of questions in AS examinations

In AS Chemistry examinations, different types of question are used to assess your abilities and skills. Unit tests mainly use structured questions requiring both short answers and more extended answers.

Short-answer questions

A short-answer question may test recall or it may test understanding. Short-answer questions normally have space for the answers printed on the question paper.

Here are some examples (the answers are shown in *italics*):

What is meant by isotopes?

Atoms of the same element with different masses.

Calculate the amount (in mol) of H_2O in 4.5 g of H_2O.

1 mol H_2O has a mass of 18 g. ∴ 4.5 g of H_2O contains 4.5/18 = 0.25 mol H_2O.

Structured questions

Structured questions are in several parts. The parts usually have a common context and they often become progressively more difficult and more demanding as you work your way through the question. A structured question may start with simple recall, then test understanding of a familiar or an unfamiliar situation.

Most of the practice questions in this book are structured questions, as this is the main type of question used in the assessment of AS Chemistry.

When answering structured questions, do not feel that you have to complete one question before starting the next. The further you are into a question, the more difficult the marks are to obtain. If you run out of ideas, go on to the next question. You need to respond to as many parts of questions on an exam paper as possible. You will not score well if you spend so long trying to perfect the first questions that you do not have time to attempt later questions.

Here is an example of a structured question that becomes progressively more demanding.

(a) Write down the atomic structure of the two isotopes of potassium: ^{39}K and ^{41}K.

    (i) ^{39}K ...19... protons; ...20... neutrons; ...19... electrons. ✓

    (ii) ^{41}K ...19... protons; ...22... neutrons; ...19... electrons. ✓ [2]

(b) A sample of potassium has the following percentage composition by mass: ^{39}K: 92%; ^{41}K: 8% Calculate the relative atomic mass of the potassium sample.

    92 × 39/100 + 8 × 41/100 = 39.16 ✓ [1]

(c) What is the electronic configuration of a potassium atom?

    $1s^2 2s^2 2p^6 3s^2 3p^6 4s^1$ ✓ [1]

(d) The second ionisation energy of potassium is much larger than its first ionisation energy.

(i) Explain what is meant by the *first ionisation energy* of potassium.

The energy required to remove an electron ✔ from each atom in 1 mole ✔ of gaseous atoms ✔

(ii) Why is there a large difference between the values for the first and the second ionisation energies of potassium?

The 2nd electron removed is from a different shell ✔ which is closer to the nucleus and experiences more attraction from the nucleus. ✔ This outermost electron experiences less shielding from the nucleus because there are fewer inner electron shells than for the 1st ionisation energy. ✔ [6]

Extended answers

In AS Chemistry, questions requiring more extended answers may form part of structured questions or may form separate questions. They may appear anywhere on the paper and will typically have between 5 and 10 marks allocated to the answers as well as several lines of answer space. These questions are also often used to assess your abilities to communicate ideas and put together a logical argument.

The correct answers to extended questions are often less well-defined than to those requiring short answers. Examiners may have a list of points for which credit is awarded up to the maximum for the question.

An example of a question requiring an extended answer is shown below.

Magnesium oxide and sulphur trioxide are both solids at 10 °C. Magnesium oxide has a melting point of 2852 °C. Sulphur trioxide has a melting point of 17 °C. Explain, in terms of structure and bonding, why these two compounds have such different melting points. [7]

Points that the examiners might look for include:

Magnesium oxide has a giant lattice structure ✔ containing ionic bonds ✔ These strong forces need to be broken during melting to allow ions to move free of the rigid lattice. ✔ This requires a large amount of energy supplied by a high temperature. ✔

Sulphur trioxide has a simple molecular structure ✔ held together by van der Waals' forces. ✔ These weak forces need to be broken during melting to allow molecules to move free of the rigid lattice. ✔ This requires a small amount of energy supplied by a low temperature. ✔

8 marking points → [7]

In this type of response, there may be an additional mark for a clear, well-organised answer, using specialist terms. In addition, marks may be allocated for legible text with accurate spelling, punctuation and grammar.

Other types of questions

Free-response and open-ended questions allow you to choose the context and to develop your own ideas. These are little used for assessing AS Chemistry but, if you decide to take the full A Level in Chemistry, you will encounter this type of question during synoptic assessment.

Multiple-choice or objective questions require you select the correct response to the question from a number of given alternatives. These are rarely used for assessing AS Chemistry, although it is possible that some short-answer questions will use a multiple-choice format.

Exam technique

Links from GCSE

Advanced Subsidiary Chemistry builds from grade CC in GCSE Science: Double Award, or equivalent in Science: Chemistry. This study guide has been written so that you will be able to tackle AS Chemistry from a GCSE Science background.

You should not need to search for important Chemistry from GCSE Science because this has been included where needed in each chapter. If you have not studied Science for some time, you should still be able to learn AS Chemistry using this text alone.

What are examiners looking for?

Examiners use instructions to help you to decide the length and depth of your answer.

If a question does not seem to make sense, you may have misread it – read it again!

State, define or list

This requires a short, concise answer, often recall of material that can be learnt by rote.

Explain, describe or discuss

Some reasoning or some reference to theory is required, depending on the context.

Outline

This implies a short response, almost a list of sentences or bullet points.

Predict or deduce

You are not expected to answer by recall but by making a connection between pieces of information.

Suggest

You are expected to apply your general knowledge to a 'novel' situation, one which you have not directly studied during the AS Chemistry course.

Calculate

This is used when a numerical answer is required. You should always use units in quantities and significant figures should be used with care.

Look to see how many significant figures have been used for quantities and give your answer to this degree of accuracy.

If the question uses three significant figures, then give your answer in three significant figures also.

Some dos and don'ts

Dos

Do answer the question

No credit can be given for good Chemistry that is irrelevant to the question.

Do use the mark allocation to guide how much you write

Two marks are awarded for two valid points – writing more will rarely gain more credit and could mean wasted time or even contradicting earlier valid points.

Do use diagrams, equations and tables in your responses

Even in 'essay-type' questions, these offer an excellent way of communicating chemistry.

Do write legibly

An examiner cannot give marks if the answer cannot be read.

Do write using correct spelling and grammar. Structure longer essays carefully

Marks are now awarded for the quality of your language in exams.

Don'ts

Don't fill up any blank space on a paper

In structured questions, the number of dotted lines should guide the length of your answer.

If you write too much, you waste time and may not finish the exam paper. You also risk contradicting yourself.

Don't write out the question again

This wastes time. The marks are for the answer!

Don't contradict yourself

The examiner cannot be expected to choose which answer is intended. You could lose a hard-earned mark, e.g.

A covalent bond is a shared pair of electrons ✓ bonded by the electrostatic attraction between ions. ✗

Don't spend too much time on a part that you find difficult

You may not have enough time to complete the exam. You can always return to a difficult part if you have time at the end of the exam.

What grade do you want?

Everyone would like to improve their grades but you will only manage this with a lot of hard work and determination. You should have a fair idea of your natural ability and likely grade in Chemistry and the hints below offer advice on improving that grade.

For a Grade A

You will need to be a very good all-rounder.

- You must go into every exam knowing the work extremely well.
- You must be able to apply your knowledge to new, unfamiliar situations.
- You need to have practised many, many exam questions so that you are ready for the type of question that will appear.

The exams test all areas of the syllabus and any weaknesses in your Chemistry will be found out. There must be no holes in your knowledge and understanding. For a Grade A, you must be competent in all areas.

For a Grade C

You must have a reasonable grasp of Chemistry but you may have weaknesses in several areas and you may be unsure of some of the reasons for the Chemistry.

- Many Grade C candidates are just as good at answering questions as the Grade A students but holes and weaknesses often show up in just some topics.
- To improve, you will need to master your weaknesses and you must prepare thoroughly for the exam. You must become a better all-rounder.

For a Grade E

You cannot afford to miss the easy marks. Even if you find Chemistry difficult to understand and would be happy with a Grade E, there are plenty of questions in which you can gain marks.

- You must memorise all definitions.
- you must practise exam questions to give yourself confidence that you do know some Chemistry. In exams, answer the parts of questions that you know first. You must not waste time on the difficult parts. You can always go back to these later.
- the areas of Chemistry that you find most difficult are going to be hard to score on in exams. Even in the difficult questions, there are still marks to be gained. Show your working in calculations because credit is given for a sound method. You can always gain some marks if you get part of the way towards the solution.

What marks do you need?

The table below shows how your average mark is transferred into a grade.

average	80%	70%	60%	50%	40%
grade	A	B	C	D	E

Four steps to successful revision

Step 1: Understand

- Study the topic to be learned slowly. Make sure you understand the logic or important concepts.
- Mark up the text if necessary – underline, highlight and make notes.
- Re-read each paragraph slowly.

GO TO STEP 2

Step 2: Summarise

- Now make your own revision note summary:
  What is the main idea, theme or concept to be learned?
  What are the main points? How does the logic develop?
  Ask questions: Why? How? What next?
- Use bullet points, mind maps, patterned notes.
- Link ideas with mnemonics, mind maps, crazy stories.
- Note the title and date of the revision notes
  (e.g. Chemistry: Atomic structure, 3rd March).
- Organise your notes carefully and keep them in a file.

This is now in **short-term memory**. You will forget 80% of it if you do not go to Step 3.
GO TO STEP 3, but first take a 10 minute break.

Step 3: Memorise

- Take 25 minute learning 'bites' with 5 minute breaks.
- After each 5 minute break test yourself:
  Cover the original revision note summary
  Write down the main points
  Speak out loud (record on tape)
  Tell someone else
  Repeat many times.

The material is well on its way to **long-term memory**.
You will forget 40% if you do not do step 4. **GO TO STEP 4**

Step 4: Track/Review

- Create a Revision Diary (one A4 page per day).
- Make a revision plan for the topic, e.g. 1 day later, 1 week later, 1 month later.
- Record your revision in your Revision Diary, e.g.
  Chemistry: Atomic Structure, 3rd March 25 minutes
  Chemistry: Atomic Structure, 5th March 15 minutes
  Chemistry: Atomic Structure, 3rd April 15 minutes
  ... and then at monthly intervals.

Chapter 1
Atomic structure

The following topics are covered in this chapter:

- Atoms and isotopes
- The electronic structure of the atom
- Experimental evidence for electronic structures

1.1 Atoms and isotopes

After studying this section you should be able to:

- recall the relative charge and mass of a proton, neutron and electron
- explain the existence of isotopes
- use atomic number and mass number to determine atomic structure

LEARNING SUMMARY

What is an atom?

AQA	M1	OCR	M1
EDEXCEL	M1	SALTERS	M1
NICCEA	M1	WJEC	CH1
NUFFIELD	M1		

An atom is the smallest part of an element that can exist on its own. Atoms are so tiny that there are more atoms in a full stop than there are people in the world. It is now possible to see individual atoms by using the most modern and powerful microscopes. However, nobody has yet seen *inside* an atom and we must devise models for the structure of an atom from experimental evidence.

Sub-atomic particles

electron shells

nucleus (protons and neutrons)

We use a model in which an atom is composed of three **sub-atomic particles**: protons, neutrons and electrons.

In this model, protons and neutrons form the nucleus at the centre of the atom with electrons orbiting in shells. Compared with the total volume of an atom, the nucleus is tiny and extremely dense. Most of an atom is empty, made up of the space between the nucleus and the electron shells.

All matter is made up from the chemical elements and 115 are now known (although others will inevitably be discovered). The number of protons in the nucleus distinguishes the atoms of each element.

An important principle of chemistry is the link between an element and the number of protons in its atoms.

> Each atom of an element has the same number of protons.
> The number of protons in an atom of an element is called the **atomic number, Z**.
>
> **KEY POINT**

All atoms of hydrogen contain 1 proton.

All atoms of carbon contain 6 protons.

All atoms of oxygen contain 8 protons.

Properties of protons, neutrons and electrons

Some properties of protons, neutrons and electrons are shown in the table below.

particle	relative mass	relative charge
proton, p	1	1+
neutron, n	1	0
electron, e	1/1840	1−

A proton has virtually the same mass as a neutron and, for most of chemistry, their masses can be assumed to be identical.

An electron has negligible mass compared with the mass of a proton or a neutron. In most of chemistry, the mass of an electron can be ignored.

The charge on a proton is opposite to that of an electron but each charge has the same magnitude. An atom is electrically neutral. To balance out the charges, an atom must have the same number of protons as electrons.

> **KEY POINT**
>
> An atom is electrically neutral.
> An atom contains the same number of protons as electrons.

All atoms of hydrogen contain:
1 proton and 1 electron.

All atoms of carbon contain:
6 protons and 6 electrons.

All atoms of oxygen contain:
8 protons and 8 electrons.

Isotopes

Without neutrons, the nucleus would just contain positively-charged protons. Like-charges repel, and a nucleus containing just protons would fly apart! Strong nuclear forces act between protons and neutrons and these hold the nucleus together. Neutrons can be thought of as 'nuclear glue'.

In each atom of an element, the number of protons in the nucleus is fixed. However, most elements contain atoms with different numbers of neutrons. These atoms are called **isotopes** and, because they have different numbers of neutrons, they also have different masses.

> **KEY POINT**
>
> Isotopes are atoms of the **same** element with **different** masses.
> Isotopes of an element have the **same** number of **protons and electrons**.
> Each isotope has a **different** number of **neutrons** in the nucleus.
> The combined number of protons and neutrons in an isotope of an element is called the **mass number**, A.

The isotopes of an element react in the same way. This is because chemical reactions involve electrons – neutrons make no difference.

Chemists represent the nucleus of an isotope in a special way.

mass number
(protons + neutrons)

$$_{Z}^{A}X$$

atomic number
(protons)

The isotopes of carbon

Carbon exists as three isotopes, $_{6}^{12}C$, $_{6}^{13}C$, $_{6}^{14}C$. In all but the most accurate work, it is reasonable to assume that the relative mass of an isotope is equal to its mass number. It is easy to work out the atomic structure of an isotope from the representation above.

isotope	atomic number	mass number	protons	neutrons	electrons
$_{6}^{12}C$	6	12	6	6	6
$_{6}^{13}C$	6	13	6	7	6
$_{6}^{14}C$	6	14	6	8	6

Because all carbon isotopes have 6 protons, it is common practice to omit the atomic number, and $_{6}^{12}C$ is often shown as ^{12}C, or even as carbon-12.

Progress check

How many protons, neutrons and electrons are in the following isotopes?

1 $_{3}^{7}Li$; 2 $_{11}^{23}Na$; 3 $_{9}^{19}F$; 4 $_{13}^{27}Al$; 5 $_{26}^{55}Fe$.

5 $_{26}^{55}Fe$: 26p, 29n, 26e.
4 $_{13}^{27}Al$: 13p, 14n, 13e;
3 $_{9}^{19}F$: 9p, 10n, 9e;
2 $_{11}^{23}Na$: 11p, 12n, 11e
1 $_{3}^{7}Li$: 3p, 4n, 3e

1.2 The electronic structure of the atom

After studying this section you should be able to:

- understand the relationship between shells, sub-shells and orbitals
- understand how atomic orbitals are filled
- describe electronic structure in terms of shells and sub-shells

LEARNING SUMMARY

Energy levels or 'shells'

AQA	M1	OCR	M1
EDEXCEL	M1	SALTERS	M1
NICCEA	M1	WJEC	CH1
NUFFIELD	M1		

energy

n = 4 ___32e⁻___

n = 3 ___18e⁻___

n = 2 ___8e⁻___

n = 1 ___2e⁻___

Electronic energy levels are like a ladder and are filled from the bottom up.

Notice that the gap between successive energy levels becomes less with increasing energy.

You can imagine a model of an atom with electrons orbiting in shells around the nucleus. The electrons in each successive shell have an orbit further away from the nucleus. The further a shell is from the nucleus, the greater the shell's energy level. Each energy level is given a number called the principal quantum number, n. The shell closest to the nucleus has an energy level with $n = 1$, then $n = 2$ and so on.

The number of electrons that can occupy the first four energy levels is shown below:

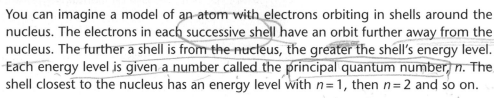

The principal quantum shell, n, is the shell number.

n	shell	electrons
1	1st shell	2
2	2nd shell	8
3	3rd shell	18
4	4th shell	32

If you look closely, you can see a pattern. The number of electrons that can occupy a shell is $2n^2$.

Using this model, electrons can be placed into available shells, starting with the lowest energy level. Each shell must be full before the next starts to fill. The table below shows how the shells are filled for the first 11 elements in the Periodic Table.

element	atomic number	electrons		
		$n = 1$	$n = 2$	$n = 3$
H	1	1		
He	2	2		
Li	3	2	1	
Be	4	2	2	
B	5	2	3	
C	6	2	4	
N	7	2	5	
O	8	2	6	
F	9	2	7	
Ne	10	2	8	
Na	11	2	8	1

This model breaks down as the $n = 3$ energy level is filled because each shell consists of sub-shells.

Sub-shells and orbitals

AQA	M1	OCR	M1
EDEXCEL	M1	SALTERS	M2
NICCEA	M1	WJEC	CH1
NUFFIELD	M1		

A more advanced model of electron structure is used in which each shell is made up of **sub-shells**.

[handwritten annotations:] The electronic structure of an atom is written using numbers to respresent the electrons in each energy level. E.g. eg. for sodium this is 2,8,1 → 2 electrons in the first energy level, 8 on 2nd energy shell, 1 or 3rd energy shell.

structure is what gives energy to atoms, and then to molecules. → many substances consist of two or more atoms joined together. shell joined

[Handwritten notes in left margin:]

What does electronic configuration?

This is the arrangement of electrons around the atom. ← They are arranged in electron shells and within each shell, there are sets of subs-shells. The sub-shells. them-selves are made up of orbit orbitals each of which can hold a maximum of two electrons, each with opposite spin.

Sub-shells

There are different types of sub-shell: s, p, d and f. Each type of sub-shell can hold a different number of electrons:

sub-shell	electrons
s	2
p	6
d	10
f	14

The table below shows the shells and sub-shells for the first four principal quantum numbers.

n	shell	sub-shell				total number of electrons	
1	1st shell	1s				2	= 2
2	2nd shell	2s	2p			2 + 6	= 8
3	3rd shell	3s	3p	3d		2 + 6 + 10	= 18
4	4th shell	4s	4p	4d	4f	2 + 6 + 10 + 14	= 32

- Each successive shell contains a new type of sub-shell.
- The 1st shell contains 1 sub-shell, the second sub-shell contains 2 sub-shells, and so on.

[Margin note:] Notice the labelling. The s sub-shell in the 2nd shell is labelled 2s.

Orbitals

How do the electrons fit into the sub-shells? Mathematicians have worked out that electrons occupy negative charge clouds called **orbitals** and these make up each sub-shell.

KEY POINT

- An orbital can hold up to two electrons.
- Each type of sub-shell has different orbitals: s, p, d and f.

The table below shows how electrons fill the orbitals in each sub-shell.

sub-shell	orbitals	electrons
s	1	1 x 2 = 2
p	3	3 x 2 = 6
d	5	5 x 2 = 10
f	7	7 x 2 = 14

s-orbitals

- An s-orbital has a spherical shape.

One s orbital

p-orbitals

- A p-orbital has a 3-dimensional dumb-bell shape.
- There are three p-orbitals, p_x, p_y and p_z, at right angles to one another.

Three p orbitals

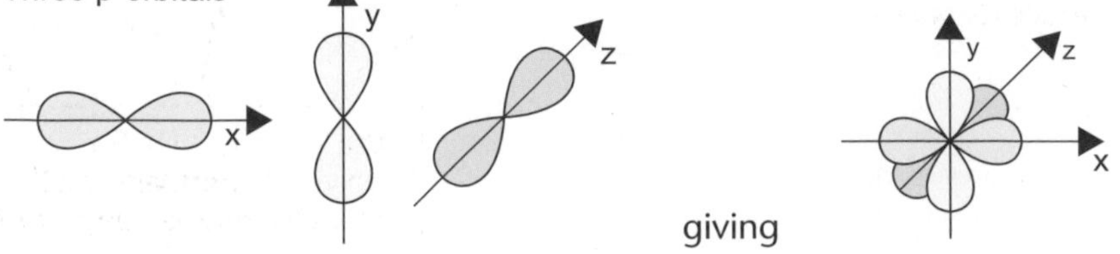

Three p-orbitals

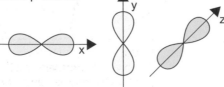

giving

d-orbitals and f-orbitals

The structures of d and f-orbitals are more complex.

Five d orbitals
Seven f orbitals

- There are five d-orbitals.
- There are seven f-orbitals.

How do two electrons fit into an orbital?

'Electrons in a box'

allowed

↑↓   ↓↑

not allowed

↑↑   ↓↓

Electrons are negatively charged and so they repel one another. An electron also has a property called **spin**. The two electrons in an orbital have opposite spins, helping to counteract the natural repulsion between their negative charges.

> **KEY POINT**
> - Chemists often represent an orbital as a box that can hold up to 2 electrons.
> - Each electron is shown as an arrow, indicating its spin: either ↑ or ↓.
> - Within an orbital, the electrons must have **opposite spins**.

to reduce the negative effect of sth by doing sth that has an opposite effect

Filling the sub-shells

AQA	M1	OCR	M1
EDEXCEL	M1	SALTERS	M2
NICCEA	M1	WJEC	CH1
NUFFIELD	M1		

Within a shell, the sub-shell energies are in the order: s, p, d and f.

Note that the 4s sub-shell is at a lower energy than the 3d sub-shell.

The 4s sub-shell fills before the 3d sub-shell.

Sub-shells have different energy levels. The diagram below shows the relative energies for the sub-shells in the first four shells.

energy

4f ——————
4d ——————
4p ——————
3d ——————   4s ——————
3p ——————
3s ——————

2p ——————
2s ——————

1s ——————

2p has a lower energy.

> The electrons occupy sub-shells in order of sub-shell energy levels.

the electrons needs to be filled for sub-shell energy

> **KEY POINT**
> - Shells and sub-shells are occupied in energy-level order.
> - Electrons occupy orbitals singly to prevent any repulsion caused by pairing.

The 2s orbital is occupied before the 2p orbitals because it is at a lower energy.

Note that the 2p orbitals are occupied singly before pairing begins.

The diagram below shows how electrons occupy orbitals from boron to oxygen.

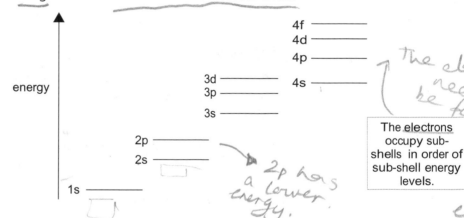

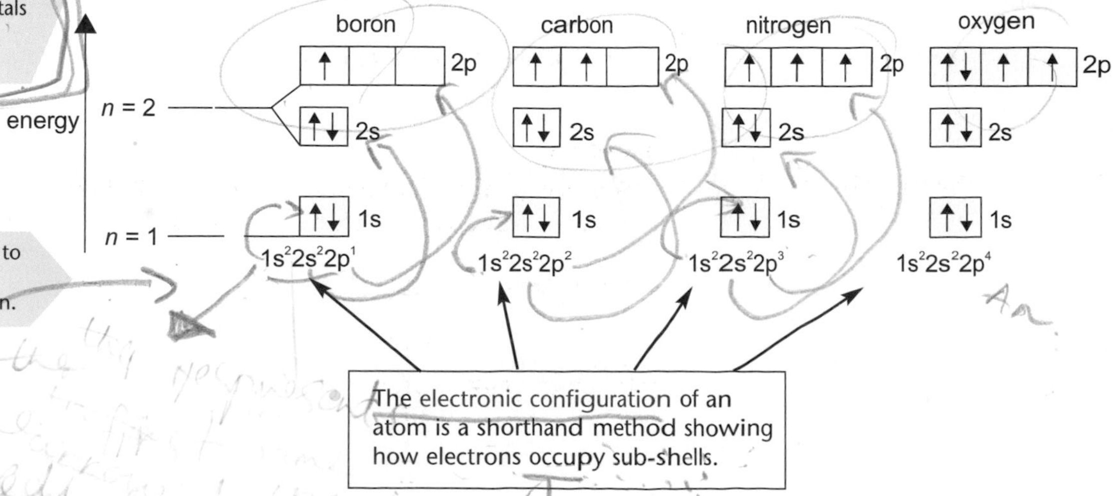

energy

Use these 4 examples to see how electronic configuration is written.

The electronic configuration of an atom is a shorthand method showing how electrons occupy sub-shells.

However, in s... small 2 represents the 2 arrows in each box only 2 can fill

The diagram below shows how the sub-shells are filled in an atom of potassium. Notice why the 4s sub-shell starts to fill before the 3d sub-shell.

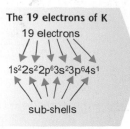

The 19 electrons of K

19 electrons

↓ ↓ ↓ ↓ ↓

$1s^2 2s^2 2p^6 3s^2 3p^6 4s^1$

↑ ↑ ↑ ↑ ↑

sub-shells

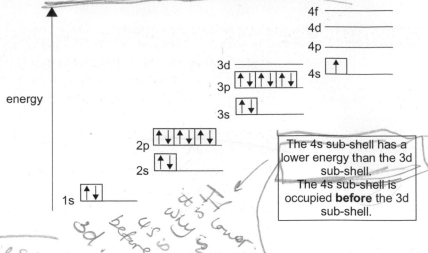

energy

4f ——————
4d ——————
4p ——————

3d ——————   4s [↑]

3p [↑↓][↑][↑][↑]

3s [↑↓]

2p [↑↓][↑][↑][↑]

2s [↑↓]

1s [↑↓]

> The 4s sub-shell has a lower energy than the 3d sub-shell.
> The 4s sub-shell is occupied **before** the 3d sub-shell.

4s sub shell occupies 3d before 4s is why is it lower.

Sub-shells and the Periodic Table

AQA	M1	OCR	M1
EDEXCEL	M1	SALTERS	M2
NICCEA	M1	WJEC	CH1
NUFFIELD	M1		

The Periodic Table is structured in blocks of 2, 6, 10 and 14, linked to sub-shells. The order of sub-shell filling can be seen by dividing the Periodic Table into blocks.

> You should be able to work out the electronic configuration for any element in the first four periods, i.e. up to krypton (Z = 36).

	1s			1s
s-block				**p-block**
2s				2p
3s		**d-block**		3p
4s		3d		4p
5s		4d		5p
6s		5d		6p
7s				

	f-block	
	4f	
	5f	

2, 3, 4, 5, 6, 7 represents the periodic number.

Simplifying electronic configurations

The similar electronic configurations within a group of the Periodic Table can be emphasised with a simpler representation in terms of the previous noble gas.

The electronic configurations for the elements in Group 1 are shown below:

Li:	$1s^2 2s^1$	*or*	$[He]2s^1$
Na:	$1s^2 2s^2 2p^6 3s^1$	*or*	$[Ne]3s^1$
K:	$1s^2 2s^2 2p^6 3s^2 3p^6 4s^1$	*or*	$[Ar]4s^1$

Progress check

1 Write the full electronic configuration in terms of sub-shells for:
(a) C; (b) Al; (c) Ca; (d) Fe; (e) Br.

1 (a) C: $1s^2 2s^2 2p^2$
(b) Al: $1s^2 2s^2 2p^6 3s^2 3p^1$
(c) Ca: $1s^2 2s^2 2p^6 3s^2 3p^6 4s^2$
(d) Fe: $1s^2 2s^2 2p^6 3s^2 3p^6 3d^6 4s^2$
(e) Br: $1s^2 2s^2 2p^6 3s^2 3p^6 3d^{10} 4s^2 4p^5$

1.3 Experimental evidence for electronic structures

After studying this section you should be able to:

- *recall the definitions for first and successive ionisation energies*
- *understand the factors affecting the sizes of ionisation energies*
- *show how ionisation energies provide evidence for shells and sub-shells*

Ions

AQA	M1	OCR	M1
EDEXCEL	M1	SALTERS	M1
NICCEA	M1	WJEC	CH1
NUFFIELD	M1		

Positive ions form when electrons are lost.

Negative ions form when electrons are gained.

An atom can either lose or gain electrons to form an **ion**: a charged atom.

A positive ion is formed when an atom loses electrons. For example, a lithium atom forms a **positive ion** with a 1+ charge by **losing** an electron:

$$Li \longrightarrow Li^+ + e^-$$
$$(3p^+, 4n, 3e^-) \quad (3p^+, 4n, 2e^-)$$

A negative ion is formed when an atom gains electrons. For example, an oxygen atom forms a **negative ion** with a 2– charge by **gaining** two electrons:

$$O + 2e^- \longrightarrow O^{2-}$$
$$(6p^+, 6n, 6e^-) \qquad\qquad (6p^+, 6n, 8e^-)$$

Ionisation energy

AQA	M1	OCR	M1
EDEXCEL	M1	SALTERS	M1
NICCEA	M1	WJEC	CH1
NUFFIELD	M1		

Ionisation energy measures the ease with which electrons are lost in the formation of positive ions. An element has as many ionisation energies as there are electrons.

> **KEY POINT**
>
> The **first** *ionisation energy* of an element is the energy required to remove **1 electron** from each atom in **1 mole** of **gaseous atoms** to form 1 mole of gaseous 1+ ions.

The equation representing the first ionisation energy of sodium is shown below.

$$Na(g) \longrightarrow Na^+(g) + e^- \qquad \text{1st ionisation energy} = +496 \text{ kJ mol}^{-1}$$

Factors affecting ionisation energy

Electrons are held in their shells by attraction from the nucleus. The first electron lost will be from the highest occupied energy level. This electron experiences least attraction from the nucleus.

Three factors affecting the size of this attraction are shown below.

These factors are very important and help to explain many chemical ideas throughout the course.

Learn them!

> **KEY POINT**
>
> **Atomic radius**
>
> The greater the distance between the nucleus and the outer electrons, the less the attractive force. Attraction falls rapidly with increasing distance and so this factor is very important and has a big effect.
>
> **Nuclear charge**
>
> The greater the number of protons in the nucleus the greater the attractive force.
>
> **Electron shielding or 'screening'**
>
> The outer shell electrons are repelled by any inner shells between the electrons and the nucleus.
>
> This repelling effect, called electron shielding or screening, reduces the overall attractive force experienced by the outer electrons.

Evidence for shells

EDEXCEL	M1	OCR	M1
NICCEA	M1	SALTERS	M1
NUFFIELD	M1	WJEC	CH1

This pattern shows that a sodium atom has 3 shells:

the first shell ($n = 1$) contains 2 electrons

the second shell ($n = 2$) contains 8 electrons

the third shell ($n = 3$) contains 1 electron.

> **KEY POINT**
>
> The **second ionisation energy** of an element is the energy required to remove **1 electron** from each atom in **1 mole of gaseous 1+ ions** to form 1 mole of gaseous 2+ ions.

The equation representing the second ionisation energy of sodium is shown below.

$$Na^+(g) \longrightarrow Na^{2+}(g) + e^-$$ 2nd ionisation energy = +4563 kJ mol^{-1}

Sodium, with 11 electrons, has 11 successive ionisation energies, shown as the graph below. Successive ionisation energies provide evidence for the different energy levels of the principal quantum numbers.

These **2** electrons are in the 1st shell ($n = 1$), closest to the nucleus. These electrons experience very **strong** attractive forces from the nucleus.

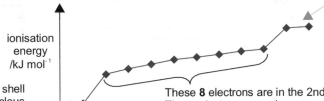

ionisation energy /kJ mol^{-1}

This **1** electron is in the 3rd shell ($n = 3$), furthest from the nucleus. This electron experiences comparatively **weak** attractive forces from the nucleus.

These **8** electrons are in the 2nd shell ($n = 2$). These electrons experience reasonably strong attractive forces from the nucleus.

ionisation number

Evidence for sub-shells

AQA	M1	NUFFIELD	M1
EDEXCEL	M1	OCR	M1
NICCEA	M1	WJEC	M1

Further details of trends in 1st ionisation energies are found on p.66.

The variation in first ionisation energies across a period in the Periodic Table provides evidence for the existence of sub-shells. The graph below shows this variation across Period 2 (Li → Ne)

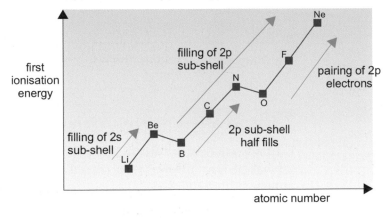

first ionisation energy

atomic number

General trend

There is a general rise in the first ionisation energy across a period.

The nuclear charge **increases** giving greater attraction.

There is little variation in atomic radius.

Across a period, increased nuclear charge is most important.

Electrons are added to the **same shell** ($n = 2$) that is the **same order of distance** from the nucleus.

All elements in the period experience the same degree of electron shielding from inner shells.

There is only one inner shell ($n = 1$) for all elements in Period 2.

> **KEY POINT**
>
> Across a period, the increased **nuclear charge** is the most important factor and the greater attractive force leads to an increased first ionisation energy.

Looking closer at first ionisation energies

Closer analysis provides evidence for sub-shells. Across the period, the graph shows three sections:

- a rise from lithium to beryllium,
- a fall to boron followed by a rise to nitrogen,
- a fall to oxygen followed by a rise to neon.

Comparing beryllium and boron

The fall in 1st ionisation energy from beryllium to boron marks the start of filling the 2p sub-shell which is at higher energy than the 2s sub-shell.

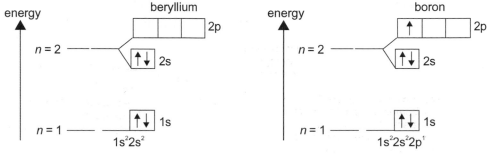

Beryllium versus boron: higher energy level.

- Beryllium's outermost electron is in the 2s sub-shell.
- Boron's outermost electron is in the 2p sub-shell.
- Boron's outermost electron is easier to remove because the 2p sub-shell has a **higher energy level** than the 2s sub-shell.

Comparing nitrogen and oxygen

The fall in 1st ionisation energy from oxygen to nitrogen marks the start of electron pairing in the p-orbitals of the 2p sub-shell.

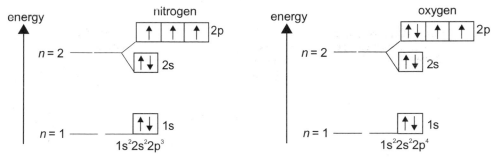

Nitrogen versus oxygen: electron pairing.

- Both nitrogen and oxygen have their outer electrons in the same 2p sub-shell with the same energy level.
- Oxygen has one 2p orbital with an electron pair – nitrogen, with one electron in each 2p orbital, has unpaired 2p electrons only.
- Because of **electron repulsion**, it is easier to remove one of oxygen's paired 2p electrons.

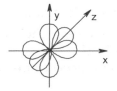

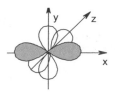

Nitrogen
2p sub-shell half-full
1 electron in each 2p orbital
parallel spins at right angles

Oxygen
2p electrons start to pair
paired electrons repel

Sample question and model answer

1

(a) Carbon has three main isotopes, ^{12}C, ^{13}C and ^{14}C. By comparing the atomic structures of these isotopes, explain what is meant by the term *isotopes*.

When comparing isotopes, it is important to stress the different numbers of neutrons.

Isotopes are atoms of the same element with different masses. ✓
The three isotopes of carbon have different numbers of neutrons. ✓
^{12}C has 6 neutrons, ^{13}C has 7 neutrons and ^{14}C has 8 neutrons. ✓

[3]

(b) The first ionisation energy of carbon is 1086 kJ mol^{-1}.
  (i) Explain the term first ionisation energy.

  The first ionisation energy of an element is the energy required to remove 1 electron ✓ from each atom in 1 mole ✓ of gaseous atoms ✓ to form 1 mole of gaseous 1+ ions.

Here, there will be 1 mark for the equation. A second mark will be available for the state symbols – important for ionisation energies. Don't forget them.

  (ii) Write an equation, including state symbols, to represent the first ionisation energy of carbon.

  $C(g) \rightarrow C^+(g) + e^-$ ✓✓

  (iii) Suggest why there is very little difference between the first ionisation energies of ^{12}C, ^{13}C and ^{14}C.

The word 'suggest' implies that you will need to use your chemical understanding to arrive at a sensible response. There are usually more marks allocated to this type of question than the maximum score possible. Here any 2 of the 3 marking points will secure the 2 marks.

  The isotopes have the same number of protons giving the same nuclear attraction ✓ on the same number of electrons. ✓ The difference in neutrons has no effect on the nuclear attraction. ✓

[7]

(c) Describe and explain the variation in first ionisation energies across Period 2 from lithium to neon.

Note the importance of nuclear charge (for trends across a period).

Across Period 2, the first ionisation energies increase ✓ because the nuclear charge increases ✓ as electrons are added to the same shell. ✓
The first ionisation energy of B is less than that of Be. ✓ The outermost electron of B is in the 2p sub-shell, ✓ at a higher energy ✓ than the outermost electron of Be, in the 2s sub-shell.

The emphasis here should be on the LOW ionisation energy of B (2p sub-shell starts) and of O (2p electron pairing starts).

It is NOT Be and N that are out of step. Be and N are NOT 'more stable'.

This is a common cause for the loss of marks in exams.

Here there are 10 marking points for 9 marks maximum.

The first ionisation energy of O is less than that of N. ✓ The three p electrons in N are unpaired in each of the three orbitals in the p sub-shell. ✓ O has pairing in one p orbital. ✓ The paired electrons in O repel ✓ and are more easily lost.

[9]

[Total: 19 marks]

Practice examination questions

1 Chemists use a model of an atom that consists of sub-atomic particles: protons, neutrons and electrons.

(a) Complete the table below to show the properties of these sub-atomic particles.

particle	relative mass	relative charge
proton		
neutron		
electron		

[3]

(b) The particles in each pair below differ **only** in the number of protons **or** neutrons or electrons. Explain what the difference is within each pair.
   (i) ^6Li and ^7Li
   (ii) ^{32}S and $^{32}S^{2-}$
   (iii) $^{39}K^+$ and $^{40}Ca^{2+}$.

[6]

[Total: 9 marks]

Cambridge Chemistry Foundation Q1 June 1999

2 (a) In terms of the numbers of sub-atomic particles, state **one** difference and **two** similarities between two isotopes of the same element. [3]

(b) Give the chemical symbol, including its mass number, for an atom which has 3 electrons and 4 neutrons. [1]

(c) (i) An element has an atomic number of 16 and it forms an ion with a charge of 2−. State the electronic configuration of this ion.
    (ii) To which block in the Periodic Table does this element belong? [2]

(d) (i) Write an equation for the process involved in the first ionisation energy of aluminium.
    (ii) Explain why the fourth ionisation energy of aluminium is much greater than the third. [6]

[Total: 12]

NEAB Atomic Structure, Bonding and Periodicity Q1 Feb 1996

3 The first ionisation energies of the elements lithium to carbon are given below:

element	Li	Be	B	C
ionisation energy / kJ mol^{-1}	519	900	799	1090

(a) Explain the meaning of the term *first ionisation energy*. [3]

(b) Explain why beryllium has a higher first ionisation energy than lithium. [2]

(c) Write the electronic configuration of a boron atom. [1]

(d) Explain why the first ionisation energy of boron is less than that for beryllium. [2]

(e) Predict a value for the first ionisation energy of nitrogen and explain how you estimated your answer. [2]

[Total: 10]

NEAB Atomic Structure, Bonding and Periodicity Q2 June 1995

Chapter 2

Atoms, moles and equations

The following topics are covered in this chapter:

- *Atomic mass*
- *The mole*

- *Formulae, equations and reacting quantities*

2.1 Atomic mass

After studying this section you should be able to:

- *define relative masses, based on the ^{12}C scale*
- *describe the basic principles of the mass spectrometer*
- *calculate the relative atomic mass of an element from its isotopic abundance or mass spectrum*
- *calculate the relative molecular mass of a compound from relative atomic masses*

LEARNING SUMMARY

Relative atomic mass

AQA	M1	OCR	M1
EDEXCEL	M1	SALTERS	M1
NICCEA	M1	WJEC	CH1
NUFFIELD	M1		

Carbon-12: an international standard

The mass of an atom is too small to be measured on even the most sensitive balance. Chemists use **relative** masses to compare the atomic masses of different elements. First, we must have an atom with which to compare other atoms and an atom of the **carbon-12 isotope** is chosen as the international standard for the measurement of atomic mass.

- The mass of an atom of carbon-12 is exactly 12 atomic mass units (amu).
- The mass of one-twelfth of an atom of carbon-12 is exactly 1 amu.

All atoms in a pure isotope have the same atomic structure and the same mass. An isotope's *relative* mass is found by comparison with carbon-12.

> **Atomic mass unit (amu)**
>
> 1 atomic mass unit (amu) is the mass of one-twelfth the mass of an atom of the carbon-12 isotope. This provides the base measurement for atomic masses:
>
> 1 amu = 1.661×10^{-23} g

> **Relative isotopic mass** is the mass of an atom of an isotope compared with one-twelfth the mass of an atom of carbon-12.
>
> **KEY POINT**

In most chemistry work, it is reasonable to:

- neglect the tiny contribution to atomic mass from electrons;
- take both the mass of a proton and a neutron as 1 amu.

Relative isotopic mass is then simply the mass number of the isotope, e.g. the relative isotopic mass of ^{16}O is 16.

Relative atomic mass

> Note that a relative mass has no units. It is simply a ratio of masses and any mass units will cancel.

Most elements consist of a mixture of isotopes, each with a different mass number. To work out the relative atomic mass of an element we must find the **weighted average mass** of the isotopes present from:

- the natural abundances of the isotopes
- the relative isotopic masses of the isotopes.

> Examples of relative atomic masses:
> H: 1.008;
> Cl: 35.45;
> Pb: 207.19

> **Relative atomic mass,** A_r, is the weighted average mass of an atom of an element compared with one-twelfth of the mass of an atom of carbon-12.
>
> **KEY POINT**

Calculating a relative atomic mass

In most chemistry work, the relative isotopic mass can be taken as a whole number.

Naturally occurring chlorine consists of 75% ^{35}Cl and 25% ^{37}Cl.

$\frac{75}{100}$ of chlorine is the ^{35}Cl isotope;

$\frac{25}{100}$ of chlorine is the ^{37}Cl isotope.

The relative atomic mass of chlorine $= \frac{75}{100} \times 35 + \frac{25}{100} \times 37 = 35.5$

Measuring relative atomic masses

A relative atomic mass is measured using a mass spectrometer.

Remember VIADD:

Vaporisation
Ionisation
Acceleration
Deflection
Detection

- A sample of the element is placed into the mass spectrometer and vaporised.
- The sample is bombarded with electrons forming positive ions.
- The positive ions are accelerated using an electric field.
- The positive ions are deflected using a magnetic field.
- Ions of lighter isotopes are deflected more than ions of heavier isotopes. This separates different isotopes.
- The ions are detected to produce a **mass spectrum**.

You can follow the stages in the diagram below:

Notice that ions are detected in the mass spectrometer. The ions are formed following electron bombardment.

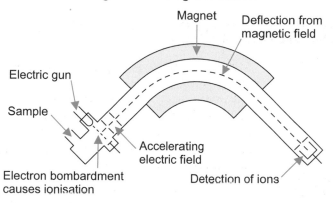

Relative atomic mass from a mass spectrum

A mass spectrum provides:

- the relative isotopic masses of the isotopes in an element
- isotopic abundances.

The diagram below shows how a relative atomic mass can be calculated from a mass spectrum.

Mass spectrum of a copper sample

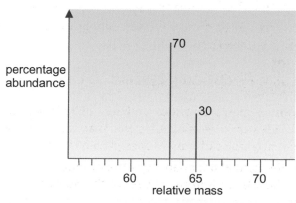

ion	relative mass	percentage abundance
$^{63}Cu^+$	63	70%
$^{65}Cu^+$	65	30%

The relative atomic mass of copper

$= \frac{70}{100} \times 63 + \frac{30}{100} \times 65$

$= 63.6$

Relative masses for compounds

AQA	M1	OCR	M1
EDEXCEL	M1	SALTERS	M1
NICCEA	M1	WJEC	CH1
NUFFIELD	M1		

You will find out more about chemical bonding and structure in Chapter 3, p.44.

Relative masses can be used to compare the masses of any chemical species. Much of matter is made up of compounds.

Relative molecular mass

A simple molecule of Cl_2, H_2O or CO_2 comprises small groups of atoms, held together by chemical bonds.

> **KEY POINT**
>
> Relative molecular mass, M_r, is the weighted average mass of a molecule of a compound compared with one-twelfth the mass of an atom of carbon-12.

The *relative molecular mass* of a compound is found by adding together the relative atomic masses of each atom in a molecule.

Examples of relative molecular masses

Cl_2: $M_r = 35.5 \times 2 = 71$;

H_2O: $M_r = 1.0 \times 2 + 16.0 = 18.0$

Some compounds, such as sand SiO_2, exist as 'giant' molecules comprising hundreds of thousands of atoms bonded together. Each molecule in these structures is the size of each crystal and, in a sense, the whole crystal is a molecule.

For giant structures, the **formula unit** of the compound is the simplest ratio of atoms or ions in the structure.

- Each 'molecule' of SiO_2 has twice as many oxygen atoms as silicon atoms.

Many compounds, such as common salt NaCl, are made up of ions and not molecules. These ionic compounds do not have single molecules and form giant structures of oppositely charged ions.

- NaCl has the same number of sodium ions as chloride ions.

Although relative **molecular** mass is often used with these compounds, a better term to use is relative **formula** mass.

Relative formula mass can be used to describe the relative mass of any compound.

> **KEY POINT**
>
> Relative formula mass is the average mass of the formula unit of a compound compared with one-twelfth the mass of an atom of carbon-12.

The *relative formula mass* of a compound is found by adding together the relative atomic masses of each atom in the formula.

Examples of relative formula masses

$CaBr_2$: relative formula mass $= 40.1 + 79.9 \times 2 = 199.9$;

Na_3PO_4: relative formula mass $= 23.0 \times 3 + 31.0 + 16.0 \times 4 = 164.0$

Progress check

1. Calculate the relative atomic mass, A_r, of the following elements from their isotopic abundances:
   (a) Gallium containing 60% [69]Ga and 40% [71]Ga.
   (b) Silicon containing 92.18% [28]Si, 4.71% [29]Si and 3.11% [30]Si.

2. Use A_r values from the Periodic Table to calculate the relative formula mass of:
   (a) CO_2
   (b) NH_3
   (c) Fe_2O_3
   (d) $C_6H_{12}O_6$
   (e) $Pb(NO_3)_2$.

1 (a) Gallium: 69.8
(b) Silicon: 28.11.
2 (a) 44
(b) 17
(c) 159.6
(d) 180
(e) 331.

2.2 The mole

After studying this section you should be able to:

LEARNING SUMMARY

- *understand the concept of the mole as the amount of substance*
- *calculate molar quantities using masses, solutions and gas volumes*

The mole

AQA	M1	OCR	M1
EDEXCEL	M1	SALTERS	M1
NICCEA	M1	WJEC	CH1
NUFFIELD	M1		

Counting and weighing atoms

To make a compound such as H_2O, it would be useful to be able to count out atoms. For H_2O, we would need twice as many hydrogen atoms as oxygen atoms. Atoms are far too small to be counted individually but they can be counted using the idea of relative mass:

element	H	C	O
relative mass	1	12	16

For 1 atom of H, C and O, the relative masses are in the ratio 1 : 12 : 16.
For 10 atoms of each, the ratio is 10 : 120 : 160 which is still 1 : 12 : 16.

> Provided we have the same number of atoms of each, the masses are in the ratio of the relative masses.

By measuring masses, chemists are able to actually count atoms.
For 1 g of H, 12 g of C and 16 g of O, the masses are still in the ratio 1 : 12 : 16.
1 g of H contains the same number of atoms as 12 g of C and 16 g of O.

Amount of substance

Chemists use a quantity called *amount of substance* for counting atoms.
Amount of substance is:

- given the symbol n
- measured using a unit called the **mole** (abbreviated as **mol**).

Note that carbon-12 is the **substance** here. So **amount of *substance*** translates to **amount of carbon-12.**

As with relative masses, amount of substance uses the carbon-12 isotope as the standard:
The amount n of carbon-12 atoms in 12 g of ^{12}C is 1 *mol*.

The actual **number** of atoms in 1 mol of carbon-12 is 6.02×10^{23}.

This can be expressed by saying:

'There are 6.02×10^{23} atoms per mole of carbon-12 atoms (or 6.02×10^{23} mol^{-1})'.

6.02×10^{23} mol^{-1} is a constant known as the **Avogadro constant**, L.

> **KEY POINT**
>
> This is a very powerful and useful idea. To make a water molecule from hydrogen and oxygen, we cannot **count** out the atoms but we can **weigh** them out in the correct ratio. To make H_2O, we need 2 atoms of hydrogen for every 1 atom of oxygen which we can get by weighing out 2 g of hydrogen and 16 g of oxygen.
>
> **2 g of H contains twice as many atoms as 16 g of O.**
> **2 g of H atoms contains 2 mol of H atoms.**
> **16 g of O atoms contains 1 mol of O atoms.**

Particle

A single particle of a substance may refer to an atom, a molecule, an ion, an electron or to any identifiable particle. Chemists refer to a collection of particles as a **chemical species**.

Amount of substance is not restricted just to atoms. We can have an amount of any chemical species: 'amount of atoms', 'amount of molecules', 'amount of ions', etc.

> **A mole** is the amount of substance that contains as many single particles as there are atoms in exactly 12 g of the carbon-12 isotope.
>
> **KEY POINT**

Using the definition of the mole above, this means that:

- 1 mole of hydrogen atoms, H, contains 6.02×10^{23} hydrogen atoms, H

- 1 mole of oxygen molecules, O_2, contains 6.02×10^{23} oxygen molecules, O_2.

- 1 mole of electrons, e^-, contains 6.02×10^{23} electrons, e^-.

It is important to always refer to what the particle is.

For example:

'1 mole of oxygen' could refer to oxygen atoms, O, or to oxygen molecules, O_2.

It is always safest to quote the formula to which the amount of substance refers.

Moles from masses

AQA	M1	OCR	M1
EDEXCEL	M1	SALTERS	M1
NICCEA	M1	WJEC	CH1
NUFFIELD	M1		

Molar mass, *M*

> **Molar mass, *M*,** is the mass of one mole of a substance.
> The units of molar mass are g mol⁻¹.
>
> **KEY POINT**

For the atoms of an element, the molar mass is simply the relative atomic mass in g mol⁻¹.

Molar mass is an extremely useful term that can be applied to any chemical species.

- Molar mass of Mg = 24.3 g mol⁻¹
- This simply means that 1 mole of Mg atoms has a mass of 24.3 g.

For a compound, the molar mass is found by adding together the relative atomic masses of each atom in the formula:

Examples of molar masses

H	1 g mol⁻¹
H_2O	18 g mol⁻¹
CO_2	44 g mol⁻¹
HNO_3	63 g mol⁻¹

- Molar mass of CO_2 = 12 + (16 x 2) = 44 g mol⁻¹
- 1 mole of CO_2 has a mass of 44 g.

> Amount of substance n, mass and molar mass are linked by the expression below:
> $$n = \frac{mass\ (in\ g)}{molar\ mass}$$
>
> **KEY POINT**

In 24 g of C, *amount of C* = $\frac{mass\ (in\ g)}{molar\ mass}$ = $\frac{24}{12}$ = 2 mol

In 11 g of CO_2, *amount of CO_2* = $\frac{mass\ (in\ g)}{molar\ mass}$ = $\frac{11}{44}$ = 0.25 mol

Progress check

1 (a) What is the amount of each substance (in mol) in the following:
   (i) 124 g P; (ii) 64 g O; (iii) 64 g O_2; (iv) 10 g $CaCO_3$;
   (v) 3.04 g Cr_2O_3?

(b) What is the mass of each substance (in g) in the following:
   (i) 0.1 mol Na; (ii) 2.5 mol NO_2; (iii) 0.3 mol Na_2SO_4;
   (iv) 0.05 mol H_2SO_4; (v) 0.025 mol Ag_2CO_3?

(b) (i) 2.3 g; (ii) 115 g; (iii) 42.6 g; (iv) 4.9 g; (v) 6.9 g.
1 (a) (i) 4 mol; (ii) 4 mol; (iii) 2 mol; (iv) 0.1 mol; (v) 0.02 mol.

Moles from solutions

AQA	M1	OCR	M1
EDEXCEL	M1	SALTERS	M1
NICCEA	M1	WJEC	CH1
NUFFIELD	M1		

KEY POINT

The concentration of a solution is the amount of solute, in mol, dissolved in 1 dm³ (1000 cm³) of solution.

If the volume of the solution is in dm³:

$$n = c \times V \text{ (in dm}^3)$$

$n =$ amount of substance, in mol.

$c =$ concentration of solution, in mol dm⁻³.

$V =$ volume of solution, in dm³.

It is more convenient to measure smaller volumes of solutions in cm³ and the expression becomes:

$$n = c \times \frac{V \text{ (in cm}^3)}{1000}$$

> Dividing a volume measured in cm³ by 1000 converts the volume automatically to dm³.

Example

What is the amount of NaCl (in mol) in 25.0 cm³ of an aqueous solution of concentration 2.00 mol dm⁻³?

$$n(\text{NaCl}) = c \times \frac{V}{1000} = 2 \times \frac{25.0}{1000} = 0.0500 \text{ mol}$$

> **Molarity**
>
> Molarity refers to the concentration in mol dm⁻³.
>
> Thus 2 mol dm³ and 2 Molar (or 2M) mean the same:
> 2 moles of solute in 1 dm³ of solution.

Standard solutions

Chemists often need to prepare **standard solutions** with an exact concentration. Using an understanding of the mole, the mass required to prepare such a solution can easily be worked out.

> Concentrations are sometimes referred to in g dm⁻³.
>
> The Na_2CO_3 here has a **molar** concentration of 0.100 mol dm⁻³ and a **mass** concentration of 10.6 g dm⁻³ of Na_2CO_3.

Example

Find the mass of sodium carbonate required to prepare 250 cm³ of a 0.100 mol dm⁻³ solution.

Find the amount of Na_2CO_3 (in mol) required in solution:

$$\text{amount, } n, \text{ of } Na_2CO_3 = c \times \frac{V}{1000} = 0.100 \times \frac{250}{1000} = 0.0250 \text{ mol}$$

Convert moles to grammes

> molar mass of Na_2CO_3 =
> $23.0 \times 2 + 12.0 + 16.0 \times 3$
> $= 106.0$ g mol⁻¹

$$n = \frac{\text{mass}}{\text{molar mass}} \quad \therefore \text{ mass} = n \times \text{molar mass}$$

mass of Na_2CO_3 required = $0.0250 \times 106.0 = 2.65$ g

Progress check

1 (a) What is the amount of each substance, in mol, in:
   (i) 250 cm³ of a 1.00 mol dm⁻³ solution;
   (ii) 10 cm³ of a 2.0 mol dm⁻³ solution?

 (b) Find the concentration, in mol dm⁻³, for:
   (i) 4 mol in 2 dm³ of solution;
   (ii) 0.0100 mol in 100 cm³ of solution.

 (c) Find the concentration, in g dm⁻³, for:
   (i) 2 mol of NaOH in 4 dm³ of solution;
   (ii) 0.500 mol of HNO_3 in 200 cm³ of solution.

(handwritten: mass = mol × molar mass)

(c) (i) 20 g dm³, (ii) 157.5 g dm⁻³.
(b) (i) 2 mol dm⁻³, (ii) 0.1 mol dm⁻³.
1 (a) (i) 0.25 mol; (ii) 0.02 mol.

Moles from gas volumes

AQA	M1	OCR	M1
EDEXCEL	M1	SALTERS	M1
NICCEA	M1	WJEC	CH1
NUFFIELD	M1		

This is sometimes summarised by Avogadro's hypothesis:

'Equal volumes of gases contain the same number of molecules under the same conditions of temperature and pressure.'

For a gas, the amount of gas molecules (in mol) is most conveniently obtained by measuring the gas volume.

Provided that the pressure and temperature are the same, equal volumes of gases contain the same number of molecules.

This means that it does not matter which gas is being measured. By measuring the volume, we are indirectly also counting the number of molecules.

> **KEY POINT**
>
> At room temperature and pressure (r.t.p), 298K (25°C) and 100 kPa, 1 mol of a gas occupies approximately 24 dm³ = 24 000 cm³.

At r.t.p., if the volume of the gas is in **dm³**: $n = \dfrac{V \text{ (in dm}^3\text{)}}{24}$

if the volume of the gas is in **cm³**: $n = \dfrac{V \text{ (in cm}^3\text{)}}{24000}$

Example

How many moles of gas molecules are in 480 cm³ of a gas at r.t.p.?

$$n = \frac{V \text{ (in cm}^3\text{)}}{24000} = \frac{480}{24000} = 0.0200 \text{ mol of gas molecules.}$$

The ideal gas equation (AQA and WJEC only)

The conditions will not always be room temperature and pressure. A gas volume depends on temperature and pressure.

The Ideal Gas Equation can be used to find the amount of gas molecules n (in mol) in a volume V (in m³) at any temperature T (in K) and pressure p (in Pa).

The Ideal Gas Equation: $pV = nRT$ ($R = 8.31$ J K⁻¹ mol⁻¹)

Before using $pV = nRT$, you must remember to convert any values to Pa, K and m³.

Conversion rules

cm³ to m³	× 10⁻⁶
dm³ to m³	× 10⁻³
°C to K	+ 273
kPa to Pa	× 10³

Note that M (molar mass) has units: g mol⁻¹

M_r (relative molecular mass) has no units.

Example

A 0.215 g sample of a volatile liquid, **X**, produces 77.5 cm³ of gas at 100°C and 100 kPa. Calculate the relative molecular mass of **X**.

Use the ideal gas equation:

$$pV = nRT \qquad \therefore n = \frac{pV}{RT}$$

$$\therefore n = \frac{1.00 \times 10^5 \times 7.75 \times 10^{-5}}{8.31 \times 373} = 0.00250 \text{ mol}$$

Find the molar mass:

$$n = \frac{\text{mass}}{\text{molar mass}}$$

$$\therefore \text{ molar mass } = \frac{\text{mass}}{n} = \frac{0.215}{0.00250} = 86.0 \text{ g mol}^{-1}$$

$$\therefore \text{ Relative molecular mass } M_r = 86.0$$

Progress check

1. (a) What is the volume, at room temperature and pressure (r.t.p.) of
   (i) 44 g CO_2(g); (ii) 7 g N_2(g); (iii) 5.1 g NH_3(g)?
   (b) What is the mass, at r.t.p. of
   (i) 1.2 dm³ O_2; (ii) 720 cm³ CO_2(g); (iii) 48 cm³ CH_4(g)?

(b) (i) 1.6 g; (ii) 1.32 g; (iii) 0.032 g.

1 (a) (i) 24 dm³; (ii) 6 dm³; (iii) 7.2 dm³.

2.3 Formulae, equations and reacting quantities

After studying this section you should be able to:

- *understand what is meant by empirical and molecular formula*
- *calculate the empirical and molecular formula of a substance*
- *write balanced equations*
- *use chemical equations to calculate reacting masses and reacting volumes (and vice versa)*
- *perform calculations involving volumes and concentrations of solutions in simple acid-base titrations*

LEARNING SUMMARY

Types of chemical formula

AQA	M1	OCR	M1
EDEXCEL	M1	SALTERS	M1
NICCEA	M1	WJEC	CH1
NUFFIELD	M1		

The **empirical formula** of ethane is CH_3.

Ethane has 1 carbon atom for each of 3 hydrogen atoms.

The molecular formula of ethane is C_2H_6

Each molecule of ethane contains 2 carbon atoms and 6 hydrogen atoms.

Empirical and molecular formula

> **Empirical formula** represents the simplest, whole-number ratio of atoms of each element in a compound.
>
> **Molecular formula** represents the actual number of atoms of each element in a molecule of a compound.

KEY POINT

Formula determination

A formula can be calculated from experimental results using the Mole Concept.

Example 1

Find the empirical formula of a compound formed when 6.75 g of aluminium reacts with 26.63 g of chlorine. [A_r: Al, 27.0; Cl, 35.5.]

Find the molar ratio of atoms:

$$
\begin{array}{ccc}
\text{Al} & : & \text{Cl} \\
= \dfrac{6.75}{27.0} & : & \dfrac{26.63}{35.5} \\
0.25 & : & 0.75
\end{array}
$$

Divide by smallest number (0.25):   1   :   3

∴ Empirical formula = $AlCl_3$

Example 2

Find the molecular formula of a compound containing carbon, hydrogen and oxygen only with the composition by mass of carbon: 40.0%; hydrogen: 6.7%. [$M_r = 180$. A_r: H, 1.00; C, 12.0; O, 16.0.]

Percentage of oxygen in compound = 100 − (40.0 + 6.7) = 53.3%
100.0 g of the compound contains 40.0 g C, 6.7 g H and 53.3 g O.

Analysis often doesn't give the oxygen content directly but it can easily be calculated.

Find the molar ratio of atoms:

$$
\begin{array}{ccccc}
\text{C} & : & \text{H} & : & \text{O} \\
= \dfrac{40.0}{12.0} & : & \dfrac{6.7}{1.00} & : & \dfrac{53.3}{16.0} \\
= 3.33 & : & 6.7 & : & 3.33
\end{array}
$$

Divide by smallest number (3.33):   1   :   2   :   1

∴ Empirical formula = CH_2O

Relate to molecular mass

Each CH_2O unit has a relative mass of $12 + (1 \times 2) + 16 = 30$
M_r, of the compound is 180 containing $180/30 = 6$ CH_2O units
∴ molecular formula is $C_6H_{12}O_6$

Atoms, moles and equations

Progress check

1 Find the empirical formula of the following:
   (a) sodium oxide (2.3 g of sodium reacts to form 3.1 g of sodium oxide).
   (b) iron oxide (11.16 g of iron reacts to form 15.96 g of iron oxide).

2 Use the following percentage compositions by mass to find the empirical formula of the compound.
   (a) A compound of sulphur and oxygen. [S, 40.0%; O, 60.0%]
   (b) A compound of iron, sulphur and oxygen. [Fe, 36.8%; S, 21.1%; O, 42.1%]

3 2.8g of a compound of carbon and hydrogen is formed when 2.4 g of carbon combines with hydrogen [M_r: 56]. What is the empirical and molecular formula of the compound formed?

3 CH_2; C_4H_8.
2 (a) SO_3; (b) $FeSO_4$.
1 (a) Na_2O; (b) Fe_2O_3.

Equations

Chemical reactions involve the **rearrangement** of atoms and ions.

Chemical equations provide two types of information about a reaction:

- qualitative – *which* atoms or ions are rearranging;
- quantitative – *how* many atoms or ions are rearranging.

Balancing equations

The formula of each substance shows the chemicals involved in the reaction.

State symbols can be added to show the physical states of each species under the conditions of the reaction.

State symbols give the physical states of each species in a reaction:

gaseous state, (g);
liquid state, (l);
solid state, (s);
aqueous solution, (aq).

The **qualitative** equation for the reaction of hydrogen with oxygen is shown below:

$$\text{hydrogen} \quad + \quad \text{oxygen} \quad \longrightarrow \quad \text{water}$$
$$H_2(g) \quad + \quad O_2(g) \quad \longrightarrow \quad H_2O(l)$$

In this equation, hydrogen is balanced, oxygen is **not** balanced.

	$H_2(g)$	+	$O_2(g)$	$\longrightarrow$	$H_2O(l)$	
hydrogen	2				2	✓
oxygen			2		1	✗

The equation must be **balanced** to give the **same number** of particles of each element on each side of the equation.

Balancing an equation often takes several stages.

- Oxygen can be balanced by placing a '2' in front of H_2O.
- However, this unbalances hydrogen:

	$H_2(g)$	+	$O_2(g)$	$\longrightarrow$	**2** $H_2O(l)$	
hydrogen	2				4	✗
oxygen			2		2	✓

- H can be re-balanced by placing a '2' in front of H_2.
- The equation is now balanced.

Notice that no formula has been changed. You balance an equation by adding numbers in front of a formula only.

	2 $H_2(g)$	+	$O_2(g)$	$\longrightarrow$	**2** $H_2O(l)$	
hydrogen	4				4	✓
oxygen			2		2	✓

More balancing numbers.

$3 Na_2SO_4$ means
Na: $3 \times 2 = 6$;
S: $3 \times 1 = 3$;
O: $3 \times 4 = 12$.

With brackets, the subscript applies to everything in the preceding bracket:

$2 Ca(NO_3)_2$ means
Ca: $2 \times 1 = 2$;
N: $2 \times (1 \times 2) = 4$;
O: $2 \times (3 \times 2) = 12$.

In a formula, a subscript applies **only** to the symbol immediately preceding,

e.g. NO_2 comprises **1** N and **2** O.

When balancing an equation, you must **not** change any formula. You can only add a balancing number in front of a formula.

The balancing number multiplies everything in the formula by the balancing number,

e.g. $2AlCl_3$ comprises Al: $2 \times 1 = 2$; Cl: $2 \times 3 = 6$.

Progress check

Balance the following equations:
(a) $Li(s) + O_2(g) \longrightarrow Li_2O(s)$
(b) $CH_4(g) + O_2(g) \longrightarrow CO_2(g) + H_2O(l)$
(c) $Al(s) + HCl(aq) \longrightarrow AlCl_3(aq) + H_2(g)$

(a) $4Li(s) + O_2(g) \longrightarrow 2Li_2O(s)$
(b) $CH_4(g) + 2O_2(g) \longrightarrow CO_2(g) + 2H_2O(l)$
(c) $2Al(s) + 6HCl(aq) \longrightarrow 2AlCl_3(aq) + 3H_2(g)$.

Reacting quantities

AQA	M1	OCR	M1
EDEXCEL	M1	SALTERS	M1
NICCEA	M1	WJEC	CH1
NUFFIELD	M1		

An understanding of molar reacting quantities provides chemists with the required recipes to make quantities of chemicals to order.

The balancing numbers give the ratio of the **amount** of each substance, in mol. Using the example above:

| *equation* | $2 H_2(g)$ | + | $O_2(g)$ | $\longrightarrow$ | $2 H_2O(l)$ |
| *moles* | 2 mol | | 1 mol | $\longrightarrow$ | 2 mol |

Chemists use this **quantitative** information to find:
• the *reacting quantities* required to prepare a particular quantity of a product
• the *quantities of products* formed by reacting particular quantities of reactants.

By working out the reacting quantities from a balanced chemical equation, quantities can be adjusted to take into account the required scale of preparation.

Example 1

These examples assume a 100% yield of products. See also percentage yield, page 117.

Calculate the masses of nitrogen and oxygen required to form 3 g of nitrogen oxide, NO (A_r: N, 14; O, 16).

equation	$N_2(g)$	+	$O_2(g)$	$\longrightarrow$	$2 NO(g)$
moles	1 mol	+	1 mol	$\longrightarrow$	2 mol
reacting masses	(14×2) g	+	(16×2) g	$\longrightarrow$	$2 (14 + 16)$ g
	28 g	+	32 g	$\longrightarrow$	60 g
for 3 g NO(g), ÷ 20:	1.4 g	+	1.6 g	$\longrightarrow$	3 g

∴ 1.4 g of $N_2(g)$ react with 1.6 g of $O_2(g)$ to form 3 g NO(g).

Example 2

You must keep the proportions the same as the reacting quantities.

Any scaling must be applied to all reacting quantities.

Calculate the volume of oxygen, at r.t.p., formed by the decomposition of an aqueous solution containing 5 g of hydrogen peroxide, H_2O_2. [A_r: H, 1; O, 16.]

equation	$2 H_2O_2(aq)$	$\longrightarrow$	$2 H_2O(l)$	+	$O_2(g)$
moles	2 mol	$\longrightarrow$	2 mol	+	1 mol
reacting quantities	$2 [(1 \times 2) + (16 \times 2)]$ g	$\longrightarrow$			1×24 dm³
	68 g	$\longrightarrow$			24 dm³
for 1 g H_2O_2, ÷ 68,	1 g	$\longrightarrow$			$\frac{24}{68}$ dm³
for 5 g H_2O_2, × 5,	5 g	$\longrightarrow$			$\frac{24}{68} \times 5$ dm³

∴ 5 g of H_2O_2 decomposes to form 1.8 dm³ of $O_2(g)$

Progress check

1 For the reaction: $S(s) + O_2(g) \longrightarrow SO_2(g)$
   (a) How many moles of S and O_2 react to form 0.5 mole of SO_2?
   (b) What mass of SO_2 is formed by reacting 64.2 g of S with O_2?
   (c) What volume of SO_2 forms when 0.963 g of S reacts with O_2 at r.t.p.?

2 Balance the equation: $Mg(s) + O_2(g) \longrightarrow MgO(s)$
   (a) What mass of MgO forms by burning 8.1 g of Mg?
   (b) What volume of O_2 at r.t.p. will react with this mass of Mg?
   (c) What masses of Mg and O_2 are needed to prepare 1 g of MgO?

2 (a) 13.4 g; (b) 4 dm³; (c) 0.6 g Mg, 0.4 g O_2.
1 (a) 0.5 mol S, 0.5 mol O_2; (b) 96.2 g; (c) 0.72 dm³.

Calculations in acid-base titrations

AQA — M1   OCR — M1
EDEXCEL — M1   SALTERS — M2
NICCEA — M1   WJEC — CH2
NUFFIELD — M1

The acid is added from the burette to the alkali.

An indicator is required to show the 'end-point' when all alkali has reacted with the added acid.

The concentration and volume of NaOH are known.

Using the equation determine the number of moles of the second reagent

Work out the concentration of $H_2SO_4(aq)$, in mol dm⁻³.

Calculations for acid-base titrations are best shown with an example:

In a titration, 25.0 cm³ of 0.100 mol dm⁻³ sodium hydroxide NaOH(aq) were found to react exactly with 10.4 cm³ of sulphuric acid, $H_2SO_4(aq)$. Find the concentration of the sulphuric acid.

To solve the problem, we must use the following pieces of information:

- the balanced equation
$$2NaOH(aq) + H_2SO_4(aq) \longrightarrow Na_2SO_4(aq) + 2H_2O (l)$$
- the concentration c_1 and reacting volume V_1 of NaOH(aq)
- the concentration c_2 and reacting volume V_2 of $H_2SO_4(aq)$.

From the titration results, the amount of NaOH (in mol) can be calculated:
$$\text{amount of NaOH} = c \times \frac{V}{1000} = 0.100 \times \frac{25.0}{1000} = 0.00250 \text{ mol}$$

From the equation, the amount of H_2SO_4 (in mol) can be determined:
$$2 NaOH (aq) + H_2SO_4(aq) \longrightarrow Na_2SO_4(aq) + 2H_2O (l)$$
$$2 \text{ mol} \qquad 1 \text{ mol } (balancing\ numbers)$$
∴ 0.00250 mol NaOH reacts with 0.00125 mol H_2SO_4
amount of H_2SO_4 that reacted = 0.00125 mol

The concentration (in mol dm⁻³) of H_2SO_4 can be calculated by scaling to 1000 cm³:
10.4 cm³ $H_2SO_4(aq)$ contains 0.00125 mol H_2SO_4
1 cm³ $H_2SO_4(aq)$ contains $\frac{0.00125}{10.4}$ mol H_2SO_4

1 dm³ (1000 cm³) $H_2SO_4(aq)$ contains $\frac{0.00125}{10.4} \times 1000 = 0.120$ mol H_2SO_4

∴ concentration of $H_2SO_4(aq)$ is 0.120 mol dm⁻³

Progress check

1 25.0 cm³ of 0.500 mol dm⁻³ NaOH(aq) react with 23.2 cm³ of $HNO_3(aq)$.
$$HNO_3(aq) + NaOH(aq) \longrightarrow NaNO_3(aq) + H_2O(l)$$
Find the concentration of the nitric acid, HNO_3.

2 25.0 cm³ of 1.00 mol dm⁻³ KOH(aq) react with 26.6 cm³ of $H_2SO_4(aq)$.
$$H_2SO_4(aq) + 2KOH(aq) \longrightarrow K_2SO_4(aq) + 2H_2O(l)$$
Find the concentration of the sulphuric acid, H_2SO_4.

2 0.470 mol dm⁻³.
1 0.539 mol dm⁻³.

Sample question and model answer

The determination of reacting quantities using the Mole Concept is one of the most important concepts in chemistry. This can be tested in exam questions almost anywhere and it underpins much of the content of a chemistry course at this level. To succeed at AS Chemistry, it is essential that you grasp this concept.

A tank contained 4000 dm^3 of waste hydrochloric acid. It was decided to neutralise the acid by adding slaked lime, $Ca(OH)_2$.

(a) The concentration of the acid was first determined by titration of a 25.0 cm^3 sample against 0.121 M sodium hydroxide of which 32.4 cm^3 were required.

(i) Calculate the concentration, in mol dm^{-3}, of the hydrochloric acid in the sample.

$$\text{no. of moles of NaOH} = \frac{0.121 \times 32.4}{1000} = 3.92 \times 10^{-3} \checkmark$$

$$NaOH + HCl \longrightarrow NaCl + H_2O$$

1 mol NaOH reacts with 1 mol HCl

$\therefore 3.92 \times 10^{-3}$ mol NaOH reacts with 3.92×10^{-3} mol HCl $\checkmark$

25.0 cm^3 HCl(aq) contains 3.92×10^{-3} mol HCl

1 cm^3 HCl(aq) contains $\dfrac{3.92 \times 10^{-3}}{25}$ mol HCl

1 dm^3 (1000 cm^3) HCl(aq) contains $\dfrac{3.92 \times 10^{-3}}{25} \times 1000 = 0.157$ mol HCl.

$\therefore$ concentration of HCl(aq) is 0.157 mol dm^{-3} $\checkmark$

Watch out for 'scaling'. This is often missed, even by the best candidates. Read all parts of a question thoroughly to see if you have missed anything.

(ii) Calculate the total number of moles of HCl in the tank.

Total number of moles of HCl in the tank = $4000 \times 0.157 = 628$ $\checkmark$

[4]

(b) Calculate the mass, in kg, of slaked lime required to neutralise the acid. Slaked lime reacts with hydrochloric acid according to the equation shown below:

$$Ca(OH)_2 + 2HCl \longrightarrow CaCl_2 + 2H_2O$$

1 mol $Ca(OH)_2$ reacts with 2 mol HCl

$\therefore \dfrac{628}{2}$ mol $Ca(OH)_2$ reacts with 628 mol HCl $\checkmark$

Molar mass of $Ca(OH)_2 = 40 + (16 + 1)2 = 74$ g mol^{-1} $\checkmark$

Mass of $Ca(OH)_2 = \dfrac{628}{2} \times 74 = 23200$ g = 23.2 kg (to 3 significant figures) $\checkmark$

Notice that the same ideas are used here.

The correct reacting quantities from the equation are essential.

[3]

(c) The slaked lime was manufactured by roasting limestone and then adding water:

$$CaCO_3 \longrightarrow CaO + CO_2$$
$$CaO + H_2O \longrightarrow Ca(OH)_2$$

Calculate the mass of limestone which is required to produce 1 kg of slaked lime.

In 1 kg $Ca(OH)_2$, number of moles of $Ca(OH)_2 = \dfrac{1 \times 10^3}{74} = 13.5$ $\checkmark$

From the equations, 1 mol $CaCO_3 \longrightarrow$ 1 mol $CaO \longrightarrow$ 1 mol $Ca(OH)_2$

$\therefore$ number of moles of $CaCO_3 = 13.5$

$\therefore$ Mass of $CaCO_3 = 13.5 \times 100 = 13500$ g = 1.35 kg $\checkmark$

It is important to always show full working. If you do make a mistake early on in a calculation, you will only be penalised once if the method that follows is sound.

[2]

[Total: 9]

NEAB Q3 Atomic structure, bonding and periodicity Feb 1996 (modified)

Practice examination questions

1 (a) When 0.25 g of sodium metal was added to 200 cm³ (an excess) of water, the following reaction occurred.

$$Na(s) + H_2O(l) \longrightarrow NaOH(aq) + \tfrac{1}{2}H_2(g)$$

    (i) Calculate the number of moles of sodium taking part in the reaction.

    (ii) Calculate the molarity of the sodium hydroxide solution which was formed.

    (iii) Calculate the volume of hydrogen gas produced at r.t.p. (298 K and 100 kPa). Assume that hydrogen is insoluble in water under these conditions.

                      [6]

(b) In another experiment 25.0 cm³ of 0.183 M sodium hydroxide were neutralised by 13.7 cm³ of sulphuric acid according to the following equation.

$$2NaOH(aq) + H_2SO_4(aq) \longrightarrow Na_2SO_4(aq) + 2H_2O(l)$$

Calculate the molarity of the sulphuric acid.

                      [3]
         [Total: 9]

NEAB Atomic Structure, Bonding and Periodicity Q3 Feb 1997 (modified)

2 50 kg of pure sulphuric acid were accidentally released into a lake when a storage vessel leaked. Two methods were proposed to neutralise it.

(a) The first proposal was to add a solution of 5 M NaOH to the lake. Sodium hydroxide reacts with sulphuric acid as follows:

$$2NaOH(aq) + H_2SO_4(aq) \longrightarrow Na_2SO_4(aq) + 2H_2O(l)$$

Calculate the volume of 5 M NaOH required to neutralise the sulphuric acid by answering the following questions.

    (i) How many moles of sulphuric acid are there in 50 kg of the acid?

    (ii) How many moles of sodium hydroxide are required to neutralise this acid?

    (iii) Calculate the volume, in dm³, of 5 M NaOH which contains this number of moles.

                      [4]

(b) The second proposal was to add powdered calcium carbonate which reacts as follows:

$$CaCO_3(s) + H_2SO_4(aq) \longrightarrow CaSO_4(s) + H_2O(l) + CO_2(g)$$

Calculate the mass of calcium carbonate required to neutralise 50 kg of sulphuric acid.

                      [3]

(c) Suggest **two** reasons why the addition of calcium carbonate to neutralise the acid is the preferred method in practice.

                      [2]
         [Total: 9]

NEAB Atomic Structure, Bonding and Periodicity Q3 Feb 1995

3 Azides are compounds of nitrogen, used mainly as detonators in explosives. However, sodium azide, NaN_3, decomposes non-explosively on heating to release nitrogen gas. This provides a convenient method of obtaining pure nitrogen in the laboratory.

$$2NaN_3(s) \longrightarrow 2Na(l) + 3N_2(g)$$

(a) A student prepared 1.80 dm³ of pure nitrogen in the laboratory by this method. This gas volume was measured at room temperature and pressure (r.t.p.).

   (i) How many moles of nitrogen, N_2, did the student prepare?
   [Assume that 1 mole of gas molecules occupies 24.0 dm³ at r.t.p.]

   (ii) What mass of sodium azide did the student heat? [3]

(b) After cooling, the student obtained 1.15 g of solid sodium. She then carefully reacted this sodium with water to form 25.0 cm³ of aqueous sodium hydroxide:

$$2Na(s) + 2H_2O(l) \longrightarrow 2NaOH(aq) + H_2(g)$$

Calculate the concentration, in mol dm⁻³, of the aqueous sodium hydroxide. [2]

(c) Liquid sodium has been used as a coolant in nuclear reactors.

   (i) Suggest **one** property of sodium that is important for this use.

   (ii) Suggest **one** particular hazard of sodium in this use. [2]

[Total: 7]

Cambridge Chemistry Foundation Q5 March 1998

Chapter 3

Bonding and structure

The following topics are covered in this chapter:

- Chemical bonding
- Shapes of molecules
- Electronegativity, polarity and polarisation
- Intermolecular forces
- Bonding, structure and properties

3.1 Chemical bonding

After studying this section you should be able to:

- appreciate that compounds often contain atoms or ions with electron structures of a noble gas
- understand the nature of an ionic bond in terms of electrostatic attraction between ions
- determine the formula of an ionic compound from the ionic charges
- understand the nature of a covalent bond and a dative covalent (coordinate) bond in terms of the sharing of an electron pair
- construct 'dot-and-cross' diagrams to demonstrate ionic and covalent bonding
- describe metallic bonding

LEARNING SUMMARY

Why do atoms bond together?

AQA	M1	OCR	M1
EDEXCEL	M1	SALTERS	M1
NICCEA	M1	WJEC	CH1
NUFFIELD	M1		

The Octet Rule

Under normal conditions, the only elements that can exist as single atoms are the Noble Gases in Group 0 of the Periodic Table. The atoms of all other elements are bonded together. To find out why, we need to look at the electronic configurations of the atoms of the Noble Gases.

The Noble Gases get their name from their unreactivity.

In the atoms of a noble gas:

- all electrons are paired,
- the bonding shells are full.

> Atoms of the Noble Gases are stable because all their electrons are paired.

He atom
$1s^2$

Ne atom
$1s^2 2s^2 2p^6$

Ar atom
$1s^2 2s^2 2p^6 3s^2 3p^6$

This arrangement of electrons is particularly stable and is often referred to as a stable 'octet' (as in neon and argon).

> This tendency to acquire a noble gas electron structure is often refer to as the 'Octet Rule'.

Apart from the Noble Gases, atoms are bonded together. Thus the elements oxygen and nitrogen exists as O_2 and N_2. Compounds exist when atoms of different elements bond together as in CO_2, CO and NaCl. In all these examples, electrons have been shared, transferred or rearranged in such a way that the atoms of the elements have acquired a noble gas electron structure. If this creates an outer shell of 8 electrons (as in all these cases), the 'Octet Rule' is obeyed. All these themes are discussed in the sections that follow.

Types of bonding

Many mistakes are made in chemistry by confusing the type of bonding.

Covalent and ionic compounds are bonded differently and they behave differently.

You cannot apply 'ionic bonding ideas' to a compound that is covalent.

Chemical bonds are classified into three main types: ionic, covalent and metallic.

- **Ionic** bonding occurs between the atoms of a **metal** and a **non-metal**. For example NaCl, MgO, Fe_2O_3.
- **Covalent** bonding occurs between the atoms of **non-metals**. For example O_2, H_2, H_2O, diamond and graphite.
- **Metallic** bonding occurs between the atoms of **metals**. For example all metals such as iron, zinc, aluminium, etc.; alloys as in brass (copper and zinc), bronze (copper and tin).

Always decide the type of bonding before:
- drawing a 'dot-and-cross' diagram
- predicting how a compound reacts
- predicting the properties of a compound.

KEY POINT

Ionic bonding

AQA	M1	OCR	M1
EDEXCEL	M1	SALTERS	M1
NICCEA	M1	WJEC	CH1
NUFFIELD	M1		

Ionic bonds

An ionic bond is formed when electrons are **transferred** from a metal atom to a non-metal atom forming **oppositely-charged ions**.

Ionic bonds are present in a compound of a metal and a non-metal.

An ionic bond is the electrical attraction between oppositely charged ions.

KEY POINT

Example: sodium chloride, NaCl

Sodium chloride, NaCl, is formed by transfer of one electron from a sodium atom ($[Ne]3s^1$) to a chlorine atom ($[Ne]3s^23p^5$) with the formation of ions:

For showing chemical bonding, 'dot-and-cross' diagrams' often only show the outer shell.

electron transfer

ionic bond forms:
attraction between ions

Na atom   Cl atom Na^+ ion Cl^- ion

A sodium atom has lost an electron to acquire the electron structure of neon:

$$Na \longrightarrow Na^+ + e^-$$
$$[Ne]3s^1 \longrightarrow [Ne] + e^-$$

The ions that are formed have stable noble gas electron structures with full outer electron shells.

A chlorine atom has gained an electron to acquire the electron structure of argon:

$$Cl + e^- \longrightarrow Cl^-$$
$$[Ne]3s^23p^5 + e^- \longrightarrow [Ne]3s^23p^6 \text{ or } [Ar]$$

Example: magnesium chloride, $MgCl_2$

A magnesium atom ($[Ne]3s^2$) has two electrons in its outer shell. One electron is transferred to each of two chlorine atoms ($[Ne]3s^23p^5$) forming ions.

The Mg^{2+} and Cl^- ions have noble gas electronic configurations.

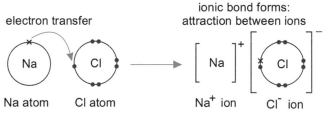

transfer of 2 electrons

ionic bond forms:
attraction between ions

Cl atom   Mg atom   Cl atom Cl^- ion Mg^{2+} ion Cl^- ion

Giant ionic lattices

Each ion is surrounded by oppositely-charged ions, forming a giant ionic lattice.

- Although it is convenient to look at ionic bonding between two ions only, each ion is able to attract oppositely charged ions in all directions.
- This results in a **giant ionic lattice** structure comprising hundreds of thousands of ions (depending upon the size of the crystal).
- This arrangement is characteristic of all ionic compounds.

Part of the sodium chloride lattice

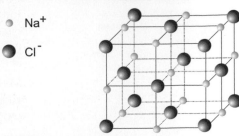

- Na$^+$
- Cl$^-$

- Each Na$^+$ ion surrounds 6 Cl$^-$ ions
- Each Cl$^-$ ion surrounds 6 Na$^+$ ions

Ionic charges and the Periodic Table

An ionic charge can be predicted from an element's position in the Periodic Table.

The diagram below shows elements in Periods 2 and 3 of the Periodic Table.

group	1	2	3	4	5	6	7	0
number of outer shell electrons	1	2	3	4	5	6	7	8
element	Li	Be	B	C	N	O	F	Ne
ion	Li$^+$				N^{3-}	O^{2-}	F$^-$	
element	Na	Mg	Al	Si	P	S	Cl	Ar
ion	Na$^+$	Mg^{2+}	Al^{3+}		P^{3-}	S^{2-}	Cl$^-$	

Remember that any metal ions will be positive and non-metal ions negative.

- Metals in Groups 1–3 **lose** sufficient electrons from their atoms to form ions with the electron configuration of the **previous** noble gas;
- Non-metals in Groups 5–7 **gain** sufficient electrons to their atoms to form ions with the electron configuration of the **next** noble gas.
- Be, B, C and Si do not normally form ionic compounds because of the large amount of energy required to transfer electrons forming their ions.

Groups of covalent-bonded atoms can also lose or gain electrons to give ions such as those shown below.

1+	1–		2–		3–	
ammonium NH$_4^+$	hydroxide	OH$^-$	carbonate	CO$_3^{2-}$	phosphate	PO$_4^{3-}$
	nitrate	NO$_3^-$	sulphate	SO$_4^{2-}$		
	nitrite	NO$_2^-$	sulphite	SO$_3^{2-}$		
	hydrogen-carbonate	HCO$_3^-$				

Learn these

Predicting ionic formulae

Although an ionic compound comprises charged ions, its overall charge is zero.

In an ionic compound,

total number of positive charges from positive ions = total number of negative charges from negative ions.

KEY POINT

Working out an ionic formula from ionic charges

calcium chloride:	ion	charge	aluminium sulphate:	ion	charge
equalise charges:	Ca²⁺	2+	equalise charges:	Al³⁺	3+
				Al³⁺	3+
	Cl⁻	1–		SO₄²⁻	2–
	Cl⁻	1–		SO₄²⁻	2–
				SO₄²⁻	2–
total charge must be zero:		0	total charge must be zero:		0
formula:		CaCl₂	formula:		Al₂(SO₄)₃

Covalent bonding

AQA	M1	OCR	M1
EDEXCEL	M1	SALTERS	M1
NICCEA	M1	WJEC	CH1
NUFFIELD	M2		

Covalent bonds

A covalent compound comprises molecules: groups of atoms held together by covalent bonds.

> **KEY POINT**
> A molecule is the smallest part of a covalent compound that can take part in a chemical reaction.

A covalent bond is formed when electrons are **shared** rather than transferred.

> **KEY POINT**
> A covalent bond is a shared pair of electrons.

> Covalent bonds are present in a compound of two non-metals.

Example: hydrogen, H₂

The covalent bond in a hydrogen molecule H_2 forms when two hydrogen atoms $(1s^1)$ bond together, each contributing 1 electron to the bond:

> A single covalent bond is written as A–B.

In forming the covalent bond, each hydrogen atom in the H_2 molecule has:

- control of 2 electrons – one of its own and the second from the other atom
- the electron configuration of the noble gas helium.

When a covalent bond forms, each atom *often* acquires a stable noble gas electronic configuration with a full outer bonding shell.

This diagram shows how two atoms are bonded together by a covalent bond.

Unlike an ionic bond, a covalent bond is *directional* and acts solely between the two atoms involved in the bond. Thus hydrogen above exists simply as H_2 molecules.

The diagrams below show examples of covalent bonding in some simple molecules.

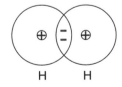

In a hydrogen molecule, the nucleus of each hydrogen atom is attracted towards the electron pair of the covalent bond.

> Notice that each atom contributes one electron to the covalent bond.

Cl₂

H₂O

NH₃

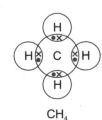

CH₄

Multiple covalent bonds

A covalent bond in which **one** pair of electrons is shared is known as a **single bond**, e.g. Cl_2, written as Cl–Cl.

Atoms can also share more than one pair of electrons to form a **multiple bond**.

- Sharing of **two** pairs of electrons forms a **double bond**, e.g. O_2, written as O=O.
- Sharing of **three** pairs of electrons forms a **triple bond**, e.g. N_2, written as N≡N.

O_2 N_2

Dative covalent bonds

> A dative covalent bond forms when the shared pair of electrons comes from just **one** of the bonded atoms.

A dative covalent or coordinate bond is one in which **one** of the atoms supplies **both** the shared electrons to the covalent bond. A dative covalent bond is written as A→B, where the direction of the arrow shows the direction in which the electron pair is donated.

Example: ammonium ion, NH_4^+

- The ammonium ion, NH_4^+, forms when ammonia NH_3 bonds with a proton H^+, with both the bonding electrons coming from NH_3.
- Note that the ammonium ion has three covalent bonds and one dative covalent bond:

> The involvement of the electron pair as a dative covalent bond is often shown as an arrow:
>
>
>
> Note that once NH_4^+ has formed, all 4 bonds are equivalent and you cannot tell which was formed by a dative covalent bond.

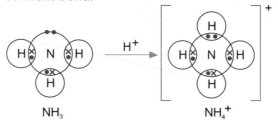

NH_3 NH_4^+

> In exams, these are easy marks provided that the facts have been learnt.

The Octet Rule doesn't always work

> The Octet Rule is useful in chemistry but it can be broken. It is probably more important that all unpaired electrons are paired.

When covalent bonds form, unpaired electrons **often** pair up to form a noble gas electronic configuration according to the Octet Rule (see page 44). Many molecules, however, form electronic configurations that do not form an octet. Some molecules have 6, 10, 12 or even 14 electrons in the outer bonding shell. These compounds form by the pairing of unpaired electrons.

Phosphorus forms two chlorides, PCl_3 and PCl_5.

PCl_3 obeys the Octet Rule.

In PCl_5,

- the outer shell of phosphorus has expanded allowing all 5 electrons to be used in bonding
- the octet rule is broken and phosphorus now has 10 electrons in its outer shell:

> In PCl_3, phosphorus is using 3 valence electrons.
>
> In PCl_5, phosphorus is using 5 valence electrons (the same as the Group number of phosphorus).

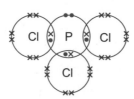   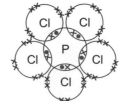

PCl_3: only 3 electrons from the outer bonding shell of phophorus are used in bonding

PCl_5: all 5 electrons from the outer bonding shell of phosphorus are used in covalent bonds.

An electron that takes part in forming chemical bonds is called a **valence electron**.

For elements in the s-block and p-block of the Periodic Table, the maximum number of valence electrons available is usually equal to the group number.

Metallic bonding

AQA	M1	OCR	M1
EDEXCEL	M1	SALTERS	M1
NICCEA	M1	WJEC	CH1
NUFFIELD	M1		

A metallic bond holds atoms together in a solid metal or alloy.

In solid metals, the atoms are ionised.

- The positive ions occupy fixed positions in a lattice.
- The outer shell electrons are **delocalised** – they are spread throughout the metallic structure and are able to move freely throughout the lattice.

> A **metallic bond** is the electrostatic attraction between the positive metal ions and delocalised electrons.

KEY POINT

A metallic lattice is held together by electrostatic attraction between positive metal ions and electrons.

In the metallic lattice, each metal atom exists as + ions by releasing its outer valence electrons to the delocalised pool of electrons (often called a 'sea of electrons').

The 'sea of electrons'

Delocalised and localised electrons

The **delocalised** electrons in metals are spread throughout the metal structure. The delocalisation is such that electrons are able to move throughout the structure and it is impossible to assign any electron to a particular positive ion.

In a covalent bond, the **localised** pair of electrons is always positioned between the two atoms involved in the bond. The electron charge is much more concentrated.

Progress check

1 Draw 'dot-and-cross' diagrams of (a) MgO; (b) Na$_2$O.

2 Using the information on page 46, predict formulae for the following ionic compounds:
   - (a) lithium chloride
   - (b) potassium sulphide
   - (c) lithium nitride
   - (d) aluminium oxide
   - (e) sodium carbonate
   - (f) calcium hydroxide
   - (g) aluminium sulphate
   - (h) ammonium phosphate.

3 Draw 'dot-and-cross' diagrams for
   (i) C$_2$H$_6$; (ii) HCN; (iii) H$_3$O$^+$.

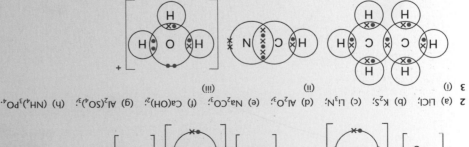

3 (i)

2 (a) LiCl; (b) K$_2$S; (c) Li$_3$N; (d) Al$_2$O$_3$; (e) Na$_2$CO$_3$; (f) Ca(OH)$_2$; (g) Al$_2$(SO$_4$)$_3$; (h) (NH$_4$)$_3$PO$_4$.

1 (a)

3.2 Shapes of molecules

After studying this section you should be able to:

- *use electron-pair repulsion to explain the shapes and bond angles of simple molecules*
- *understand that lone pairs have a larger repulsive effect than bonded pairs leading to distortion of the molecular shape*
- *understand how multiple bonds affect the shape of a molecule*

LEARNING SUMMARY

Electron-pair repulsion theory

AQA	M1	OCR	M1
EDEXCEL	M1	SALTERS	M1
NICCEA	M1	WJEC	CH1
NUFFIELD	M2		

Electron-pair repulsion theory states that:

- the shape of a molecule depends upon the number of electron pairs surrounding the central atom
- electron pairs repel one another and move as far apart as possible.

The shape of any molecule can be predicted by applying this theory.

Molecules with bonded pairs

> Learn these shapes and bond-angles.

You can predict the shape of a molecule from its '**dot-and-cross**' diagram. Try to relate the number of electron pairs in the examples below with the three-dimensional shape of each molecule.

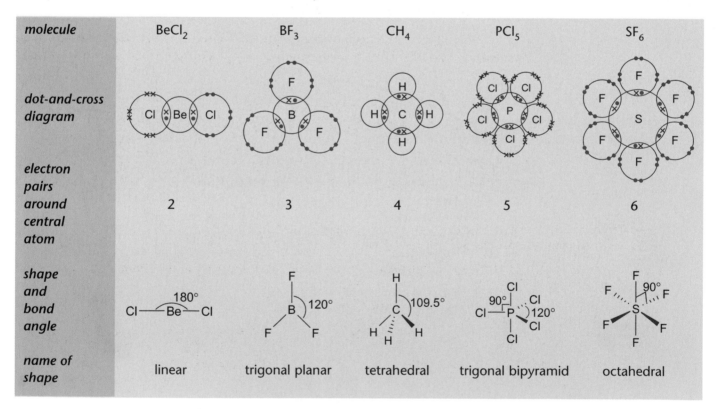

molecule	$BeCl_2$	BF_3	CH_4	PCl_5	SF_6
dot-and-cross diagram	Cl–Be–Cl	F,B,F	H,C,H	Cl,P,Cl	F,S,F
electron pairs around central atom	2	3	4	5	6
shape and bond angle	180° Cl—Be—Cl	120°	109.5°	90° / 120°	90°
name of shape	linear	trigonal planar	tetrahedral	trigonal bipyramid	octahedral

Molecules with lone pairs

Each of the examples below are molecules with four electron pairs surrounding the central atom. The molecular shape will therefore be based upon a tetrahedron. However, a lone pair is closer to an atom than a bonded pair of electrons and has a larger repulsive effect than a bonded pair.

> **KEY POINT**
>
> The relative magnitudes of electron-pair repulsions are:
> **lone-pair/lone-pair > bonded-pair/lone-pair > bonded-pair/bonded-pair.**

Lone pairs distort the shape of a molecule and reduce the bond angle. Look at the examples below and notice how each lone pair decreases the bond-angle by 2.5°.

Always draw 3-D shapes – you may be penalised in exams otherwise.

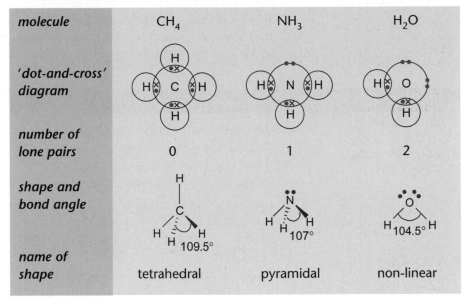

molecule	CH_4	NH_3	H_2O
'dot-and-cross' diagram			
number of lone pairs	0	1	2
shape and bond angle	109.5°	107°	104.5°
name of shape	tetrahedral	pyramidal	non-linear

Molecules with double bonds

In molecules containing multiple bonds, each double bond is treated in the same way as a bonded pair. In the diagram of carbon dioxide below, each double bond is treated as a *'bonding region'*.

molecule	'dot-and-cross' diagram	number of bonding regions	shape and bond angle	name of shape
CO_2	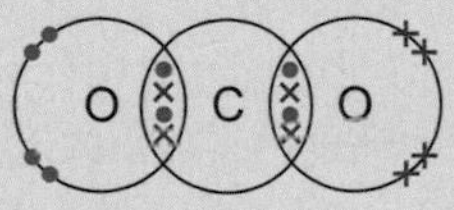	2	 180° $O=C=O$	linear

Progress check

1 For each of the following molecules predict the shape and bond angles.
  (a) BeF_2
  (b) $AlCl_3$
  (c) SiH_4
  (d) H_2S
  (e) PH_3
  (f) CS_2
  (g) SO_3
  (h) SO_2

1 (a) linear, 180°
(b) trigonal planar, 120°
(c) tetrahedral, 109.5°
(d) non-linear, 104.5°
(e) pyramidal, 107°
(f) linear, 180°
(g) trigonal planar, 120°.
(h) non-linear, 120°.

3.3 Electronegativity, polarity and polarisation

LEARNING SUMMARY

After studying this section you should be able to:

- appreciate that many chemical bonds have bonding intermediate between ionic and covalent bonding
- describe electronegativity in terms of attraction for bonding electrons
- understand the nature of polarity in molecules of covalent compounds
- understand the nature of polarisation in ionic compounds

Ionic or covalent?

AQA	M1	OCR	M1
EDEXCEL	M1	SALTERS	M2
NICCEA	M1	WJEC	CH1
NUFFIELD	M2		

> Between the extremes of ionic and covalent bonding, there is a whole range of intermediate bonds, which have both ionic and covalent contributions.

An ionic bond with 100% ionic character would require the complete transfer of an electron from a metal atom to a non-metal atom. In practice, this never completely happens, but a compound such as caesium fluoride Cs^+F^- forms an ionic bond in which the electron transfer is nearly complete with close to 100% ionic character.

In a hydrogen molecule H_2, the two hydrogen atoms are identical and form a covalent bond with 100% covalent character by equally sharing the bonded pair of electrons. However, in molecules such as HCl, the bonded pair of electrons is not shared equally.

An ionic bond is often formed with some covalent character:
- the electron transfer is incomplete
- there is a degree of electron sharing.

A covalent bond is often formed with some ionic character:
- the electrons are not equally shared
- there is a degree of electron transfer.

The sections that follow discuss the factors that lead to intermediate bonding.

Electronegativity

AQA	M1	OCR	M1
EDEXCEL	M1	SALTERS	M2
NICCEA	M1	WJEC	CH1
NUFFIELD	M2		

The nuclei of the atoms in a molecule attract the electron pair in a covalent bond.

> **KEY POINT**
>
> Electronegativity is a measure of the attraction of an atom in a molecule for the pair of electrons in a covalent bond.

> The most electronegative atoms attract bonding electrons most strongly.

- In general, small atoms are electronegative atoms.
- The most electronegative atoms are those of highly reactive non-metallic elements (such as O, F and Cl).
- Reactive metals (such as Na and K) have the least electronegative atoms.

How is electronegativity measured?

> Notice that the Noble Gases are not included. Although neon and helium atoms are smaller than those of fluorine, they do not form bonds and so they have no affinity for bonded electrons.

> See how the Pauling values of electronegativity relate to the element's position in the Periodic Table.

The 'Pauling scale' is often used to compare the electronegativities of different elements. The diagram below shows how the electronegativity of an element relates to its position in the Periodic Table. The numbers give the Pauling electronegativity values.

electronegativity increases						
Li	Be	B	C	N	O	F
1.0	1.5	2.0	2.5	3.0	3.5	4.0
Na						Cl
0.9						3.0
K						Br
0.8						2.8

Fluorine, the most electronegative element has:

- an electronegativity of 4.0
- small atoms with a greater attraction than larger atoms for the pair of electrons in a covalent bond.

> The greater the **difference** between electronegativities, the greater the **ionic** character of the bond.
>
> The greater the **similarity** in electronegativities, the greater the **covalent** character of the bond.

KEY POINT

Polar and non-polar molecules

AQA	M1	OCR	M1
EDEXCEL	M1	SALTERS	M2
NICCEA	M1	WJEC	CH1
NUFFIELD	M2		

Non-polar bonds

A covalent bond is non-polar when:

- the bonded electrons are shared **equally** between both atoms
- the bonded atoms are the same
- the bonded atoms have similar electronegativities.

When bonding atoms are the same, the attraction for the bonded pair of electrons is the same and the bond is non-polar.

A covalent bond **must** be non-polar if the bonded atoms are the same, as in molecules of H_2 and Cl_2 shown here:

H_2 molecule Cl_2 molecule

Polar bonds

A covalent bond is polar when:

- the electrons in the bond are shared **unequally** making a *polar bond*
- the bonded atoms are different, each has a different electronegativity.

In a polar covalent bond, the bonded electrons are attracted towards the more electronegative of the atoms.

The bond in hydrogen chloride, H–Cl, is predominantly covalent with one pair of electrons shared between the two atoms. However, the chlorine atom is **more electronegative** than the hydrogen atom. The chlorine atom has a greater attraction for the bonded pair of electrons than the hydrogen atom making the covalent bond polar.

The electronegativities of hydrogen halides are discussed in more detail in p.76.

bonded pair of electrons
attracted closer to chlorine

The molecule is *polarised* with a small positive charge $\delta+$ on the hydrogen atom and a small negative charge $\delta-$ on the chlorine atom.

The hydrogen chloride molecule is *polar* with a *permanent dipole*.

Symmetrical and unsymmetrical molecules

If the molecule is symmetrical, any dipoles will cancel and the molecule will not have a permanent dipole. The diagram below shows why the symmetrical molecule, CCl_4, is non-polar.

- Each C–Cl bond is polar but the dipoles act in different directions.
- The overall effect is for the dipoles to cancel each other.
- ∴ CCl_4 is a non-polar molecule.

Dipoles in symmetrical molecules cancel.

CCl_4 is non-polar

Although each C–Cl bond is polar, the dipoles cancel

Polarisation in ionic compounds

| AQA | M1 | OCR | M1 |
| EDEXCEL | M2 | WJEC | CH1 |

A 100% ionic bond would only form by complete transfer of electrons from a metallic atom to a non-metallic atom. Many ionic compounds have a degree of covalency resulting from incomplete transfer of electrons.

This effect, called *polarisation*, occurs in ionic compounds with:

- a small + ion (cation) with a high charge density
- a large – ion (anion) with a low charge density.

Charge/size ratio increases

Cations and Anions

During electrolysis, the negative electrode or *cathode* attracts positive ions, called **cations**.

The positive electrode or *anode* attracts negative ions, called **anions**.

Polarisation will be greatest when a small, highly-charged cation polarises a large anion.

Polarisation is most apparent with a small, densely charged 3+ cation.

The electric field around a **small cation** distorts the electron shells around a **large anion**. As a result, electrons are attracted towards the cation, giving a degree of covalency (electron sharing).

$$\text{charge density} = \frac{\text{charge}}{\text{ionic radius}}$$

Notice how the charge density increases as the ionic charge increases:

ion	Na^+	Mg^{2+}	Al^{3+}
charge	1+	2+	3+
radius/nm	0.102	0.072	0.053
charge density	10	28	57

Note the polarisation between Al^{3+} and O^{2-} – a 'perfect' ionic compound would contain spherical ions. Here, the O^{2-} ion is distorted.

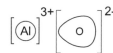

Small Al^{3+} ion is highly charged.
O^{2-} ion is distorted and polarised resulting in partial covalent character.

Progress check

1 Decide whether each molecule is polar and state which atom, if any, has the δ– charge.
   (a) Br_2;   (b) H_2O;   (c) O_2;   (d) HBr;   (e) NH_3.

2 (a) Predict the shape of a molecule of BF_3 and of PF_3.
   (b) Explain why BF_3 is non-polar whereas PF_3 is polar.

2 (a) BF_3: trigonal planar; PF_3: pyramidal.
 (b) BF_3 has a symmetrical shape. Although each bond is polar ($F^{\delta-}$), the dipoles cancel. PF_3 is not symmetrical. The F atoms are all on the same side of the molecule resulting in a permanent dipole.

1 (a) non-polar; (b) polar, $O^{\delta-}$; (c) non-polar; (d) polar, $Br^{\delta-}$; (e) polar, $N^{\delta-}$.

3.4 Intermolecular forces

After studying this section you should be able to:

- *understand the nature of van der Waals' forces (induced dipole-dipole interactions)*
- *understand the nature of permanent dipole-dipole interactions and hydrogen bonds*
- *describe the anomalous properties of water arising from hydrogen bonding.*

LEARNING SUMMARY

Van der Waals' forces

AQA	M1	OCR	M1
EDEXCEL	M1	SALTERS	M2
NICCEA	M1	WJEC	CH1
NUFFIELD	M2		

Van der Waals' forces (induced dipole-dipole interactions) exist between all molecules whether polar or non-polar.

Induced dipole-dipole interactions are often referred to as van der Waals' forces.

What causes van der Waals' forces?

Van der Waals' forces are:

- weak intermolecular forces
- caused by attractions between very small dipoles in molecules.

Ionic and covalent bonds are of comparable strength.
Intermolecular forces are far weaker.

The diagram below shows how these forces arise between single atoms in a noble gas.

Movement of electrons produces an oscillating dipole

dipole oscillates and continually changes with time

An oscillating or instantaneous dipole is caused by an uneven distribution of electrons at an instant of time.

Oscillating dipole induces a dipole in a neighbouring molecule which is induced onto further molecules

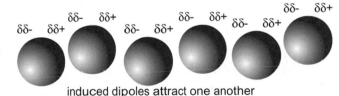

induced dipoles attract one another

What affects the strength of van der Waals' forces?

Van der Waals' forces result from interactions of electrons between molecules.

The greater the number of electrons in each molecule:

- the larger the oscillating and induced dipoles
- the greater the attractive forces between molecules
- the greater the van der Waals' forces.

Van der Waals' forces increase in strength with increasing number of electrons

This can be seen by comparing the boiling points of the Noble Gases.

Noble gas	b. pt./°C	number of electrons	trend
He	−269	2	
Ne	−246	10	• easier to distort electron clouds
Ar	−186	18	
Kr	−152	36	• induced dipoles increase
Xe	−107	54	
Rn	− 62	86	• boiling point increases

Permanent dipole-dipole interactions

AQA	M1	OCR	M1
EDEXCEL	M1	SALTERS	M2
NICCEA	M1	WJEC	CH1
NUFFIELD	M2		

Permanent dipole-dipole interactions are 'non-directional'.

They are simply weak attractions between dipole charges on different molecules.

The small δ+ and δ– charges on a polar molecule attract oppositely charged dipoles on another polar molecule.

This gives a weak *intermolecular force* called a permanent dipole-dipole interaction.

Example: Intermolecular forces between HCl molecules

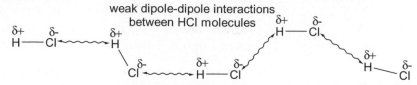

Between hydrogen chloride molecules, there will be both:

- van der Waals' forces and
- permanent dipole-dipole interactions.

Although both are weak forces, the permanent dipole-dipole interactions are still stronger than the van der Waals' forces.

Hydrogen bonds

AQA	M1	OCR	M1
EDEXCEL	M1	SALTERS	M2
NICCEA	M1	WJEC	CH1
NUFFIELD	M2		

A hydrogen bond is a special type of permanent dipole-dipole interaction found between molecules containing the following groups:

A hydrogen bond is a comparatively strong intermolecular attraction between:

- an electron deficient hydrogen atom, $H^{\delta+}$, on one molecule and
- a lone pair of electrons on a highly electronegative atom of F, O or N on another molecule.

Hydrogen bonding occurs between molecules such as H_2O:

Note the role of the lone pair – this is essential in hydrogen bonding.

A hydrogen bond is shown between molecules as a dashed line.

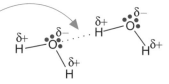
hydrogen bond formed by attraction between dipole charges on different molecules

Hydrogen bonding is especially important in organic compounds containing –OH or –NH bonds: e.g. alcohols, carboxylic acids, amines, amino acids.

Special properties of water arising from hydrogen bonding

A hydrogen bond has only one-tenth the strength of a covalent bond. However, hydrogen bonding is strong enough to have significant effects on physical properties, resulting in some unexpected properties for water.

The solid (ice) is less dense than the liquid (water)

Solids are usually denser than liquids – but ice is less dense than water.

- Particles in solids are **usually** packed closer together than in liquids.
- Hydrogen bonds hold water molecules apart in an open lattice structure.

Therefore ice is less dense than water.

When water changes state, the covalent bonds between the H and O atoms in an H_2O molecule are strong and do **not** break.

It is the intermolecular forces that break.

The diagram below shows how the open lattice of ice collapses on melting.

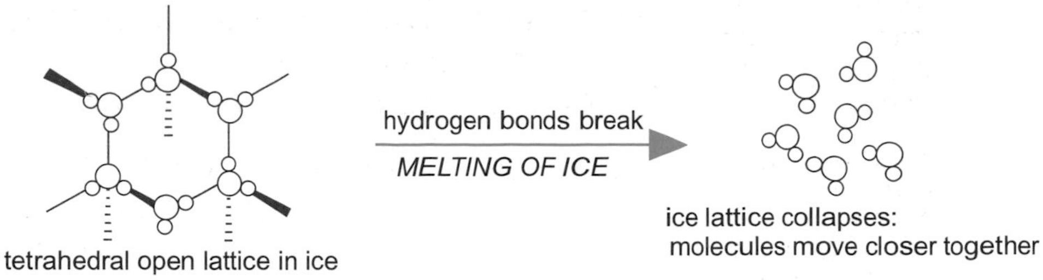

MELTING OF ICE
hydrogen bonds break

tetrahedral open lattice in ice

ice lattice collapses: molecules move closer together

Ice has a relatively high melting point, and water a relatively high boiling point

- There are relatively strong hydrogen bonds between H_2O molecules.
- The hydrogen bonds are extra forces over and above van der Waals' forces.
- These extra forces result in higher melting and boiling points than would be expected from just van der Waals' forces.
- When the ice lattice breaks, hydrogen bonds are broken.

Other properties

The extra intermolecular bonding from hydrogen bonds also explains the relatively high surface tension and viscosity in water.

Progress check

1 Which of the following molecules have hydrogen bonding: H_2O, H_2S, CH_4, CH_3OH, NO_2?

2 Draw diagrams showing hydrogen bonding between:
(a) 2 molecules of ammonia
(b) 1 molecule of water and 1 molecule of ethanol.

(b)

(a) 2

1 H_2O, CH_3OH.

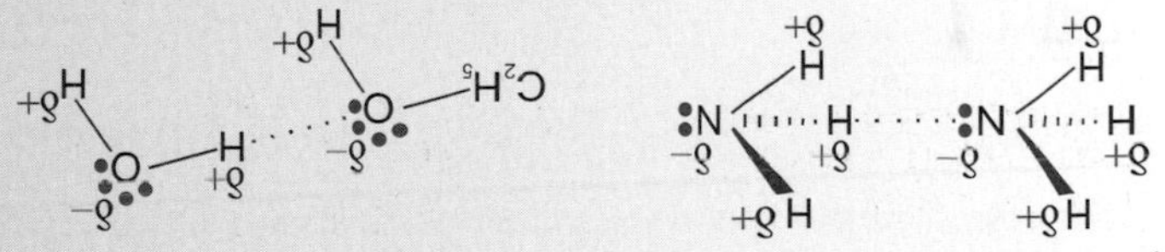

3.5 Bonding, structure and properties

After studying this section you should be able to:

LEARNING SUMMARY

- describe the typical properties of an ionic compound in terms of its structure
- describe the typical properties of a covalent compound in terms of simple molecular and giant molecular structures
- describe the structure and associated properties of diamond and graphite
- understand the properties of metals in terms of metallic bonding

Bonds and forces

AQA	M1	OCR	M1
EDEXCEL	M1	SALTERS	M2
NICCEA	M1	WJEC	CH1
NUFFIELD	M2		

These are very important and are needed to understand the links between bonding, structure and properties.

Bonding, structure and properties are all related.

A structure is held together by bonds and forces. The different types of bonds introduced in the last section are shown below.

> **KEY POINT**
>
> Covalent bonds act between atoms.
> Ionic bonds act between ions.
> Metallic bonds act between positive ions and electrons.
> Hydrogen bonds act between polar molecules.
> Dipole-dipole interactions act between polar molecules.
> Van der Waals' forces act between molecules.

Ionic, covalent and metallic bonds are of comparable strength.

Intermolecular forces are much weaker, and their relative strengths are compared with the strength of covalent bonds in the table below.

type of bond	bond enthalpy / kJ mol^{-1}
covalent bond	200–500
hydrogen bond	5–40
van der Waals' forces	~2

The properties of a substance depend upon its bonding and structure.

Properties of ionic compounds

AQA	M1	OCR	M1
EDEXCEL	M1	SALTERS	M2
NICCEA	M1	WJEC	CH1
NUFFIELD	M1		

Ionic compounds are solids at room temperature.

The strong forces between ions result in high melting and boiling points.

Ionic compounds conduct only when ions are free to move – when molten or in aqueous solution.

Ionic compounds form giant ionic lattices with each ion surrounded by ions of the opposite charge. The ions are held together by **strong** electrostatic attraction between positive and negative ions. (See page 46.)

High melting point and boiling point
- High temperatures are needed to break the strong electrostatic forces holding the ions rigidly in the solid lattice.

Therefore ionic compounds have high melting and boiling points.

Electrical conductivity
*In the **solid** lattice,*
- the ions are in a fixed position and there are no mobile charge carriers.

Therefore an ionic compound is a **non-conductor** of electricity in the solid state.

*When **melted** or **dissolved** in water,*
- the solid lattice breaks down,
- the ions are now free to move as mobile charge carriers.

Therefore an ionic compound is a **conductor** of electricity in liquid and aqueous states.

Solubility

- The ionic lattice often dissolves in **polar** solvents (e.g. water).
- Polar water molecules break down the lattice and surround each ion in solution as shown below for sodium chloride.

The structure of an aqueous cation such as Na+(aq) is often shown as a *complex ion*, e.g. [Na(H$_2$O)$_6$]+.

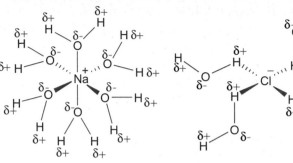

- Water molecules attract Na+ and Cl- ions.
- Lattice breaks down as it dissolves.
- Water molecules surround ions.

Na+ attracts δ− charges on the O atoms of water molecules.

Cl- attracts δ+ charges on the H atoms of water molecules.

Properties of covalent compounds

AQA	M1	OCR	M1
EDEXCEL	M1	SALTERS	M2
NICCEA	M1	WJEC	CH1
NUFFIELD	M2		

Elements and compounds with covalent bonds have either of two structures:
- a simple molecular structure
- a giant molecular structure.

Intermolecular forces are weak

Only the intermolecular forces break when a simple molecule melts or boils

A common mistake in exams is to confuse intermolecular forces with covalent bonds.

Simple molecular structures, e.g. iodine, I$_2$

Simple molecular structures form solid lattices with small molecules, such as Ne, H$_2$, O$_2$, N$_2$, held together by **weak** intermolecular forces.

Low melting point and boiling point

- Low temperatures provide sufficient energy to break the weak intermolecular forces between molecules.

Therefore simple molecular structures have low melting and boiling points.

The simple molecular structure of solid I$_2$

Giant structure:

high melting point.

Simple molecular structure:

low melting point.

- When the I$_2$ lattice breaks down, only the weak van der Waals' forces between the I$_2$ molecules break.
- In the I$_2$ molecule, the covalent bond, I–I, is strong and does **not** break when the lattice breaks down.

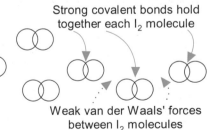

Strong covalent bonds hold together each I$_2$ molecule

Weak van der Waals' forces between I$_2$ molecules

Electrical conductivity

- There are no free charged particles.

Therefore simple molecular structures are **non-conductors** of electricity.

Solubility

- Van der Waals' forces form between a simple molecular structure and a non-polar solvent, such as hexane. These weaken the structure.

Therefore simple molecular structures are often soluble in **non-polar** solvents (e.g. hexane).

Giant molecular structures, e.g. carbon (diamond and graphite)

Diamond, graphite and SiO_2 are examples of giant molecular lattices.

Covalent bonds are strong.

The covalent bonds break when a giant molecular structure melts or boils – this only happens at high temperatures.

Giant molecular structures have thousands of atoms bonded together by **strong** covalent bonds. This type of structure is known by a variety of names: a giant **molecular** lattice, a giant **covalent** lattice or a giant **atomic** lattice.

High melting point and boiling point

- High temperatures are needed to break the strong covalent bonds in the lattice.

Therefore giant molecular structures have high melting and boiling points.

Electrical conductivity

- Except for graphite (see below), there are no free charged particles.

Therefore giant molecular structures are **non-conductors** of electricity.

Solubility

- The strong covalent bonds in the lattice are too strong to be broken by either polar **or** non-polar solvents.

Therefore giant molecular structures are insoluble in polar *and* non-polar solvents.

Comparison between the properties of diamond and graphite

property	diamond	graphite
structure	symmetrical structure held together by strong covalent bonds throughout lattice tetrahedral	strong layer structure but with weak bonds between the layers hexagonal layers
electrical conductivity	*poor* • There are no delocalised electrons as all outer shell electrons are used for covalent bonds.	*good* • Delocalised electrons between layers. • Electrons are free to move parallel to the layers when a voltage is applied.
hardness	*hard* • Tetrahedral shape enables external forces to be spread out throughout the lattice.	*soft* • Bonding within each layer is strong but weak forces between layers easily allow layers to slide.

Properties of metals

Metals have a giant metallic lattice structures held together by **strong** electrostatic attractions between positive ions and negative electrons.

High melting point and boiling point

- Generally high temperatures are needed to separate the ions from their rigid positions within the lattice.

Therefore most metals have high melting and boiling points.

Good thermal and electrical conductivity

When a metal conducts electricity, only the electrons move.

- The existence of mobile, delocalised electrons allows metals to conduct heat and electricity well, even in the solid state.
- The electrons are free to flow between positive ions.
- The positive ions do **not** move.

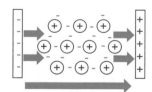

drift of delocalised electrons across potential difference

Summary of properties from structure and bonding

structure	bonding	melting pt / boiling pt	reason	electrical conductivity	reason	solubility	reason
giant ionic	ionic bonds	high	strong electrostatic attraction, between oppositely charged ions	poor when solid good when aqueous or molten	ions in a fixed position in lattice lattice has broken down: mobile ions	good in polar solvents, e.g. water	attraction between ionic lattice and polar solvent
simple molecular	covalent bonds within the molecules	low	weak van der Waals' forces between molecules	poor	no mobile charged particles (electrons or ions)	good in non-polar solvents	van der Waals' forces between molecular structure and solvent
giant molecular	covalent bonds	high	strong covalent bonds between atoms	poor	no mobile charged particles (electrons or ions)	poor	forces within lattice too strong to be broken by solvents
hydrogen bonded	hydrogen bonds	low *but* higher than expected	dipole-dipole attractions between molecules	poor	no mobile charged particles (electrons or ions)	good in polar solvents, e.g. water	attraction between dipoles
giant metallic	metallic bonds	usually high	strong electrostatic attractions between ions and electrons	good	mobile electrons, even in solid state	poor	forces within lattice too strong to be broken by solvents

> A high melting point is the result of any giant structure, bonded together with strong forces.
>
> Giant structures can be ionic, covalent or metallic.

KEY POINT

Progress check

1 For (a) MgO; (b) CH_4; (c) SiO_2, state the structure and explain the following physical properties: melting and boiling points, electrical conductivity, solubility.

1 (a) giant ionic: high m.pt/b.pt. – strong forces between ions.
Non-conductor when solid – ions fixed in lattice. Conductor when molten or dissolved in water – ions are able to move.
Soluble in water – dipole on water attracted to ions.
(b) simple molecular: low m. pt/b. pt. – weak van der Waals' forces between molecules.
Non-conductor – no mobile charge carriers, electrons localised in covalent bonds.
Soluble in non-polar solvents – van der Waals' forces in solvent interact with van der Waals' forces in CH_4.
(c) giant molecular: high m. pt. – strong covalent bonds between atoms.
Non-conductor – no mobile charge carriers, electrons localised in covalent bonds.
Insoluble in all solvents – strong covalent bonds are too strong to be broken by either polar or non-polar solvents.

Sample question and model answer

Chemical bonding helps to explain different properties of materials.

The phrase 'electrostatic attraction between ions' is essential when describing an ionic bond.

(a) Using suitable diagrams, explain what is meant by *ionic*, *covalent* and *metallic* bonding.

An ionic bond is the electrostatic attraction between ions. ✓
An example is sodium chloride:

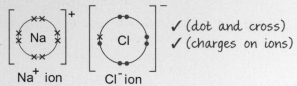

Na⁺ ion   Cl⁻ ion

✓ (dot and cross)
✓ (charges on ions)

DO use diagrams in your answers. Here, dot and cross diagrams are essential for showing ionic and covalent compounds

A covalent bond is a **shared** ✓ **pair** of electrons. ✓
An example is hydrogen chloride.

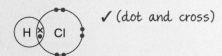

✓ (dot and cross)

The word 'pair' is essential when describing a covalent bond. Omit it and you risk losing the mark.

Metallic bonding occurs between positive centres/ions ✓
surrounded by mobile or delocalised electrons: ✓

There are plenty of examples. Keep them simple and make sure that you choose correctly. Don't use NaCl as a covalent example – it is ionic!

Many candidates do actually make this mistake.

Metallic bonding is the attraction between positive ions and mobile electrons. ✓

(b) Three substances have ionic, covalent and metallic bonding respectively. Compare and explain the electrical conductivity of the three materials.

The ionic compound does not conduct electricity when solid ✓
because the ions are fixed ✓ in a lattice.
The ionic compound does conduct electricity when aqueous or molten ✓
because the ions are mobile. ✓
The covalent compound does not conduct ✓ at all because there are no free charge carriers ✓ (electrons or ions).
The metallic compound does conduct electricity ✓ because the delocalised electrons are able to move ✓ across a potential difference.

17 marking points ⟶ maximum of [15]

Special care needed throughout with language. A-grade students will score all of these marks.

Ionic bonding: positive ions and negative ions – **both** carry electricity.

Metallic bonding: positive ions and delocalised electrons – **only** the electrons move and carry electricity.

Practice examination questions

1 (a) An ammonia molecule comprises nitrogen and hydrogen atoms held together by covalent bonds. What is meant by the term *covalent bond*? [2]

(b) Use the formation of the ammonium ion from ammonia to give the meaning of the term *co-ordinate bond*. [2]

(c) Give the bond angle in the ammonium ion and predict, with an explanation, an appropriate value for the bond angle in the ammonia molecule. [3]

(d) Name the major force of attraction existing between molecules in liquid ammonia and explain how this type of force arises. [3]

[Total: 10]

NEAB Atomic Structure, Bonding and Periodicity Q5 Feb 1995 (modified)

2 Describe how the principle of *electron pair repulsion* helps to explain the shapes of simple molecules. Using diagrams, illustrate how this determines the molecular shapes and bond angles in molecules of (a) CH_4; (b) NH_3; (c) BF_3; (d) SF_2. [17]

[Total: 17]

Cambridge Chemistry Foundation Q1 June 1995 (modified)

3 A crystal of iodine, heated gently in a test tube, gave off a purple vapour. A crystal of sodium chloride, heated to the same temperature, remained unchanged.

(a) Name the type of bonding or interaction which occurs:

  (i) between iodine atoms in a molecule of iodine;

  (ii) between iodine molecules in a crystal of iodine. [2]

(b) Give the formula for the species that is responsible for the purple colour in the vapour. [1]

(c) Explain why iodine vapourises at a relatively low temperature. [1]

(d) Explain, in terms of its bonding, why the sodium chloride crystal did not melt or vapourise. [2]

(e) Describe what happens to the particles in sodium chloride when the solid is heated above room temperature but below its melting point. [1]

(f) Explain why the heat energy required to vapourise sodium chloride (171 kJ mol^{-1}) is greater than the heat energy required to melt it (29 kJ mol^{-1}). [2]

[Total: 9]

NEAB Atomic Structure, Bonding and Periodicity Q4 Feb 1996

4 Suggest reasons for the different boiling points of (a) water, (b) trichloromethane and (c) argon (given below) by considering the nature and strength of the intermolecular forces in each case.

liquid	H_2O	$CHCl_3$	Ar
boiling point /°C	100	62	−186

[Total: 10]

Cambridge Chemistry Foundation Q6 March 1996 (modified)

Chapter 4

The Periodic Table

The following topics are covered in this chapter:

- *The modern Periodic Table*
- *Redox reactions*
- *The s-block elements: Group 1 and Group 2*
- *Extraction of metals*

4.1 The modern Periodic Table

After studying this section you should be able to:

- *describe the Periodic Table in terms of atomic number, periods and groups*
- *classify the Periodic Table into s, p, d and f blocks*
- *describe and explain periodic trends in atomic radii, ionisation energies, melting and boiling points, and electrical conductivities*

LEARNING SUMMARY

Arranging the elements

AQA	M1	OCR	M1
EDEXCEL	M1	SALTERS	M1
NICCEA	M1	WJEC	CH1
NUFFIELD	M1		

In the Periodic Table:

- elements are arranged in order of increasing atomic number
- Groups are vertical columns including elements with similar properties
- Periods are horizontal rows across which there is a trend in properties.

Key areas in the Periodic Table

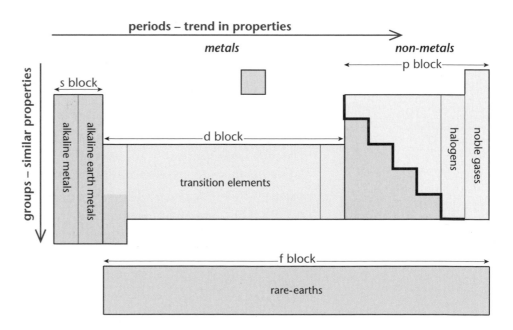

- The diagonal stepped line separates metals (to the left) from non-metals (to the right).
- Elements close to this line, such as silicon and germanium, are called **semi-metals** or **metalloids**. They show properties intermediate between those of a metal and a non-metal.
- The four blocks (s, p ,d and f) show the sub-shell being filled.

The trend in properties across a period is repeated across each period – this is called *periodicity*.

Periodicity

Periodicity is the periodic trend in properties, repeated across each period.

E.g. Period 2   METAL $\longrightarrow$ NON-METAL
      Period 3   METAL $\longrightarrow$ NON-METAL

This periodicity of properties means that predictions can be made about the likely properties of an element and its compounds.

Note, however, the metal/non-metal divide. On descending the Periodic Table, the changeover from metal to non-metal takes place further to the right. For example at the top of Group 4 carbon is a non-metal whereas, at the bottom of Group 4, tin and lead are metals. This means that subtle trends in properties take place down a group.

Group 4
C
Si
Ge   non-metal
Sn   to metal
Pb

Trends in atomic radii

AQA	M1	OCR	M1
EDEXCEL	M1	SALTERS	M1
NICCEA	M1	WJEC	CH1
NUFFIELD	M1		

Across a period

Across a period, the atomic radius decreases. The trend in atomic radius across Period 2 is shown below:

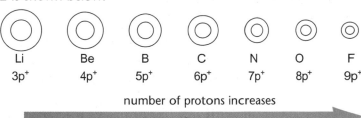

Li	Be	B	C	N	O	F
$3p^+$	$4p^+$	$5p^+$	$6p^+$	$7p^+$	$8p^+$	$9p^+$

number of protons increases

radius decreases

Atomic radius decreases across a period.

Across a period:

- the **nuclear charge increases**
- outer electrons are being added to the same shell
- the attraction between the nucleus and outer electrons increases.

Therefore the atomic radius **decreases** across a period.

Down a group

Atomic radius increases down a group.

Down a group:

- extra shells are added that are **further** from the nucleus
- there are more shells between the outer electrons and the nucleus leading to **greater shielding** of the nuclear charge
- the attraction between the nucleus and the outer electrons decreases.

shells increase

shielding increases

atomic radius increases

Therefore the atomic radius **increases** down a group.

Down a group, the nuclear charge also increases but this is more than compensated by the increase in atomic radius and shielding.

> **KEY POINT**
>
> Across a Period, nuclear change is the key factor.
>
> Down a Group, atomic radius and shielding are the key factors (see also ionisation energies, page 25).

Trends in first ionisation energies

AQA	M1	OCR	M1
EDEXCEL	M1	SALTERS	M1
NICCEA	M1	WJEC	CH1
NUFFIELD	M1		

The graph below shows the variation of first ionisation energy with increasing atomic number from hydrogen to calcium:

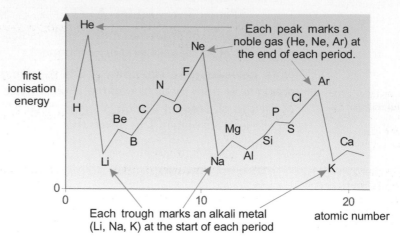

The graph shows:

* a general **increase** in first ionisation energy across a period
  (see H→He; Li→Ne; Na→Ar).
  This results from the **increase** in nuclear charge as electrons are added to the same shell across each period.

> Across a period, increased nuclear charge is most important.

* a sharp **decrease** in first ionisation energy between the end of one period and the start of the next period (see He→Li; Ne→Na; Ar→K).
  This reflects the addition of a new shell with the resulting increase in **distance** and **shielding**.

* a **decrease** in first ionisation energy down a group
  (see He→Ne→Ar and other groups).
  This reflects the presence of extra shells down the group.

> Down a group, increased distance and shielding are most important.

These trends are discussed in more detail elsewhere in this guide. (See pages 25–27). The reasons for the variation in ionisation energies are similar to those explaining the trend in atomic radii (page 65).

Trends in boiling points

AQA	M1	OCR	M1
EDEXCEL	M1	SALTERS	M1
NICCEA	M1	WJEC	CH1
NUFFIELD	M1		

The graphs below show the variation in boiling points across Period 2 and Period 3 of the Periodic Table.

> Trends in boiling point provide information about structure: giant or simple molecular.

Period 2

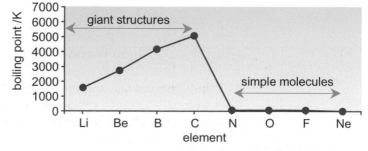

> **Giant structure**
> On boiling, strong forces are broken:
> high boiling point.
> **Simple molecular structure**
> On boiling, weak forces are broken:
> low boiling point.

Period 3

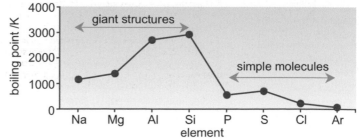

Across a period:

- the boiling point increases from Group 1 to Group 4
- there is a sharp decrease in boiling point between Group 4 and Group 5 the boiling points are comparatively low from Group 5 to Group 8.

> **KEY POINT**
>
> The sharp decrease in boiling point marks a change from giant to simple molecular structures.

Note the periodicity in boiling point: the trend for Period 2 is repeated across Period 3.

More details of the structures of these elements are shown below.

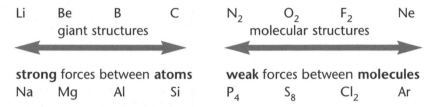

Li   Be   B   C N_2 O_2 F_2 Ne
                giant structures   molecular structures

⟷ **strong** forces between **atoms**   ⟷ **weak** forces between **molecules**
Na   Mg   Al   Si P_4 S_8 Cl_2 Ar

Notice the changes in the molecular formulae of P_4, S_8, Cl_2 and Ar. This shows up well in the graph of boiling points across Period 3.

With more atoms (and consequently electrons) in each molecule, the forces between molecules (van der Waals' forces) increase (See also p.55).

Comparison of the boiling points of the metals

Note the increase in boiling points of the metals Na → Al in Period 3.

- The number of delocalised electrons in the lattice increases.
- The charge on each cation increases.
- This results in increasing attractive forces within the metallic lattice.

The diagram below compares the attractive forces between atoms of sodium, magnesium and aluminium.

See also Metallic bonding, p.49, 60.

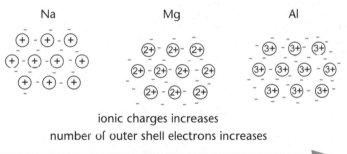

Na   Mg   Al

ionic charges increases
number of outer shell electrons increases

⟶

attraction increases: boiling point increases

The increasing number of delocalised electrons in the lattice also explains the increase in electrical conductivity from Na → Al.

Progress check

1 Using ideas about nuclear charge, attraction and shells, explain the trend in atomic radii across a period and down each group.
2 Why is the melting point of carbon much higher than that of nitrogen?

2 Carbon has a giant molecular structure: between the atoms there are strong covalent bonds which break on melting. Nitrogen has a simple molecular structure; between the molecules there are weak van der Waals' forces that break on melting.

1 Period: the nuclear charge increases as electrons are being added to the same shell. The attraction between the nucleus and outer electrons increases; ∴ the atomic radius decreases across a period. Group: extra shells are added that are further from the nucleus leading to an increased shielding of the outer electrons from the nucleus. Both these factors lead to less attraction between the nucleus and the outer electrons; ∴ the atomic radius increases down a group.

4.2 Redox reactions

After studying this section you should be able to:

- explain the terms *reduction* and *oxidation* in terms of electron transfer
- explain the terms *oxidising agent* and *reducing agent*
- apply the rules for assigning oxidation states
- identify changes in oxidation numbers from an equation
- construct an overall equation for a redox reaction from half-equations

LEARNING SUMMARY

Oxidation and Reduction

AQA	M2	OCR	M1
EDEXCEL	M1	SALTERS	M2
NICCEA	M1	WJEC	CH1
NUFFIELD	M2		

Redox reactions

Oxidation and *reduction* were originally used for reactions involving oxygen.

> Oxidation is the gain of oxygen.
> Reduction is the loss of oxygen.
>
> KEY POINT

Nowadays, oxidation and reduction have a much broader definition in terms of electron transfer in a *redox* reaction (**red**uction and **ox**idation).

> **Reduction** is the **gain** of electrons.
> **Oxidation** is the **loss** of electrons.
>
> KEY POINT

OIL
RIG

Oxidation
Is
Loss of electrons
Reduction
Is
Gain of electrons

Reduction and oxidation must take place together:
- if one species **gains** electrons
- another species **loses** the same number of electrons

Half-equations

Half-equations are useful to identify the species being oxidised or reduced. Notice that the number of electrons lost and gained must balance.

The formation of magnesium chloride from its elements is a redox reaction:

$$\text{overall reaction} \qquad Mg + Cl_2 \longrightarrow MgCl_2$$

The overall equation conceals the electron transfer that has taken place. This can be shown by writing half-equations:

$$\text{electron transfer} \qquad Mg \longrightarrow Mg^{2+} + 2e^- \quad \text{oxidation (loss of electrons)}$$
$$Cl_2 + 2e^- \longrightarrow 2Cl^- \quad \text{reduction (gain of electrons)}$$

Oxidising and reducing agents

Non-metals are oxidising agents.

Metals are reducing agents.

> An **oxidising agent removes** electrons from another reactant.
> Non-metals are oxidising agents, e.g. F_2, Cl_2, O_2.
> A **reducing agent adds** electrons to another reactant.
> Metals are reducing agents, e.g. Na, Fe, Zn.
>
> KEY POINT

In the example above:
- Mg is the reducing agent – it has *reduced* the Cl_2 to $2Cl^-$ by *adding* electrons
- Cl_2 is the oxidising agent – it has *oxidised* Mg to Mg^{2+} by *removing* electrons.

Using oxidation numbers

AQA	M2	OCR	M1
EDEXCEL	M1	SALTERS	M2
NICCEA	M1	WJEC	CH1
NUFFIELD	M2		

Chemists use the concept of **oxidation number** as a means of accounting for electrons.

The oxidation number of a species is assigned by applying a set of rules.

Oxidation number rules

Exceptions to rules

In compounds with fluorine and in peroxides, the oxidation number of oxygen is not −2 and must be calculated from other oxidation numbers.

In metal hydrides, the oxidation number of hydrogen is −1.

species	oxidation number	examples
uncombined element	0	C, zero; Na, zero; O_2, zero.
combined oxygen	−2	H_2O; CaO.
combined hydrogen	+1	NH_3, H_2S.
simple ion	charge on ion	Na^+, +1; Mg^{2+}, +2; Cl^-, −1
combined fluorine	−1	NaF, CaF_2.

> **KEY POINT**
>
> When applying oxidation numbers to elements, compounds and ions, **the sum of the oxidation numbers must equal the overall charge.**

Examples of applying oxidation numbers

Oxidation Number rules can be applied to any compound, whether ionic or covalent.

In CO_2, the overall charge is zero.

* There are **2** oxygen atoms, each with an oxidation number of **−2**, giving a total contribution of **−4**.
* The oxidation number of carbon must be **+4** to give the overall charge of zero.

In NO_3^-, the overall charge is 1−.

* There are **3** oxygen atoms, each with an oxidation number of **−2**, giving a total contribution of **−6**.
* The oxidation number of nitrogen must be +5 to give the overall charge of −1.

Oxidation number applies to each atom in a species.

Notice how this has been shown in the examples.

In a compound:	CO_2
• the total of all the oxidation numbers is zero:	+4
	−2
	−2
	+4 −4
CO_2: overall charge = 0	0

In an ion:	NO_3^-
• the total of all the oxidation numbers is equal to the overall charge on the ion:	+5
	−2
	−2
	−2
	+5 −6
NO_3^-: overall charge = −1	−1

Using oxidation number with equations

Oxidation numbers can be used to identify redox reactions in which electron loss and electron gain are not easy to see.

By applying oxidation numbers to an equation:

* the species being oxidised and reduced can be identified
* the number of electrons on both sides of the equation can be checked.

The sum of the oxidation numbers on both sides of a chemical equation must be the same.

	$Cr_2 O_3$ (s)	+	2 Al (s)	⟶	$Al_2 O_3$ (s)	+	2 Cr (s)
oxidation	+3 −2		0		+3 −2		0
numbers	+3 −2		0		+3 −2		0
	−2				−2		
sum of oxidation *numbers:*	0		0		0		0

In this reaction, the changes in oxidation number are:

Cr: $+3 \longrightarrow 0$ reduction   *oxidation number decreases*
Al: $0 \longrightarrow +3$ oxidation   *oxidation number increases*

> **KEY POINT**
>
> **Oxidation** is an **increase** in oxidation number.
> **Reduction** is a **decrease** in oxidation number.

Combining half-equations

An overall equation for a redox reaction can be constructed by combining the half-equations showing the transfer of electrons in the reaction.

Example: The reaction of Ag+ ions react with Zn metal.

electron gain: reduction

electron loss: oxidation

The half-equations are:

$$Ag^+(aq) \; + \; e^- \; \longrightarrow \; Ag \, (s)$$
$$Zn(s) \; \longrightarrow \; Zn^{2+}(aq) \; + \; 2e^-$$

To combine the half-equations in an overall equation, the number of electrons transferred must be the same in each half-equation. This will ensure that every electron lost by $Zn(s)$ is gained by Ag^+.

Balance the electrons:

Ag+ reaction × 2.

- The Ag+ half-equation is multiplied by '2' to balance the electrons.

$$2Ag^+(aq) \; + \; 2e^- \; \longrightarrow \; 2Ag(s)$$

- The half-equations are now added:

$$2Ag^+(aq) \; + \; 2e^- \; + \; Zn(s) \; \longrightarrow \; 2Ag(s) + Zn^{2+}(aq) + 2e^-$$

Cancel the electrons to give the overall equation.

- Any species appearing on both sides are cancelled to give the overall equation.

$$2Ag^+(aq) \; + \; Zn(s) \; \longrightarrow \; 2Ag(s) + Zn^{2+}(aq)$$

- Finally, check that the oxidation numbers balance on either side of the equation:

	$2Ag^+(aq)$	$+$	$Zn(s)$	$\longrightarrow$	$2Ag(s)$	$+$	$Zn^{2+}(aq)$
oxidation numbers:	+1		0		0		+2
	+1				0		

oxidation number check: $+2$ $+2$

Progress check

1   What is the oxidation state of the elements in the following:
    (a) Ag^+; (b) F_2; (c) N^{3-}; (d) Fe; (e) MgF_2; (f) $NaClO_3$?

2   Write down the oxidation number of sulphur in the following:
    (a) H_2S; (b) SO_2; (c) SO_4^{2-}; (d) SO_3^{2-}; (e) $Na_2S_2O_3$.

3   The following reaction is a redox process:
    $Mg + 2HCl \longrightarrow MgCl_2 + H_2$
    (a) Identify the changes in oxidation number.
    (b) Which species is being oxidised and which is being reduced?
    (c) Identify the oxidising agent and the reducing agent.

3 (a) Mg, $0 \longrightarrow +2$, H, $+1 \longrightarrow 0$; (b) Mg oxidised, H reduced;
   (c) Oxidising agent: HCl, Reducing agent: Mg.
2 (a) -2; (b) $+4$; (c) $+6$; (d) $+4$; (e) $+2$.
1 (a) $+1$; (b) 0; (c) -3; (d) 0; (e) Mg: $+2$, F: -1; (f) Na: $+1$, Cl: $+5$, O: -2.

4.3 The s-block elements: Group 1 and Group 2

General properties

AQA	M1	OCR	M1
EDEXCEL	M1	SALTERS	M1
NICCEA	M2	WJEC	CH1
NUFFIELD	M1		

The elements in Group 1 and Group 2 have hydroxides that are alkaline and their common names reflect this.
- The Group 1 elements are the **Alkali Metals**.
- The Group 2 elements are the **Alkaline Earth Metals**.

Electronic configuration

The elements in Group 1 and Group 2 have their highest energy electrons in an s sub-shell and these two groups are known as the *s-block elements*.

The s-block elements
Group 1
Group 2

Each Group 1 element has:
- **one** electron more than the electronic configuration of a noble gas
- an outer **s** sub-shell containing **one** electron.

Each Group 2 element has:
- **two** electrons more than the electronic configuration of a noble gas
- an outer **s** sub-shell containing **two** electrons.

Electronic configuration of the s-block elements			
Group 1		**Group 2**	
Li	[He] $2s^1$	Be	[He] $2s^2$
Na	[Ne] $3s^1$	Mg	[Ne] $3s^2$
K	[Ar] $4s^1$	Ca	[Ar] $4s^2$
Rb	[Kr] $5s^1$	Sr	[Kr] $5s^2$
Cs	[Xe] $6s^1$	Ba	[Xe] $6s^2$
Fr	[Rn] $7s^1$	Ra	[Rn] $7s^2$

Physical properties

Group 1
- They are soft metals and can be cut with a knife.
- They have low melting and boiling points.
- They have low densities: Li, Na and K all float on water.
- They have colourless compounds.

Melting points of s-block elements			
Melting point/°C		Melting point/°C	
Li	181	Be	1528
Na	98	Mg	649
K	63	Ca	839
Melting points for comparison: Fe: 1535°C; Cu 1083°C			

Group 2
- They have reasonably high melting and boiling points.
- They have low densities although not as low than those in Group 1.
- They have colourless compounds.

Densities of s-block elements			
Density/g cm^{-3}		Density/g cm^{-3}	
Li	0.53	Be	1.85
Na	0.97	Mg	1.74
K	0.86	Ca	1.54
Densities for comparison: H_2O, 1.00 g cm^{-3}; Fe: 7.86 g cm^{-3}			

Flame tests

EDEXCEL	M1	SALTERS	M1
NICCEA	M1	WJEC	CH1
NUFFIELD	M1		

When heated in a flame, compounds of Groups 1 and 2 show characteristic flame colours, allowing different cations to be identified.

cation	Li^+	Na^+	K^+	Ca^{2+}	Ba^{2+}	Sr^{2+}
flame colour	red	orange	lilac	brick-red	green	red

The characteristic flame colours arise from electronic transitions within the atom.

> **KEY POINT**
>
> Heat energy from the flame excites an electron to a higher energy level.
> The electron returns to a lower energy level.
> The excess energy is released as light energy.
> The frequency of the flame colour depends upon the difference between the energy levels involved.

Reactivity of the s-block elements

AQA	M1	OCR	M1
EDEXCEL	M1	SALTERS	M1
NICCEA	M2	WJEC	CH1
NUFFIELD	M1		

The elements in the s-block are the most reactive metals and are strong reducing agents.

Group 1 elements are oxidised in reactions, each atom losing one electron from its outer s sub-shell to form a 1+ ion (+1 oxidation state):

$$M \longrightarrow M^+ + e^-$$

Group 2 elements are oxidised in reactions, each atom losing two electrons from its outer s sub-shell to form a 2+ ion (+2 oxidation state):

$$M \longrightarrow M^{2+} + 2e^-$$

Reactivity **increases** down each group reflecting the increasing ease of losing electrons. The ionisation energy of the metal is an important factor in this process (see also pages 25, 26).

> **KEY POINT**
>
> Within each group, the elements become **more** reactive as the group descends.
> Within each period, the Group 1 element is a **more** reactive metal than the Group 2 element.

The Group 1 elements

| EDEXCEL | M1 |

Reaction with oxygen

The Group 1 elements react vigorously with oxygen. Each element forms the expected ionic oxide with the general formula $(M^+)_2O^{2-}$. However, on descending the group, further oxidation takes place resulting in a mixture of metal oxides.

Lithium forms the expected oxide Li_2O:

$$4Li(s) + O_2(g) \longrightarrow 2Li_2O(s)$$

Sodium forms a mixture of Na_2O and the peroxide Na_2O_2:

$$4Na(s) + O_2(g) \longrightarrow 2Na_2O(s)$$
$$2Na(s) + O_2(g) \longrightarrow Na_2O_2(s)$$

> When sodium reacts with an excess of oxygen, a peroxide Na_2O_2 is formed.

Potassium forms a mixture of K_2O, the peroxide K_2O_2 and the superoxide KO_2:

$$4K(s) + O_2(g) \longrightarrow 2K_2O(s)$$
$$2K(s) + O_2(g) \longrightarrow K_2O_2(s)$$
$$K(s) + O_2(g) \longrightarrow KO_2(s)$$

> When potassium reacts with an excess of oxygen, a superoxide KO_2 is formed.

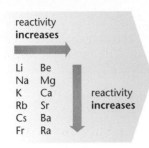

Reaction with water

The Group 1 elements react vigorously with water, forming alkaline solutions of the metal hydroxide and hydrogen.

E.g. $$2Na(s) + 2H_2O(l) \longrightarrow 2NaOH(aq) + H_2(g)$$

The reactivity increases down the group, reflecting the decrease in the first ionisation energies of the elements.

The Group 1 oxides

Reaction with water

The Group 1 oxides react with water forming strongly alkaline solutions.

E.g. $$Na_2O(s) + H_2O(l) \longrightarrow 2NaOH(aq)$$

Reaction with acids

The oxides behave as bases and are neutralised by acids, such as HCl(aq). In each reaction, a salt and water is formed. (See also page 110.)

E.g. $$Na_2O(s) + 2HCl(aq) \longrightarrow 2NaCl(aq) + H_2O(l)$$

The Group 2 elements

AQA	M1	OCR	M1
EDEXCEL	M1	SALTERS	M1
NICCEA	M2	WJEC	CH1
NUFFIELD	M1		

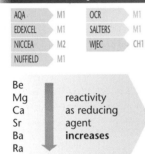

Reaction with oxygen

The Group 2 elements react vigorously with oxygen. Each element forms the expected ionic oxide with the general formula, $M^{2+}O^{2-}$.

E.g. $$2Ca(s) + O_2(g) \longrightarrow 2CaO(s)$$

Action of water

Reactivity increases down the group reflecting the increasing ease with which electrons can be lost.

* Mg reacts very slowly with water, forming the **hydroxide** and hydrogen:
$$Mg(s) + 2H_2O(l) \longrightarrow Mg(OH)_2(aq) + H_2(g)$$

Mg forms MgO with steam.

* With steam, reaction is much quicker forming the **oxide** and hydrogen:
$$Mg(s) + H_2O(g) \longrightarrow MgO(s) + H_2(g)$$

* Further down the group from calcium, each metal reacts vigorously with water:
$$Ca(s) + 2H_2O(l) \longrightarrow Ca(OH)_2(aq) + H_2(g)$$

Group 2 oxides and hydroxides

Reaction with water

Mg(OH)$_2$
Ca(OH)$_2$ — solubility increases
Sr(OH)$_2$ — alkalinity increases
Ba(OH)$_2$

Thus solid barium hydroxide is reasonably soluble in water to form a strong alkaline solution with greater OH⁻(aq) concentration.

The Group 2 oxides form weakly alkaline solutions with water:

e.g. $$MgO(s) + H_2O(l) \longrightarrow Mg^{2+}(aq) + 2OH^-(aq)$$

* Magnesium hydroxide, $Mg(OH)_2(s)$ is only slightly soluble in water and so the resulting solution is only a weak alkali.

* The solubility in water increases down the group, and resulting solutions Tare more alkaline:
$$Ba(OH)_2(s) + aq \longrightarrow Ba^{2+}(aq) + 2OH^-(aq)$$

The alkalinity of the Group 2 hydroxides is exploited commercially. Some indigestion tablets contain $Mg(OH)_2$ to neutralise excess acid in the stomach. Farmers add $Ca(OH)_2$ as 'lime' to neutralise acid soils.

MgSO$_4$
CaSO$_4$
SrSO$_4$
BaSO$_4$

solubility **decreases**

Reaction with acids

- The Group 2 oxides behave as bases and are neutralised by acids, such as HCl(aq), forming salts and water.

  E.g. $MgO(s) + 2HCl(aq) \longrightarrow MgCl_2(aq) + H_2O(l)$

In medicine, barium sulphate is used as a 'barium meal' which shows up imperfections in the gut when exposed to X-rays. Barium compounds are extremely poisonous in solution but the insolubility of BaSO$_4$ is such that no harm is caused to the patient by this treatment.

Solubility of Group 2 sulphates

Magnesium sulphate MgSO$_4$ is very soluble in water but the solubility decreases as the group is descended. The trend is so marked that barium sulphate BaSO$_4$ is virtually insoluble.

In the laboratory, the precipitation of BaSO$_4$ is used to test for the sulphate ion. A solution of a soluble barium salt (usually BaCl$_2$(aq) or Ba(NO$_3$)$_2$) is added to a solution of a substance in dilute nitric acid. In the presence of aqueous sulphate ions, a dense white precipitate of barium sulphate is formed.

$$Ba^{2+}(aq) + SO_4^{2-}(aq) \longrightarrow BaSO_4(s)$$

The thermal stability of s-block compounds

EDEXCEL M1   NUFFIELD M1
NICCEA M2

The s-block carbonates

Group 2 carbonates

The Group 2 carbonates are decomposed by heat:

e.g. $MgCO_3(s) \longrightarrow MgO(s) + CO_2(g)$

This is caused by the polarising effect of the Group 2 cation:

carbon–oxygen bond is weakened

MgCO$_3$
CaCO$_3$
SrCO$_3$
BaCO$_3$

ease of thermal decomposition **decreases**

Polarisation occurs when a small cation distorts the electron shells of a larger anion.

The polarising effect is greatest between a small densely charged cation and a large anion.

It is increasingly more difficult to decompose the carbonate as the group is descended. The polarising power of the metal cation decreases as the charge : size ratio decreases (see also page 54).

- As the size of the cation increases, the polarising effect becomes less.
- There is less weakening of the C–O bond.
- It becomes harder to decompose the carbonate.

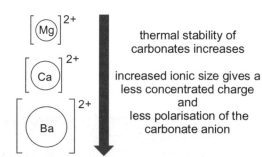

thermal stability of carbonates increases

increased ionic size gives a less concentrated charge and less polarisation of the carbonate anion

Group 1 carbonates

The Group 1 carbonates are thermally stable (except Li$_2$CO$_3$)

The cations in the Group 1 carbonates are larger than those of the Group 2 carbonates and their polarising power is less. The Li$^+$ ion, however, is comparable in size with the Mg^{2+} ion and lithium carbonate is also decomposed by heat:

$$Li_2CO_3(s) \longrightarrow Li_2O(s) + CO_2(g)$$

The remaining Group 1 carbonates are **not** decomposed by heat.

The s-block nitrates

Group 2 nitrates

The Group 2 nitrates are decomposed by heat, releasing NO_2 and O_2 gases:

e.g. $$Ca(NO_3)_2(s) \longrightarrow CaO(s) + 2NO_2(g) + \tfrac{1}{2}O_2(g)$$

As with the Group 2 carbonates, the nitrates become more thermally stable as the group is descended. This again reflects the increasing size of the cations and the resultant decrease in polarising power.

> With heat, Li_2CO_3 and $LiNO_3$ behave similarly to the carbonates and nitrates of magnesium. This is known as a *diagonal relationship*. You can see why by comparing the positions of Li and Mg in the Periodic Table.

Group 1 nitrates

The Group 1 nitrates decompose with heat releasing only oxygen gas:

e.g. $$NaNO_3(s) \longrightarrow NaNO_2(s) + \tfrac{1}{2}O_2(g)$$

However, as with the carbonates above, the increasing polarising power of the smaller Li^+ ion makes lithium nitrate less thermally stable and lithium nitrate is decomposed by heat in a similar way to the Group 2 nitrates:

$$2LiNO_3(s) \longrightarrow Li_2O(s) + 2NO_2(g) + \tfrac{1}{2}O_2(g)$$

Calcium carbonate

Present in limestone and chalk, calcium carbonate is by far the most important calcium compound. It is used in the building industry and in the manufacture of glass and steel.

Much of the calcium carbonate is used in the production of quicklime, CaO.

$$CaCO_3(s) \xrightarrow{heat} CaO(s) + CO_2(g)$$
$$\text{quicklime}$$

> *Quicklime* is used by the building industry to make cement and mortar.
>
> *Slaked lime* is used in mortar and used by farmers to neutralise acid soils.

Addition of water to calcium oxide, produces 'slaked lime', $Ca(OH)_2$.

$$CaO(s) + H_2O(l) \longrightarrow Ca(OH)_2(s)$$
$$\text{slaked lime}$$

In the laboratory, a solution of calcium hydroxide is used as 'lime water' to test for carbon dioxide. The *milky* appearance is a fine suspension of calcium carbonate.

$$\underset{\text{lime water}}{Ca(OH)_2(aq)} + CO_2(g) \longrightarrow CaCO_3(s) + H_2O(l)$$

If an excess of carbon dioxide is bubbled through lime water, the precipitate disappears with the formation of calcium hydrogencarbonate:

$$CaCO_3(s) + CO_2(g) + H_2O(l) \xrightarrow{\text{excess } CO_2} Ca^{2+}(aq) + 2HCO_3^-(aq)$$

Calcium hydrogencarbonate is responsible for 'hardness' in water. On being heated, the reverse reaction takes place and calcium carbonate is precipitated as 'scale'.

$$\underset{\text{hard water}}{Ca^{2+}(aq) + 2HCO_3^-(aq)} \xrightarrow{heat} \underset{\text{scale}}{CaCO_3(s)} + CO_2(g) + H_2O(l)$$

Progress check

1 Write down equations for the following reactions:
(a) barium with water
(b) calcium oxide with nitric acid.

2 Identify the oxidation number changes taking place during the thermal decomposition of sodium nitrate:
$$2NaNO_3(s) \longrightarrow 2NaNO_2(s) + O_2(g)$$

2 N, +5 $\longrightarrow$ +3, O, −2 $\longrightarrow$ 0.
1 (a) Ba(s) + 2H$_2$O(l) $\longrightarrow$ Ba(OH)$_2$(aq) + H$_2$(g)
(b) CaO(s) + 2HNO$_3$(aq) $\longrightarrow$ Ca(NO$_3$)$_2$(aq) + H$_2$O(l)

4.4 The Group 7 elements and their compounds

After studying this section you should be able to:

LEARNING SUMMARY

- describe the characteristic properties of the Group 7 elements
- recall the relative reactivity of the halogens as oxidising agents
- recall the characteristic tests for halide ions
- describe the characteristics of hydrogen halides
- describe the reactions of halides with concentrated sulphuric acid
- describe the use of thiosulphate titrations.

General properties

AQA	M2	OCR	M1
EDEXCEL	M1	SALTERS	M2
NICCEA	M1	WJEC	CH1
NUFFIELD	M2		

The common name for the elements in Group 7 is the **Halogens**.

Electronic configuration

Each halogen has **seven** outer shell electrons, just one electron short of the electronic configuration of a noble gas. The outer **p** sub-shell contains **five** electrons.

Electronic configuration of the halogens

F	[He] $2s^2 2p^5$
Cl	[Ne] $3s^2 3p^5$
Br	[Ar] $3d^{10}4s^2 4p^5$
I	[Kr] $4d^{10}5s^2 5p^5$
At	[Xe] $4f^{14}5d^{10}6p^2 6p^5$

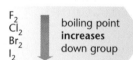

F_2
Cl_2
Br_2
I_2

boiling point **increases** down group

Trend in physical states

The halogens exist as diatomic molecules, X_2. The boiling points of the halogens increase in descending the group.

- The physical states of the halogens at r.t.p. show the classic trend of gas $\longrightarrow$ liquid $\longrightarrow$ solid.
- On descending the group the number of electrons increases leading to an increase in the van der Waals' forces between molecules.
- Therefore the boiling point increases.

Boiling points of the Halogens

	Boiling point/°C	State at r.t.p.
F_2	−188	gas
Cl_2	−35	gas
Br_2	59	liquid
I_2	184	solid
At_2	337	solid

Trend in electronegativity

AQA	M2
NUFFIELD	M2

Electronegativity is a measure of the attraction of an atom for the pair of electrons in a covalent bond (see also page 52).

The hydrogen halides have polar molecules: $H^{\delta+}-X^{\delta-}$. The polarity decreases on descending the halogens and the order of polarity is:

most polar   H–F > H–Cl > H–Br > H–I   *least polar*

This trend in polarity results from the **decreasing** electronegativity of the halogen atom on descending the group.

- The atomic radius increases from F → Cl → I resulting in less nuclear attraction on the bonding electrons (this is despite the increase in nuclear charge!).
- There are more electron shells between the nucleus and the bonding electrons to shield the nuclear charge.

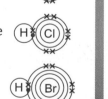

electronegativity of halogen **decreases**

polarity of H–X bond **decreases**

Electronegativity of the Halogens

	Pauling value
F	4.0
Cl	3.0
Br	2.8
I	2.5

The overall effect is that the smaller the halogen atom, the greater the nuclear attraction experienced by the bonding electrons. Thus the large electronegativity of fluorine results in a bond between hydrogen and fluorine that is more polar than between hydrogen and other halogens.

The relative reactivity of the halogens as oxidising agents

> **KEY POINT**
>
> The halogens are the most reactive non-metals and are strong oxidising agents.
>
> The halogens become **less** reactive as the group descends as their oxidising power decreases.

The halogens are reduced in reactions, each atom gaining one electron into a p sub-shell to form a –1 ion (–1 oxidation state):

e.g.
$$\tfrac{1}{2}F_2(g) + e^- \longrightarrow F^-(g)$$
[He] $2s^2 2p^5$ [He] $2s^2 2p^6$ (or [Ne])

- An extra electron is captured by being attracted to the outer shell of an atom by the nuclear charge of an atom.

F
Cl
Br
I
At

reactivity as oxidising agent **decreases**

On descending the halogens:

- the atomic radii increase resulting in less nuclear attraction at the edge of the atom (despite the increase in nuclear charge).
- there are more electron shells between the nucleus and the edge of the atom to shield the nuclear charge.

F

Cl

Br

The overall effect is that most nuclear attraction is experienced at the edge of the small fluorine atoms. This attraction decreases down the group as the atoms get bigger.

Therefore the oxidising power of the halogens decreases down the group.
Fluorine is the strongest oxidising agent and is able to attract an extra electron more strongly than other halogens.

Displacement reactions of the halogens

The decrease in reactivity as the group is descended can be demonstrated by displacement reactions of aqueous halides using Cl_2, Br_2 and I_2.

chlorine oxidises both Br^- and I^-:
$$Cl_2(aq) + 2Br^-(aq) \longrightarrow 2Cl^-(aq) + Br_2(aq)$$
$$Cl_2(aq) + 2I^-(aq) \longrightarrow 2Cl^-(aq) + I_2(aq)$$

bromine oxidises I^- only:
$$Br_2(aq) + 2I^-(aq) \longrightarrow 2Br^-(aq) + I_2(aq)$$

iodine does not oxidise either Cl^- or Br^-

Industrial extraction of bromine

The main source of bromine is as bromide ions, Br^- in sea water. Bromine is extracted by oxidising sea water with chlorine. Because chlorine is a stronger oxidising agent, it displaces the bromide ions using the principles of the displacement reaction.

Testing for halide ions

Addition of aqueous silver ions (using $AgNO_3(aq)$) to aqueous halide ions produces coloured precipitates that have different solubilities in aqueous ammonia.

chloride: $Ag^+(aq) + Cl^-(aq) \longrightarrow AgCl(s)$ white precipitate, soluble in dilute $NH_3(aq)$

bromide: $Ag^+(aq) + Br^-(aq) \longrightarrow AgBr(s)$ cream precipitate, soluble in conc $NH_3(aq)$

iodide: $Ag^+(aq) + I^-(aq) \longrightarrow AgI(s)$ yellow precipitate, insoluble in conc $NH_3(aq)$

Hydrogen halides

The hydrogen halides, HX, are colourless gases, very soluble in water, forming strongly acidic solutions. For example, hydrochloric acid forms when hydrogen chloride gas dissolves in water:

$$HCl(g) + aq \longrightarrow H^+(aq) + Cl^-(aq)$$

- As the group is descended, the H–X bond enthalpy decreases and the bond breaks more readily, resulting in an increase in acidity down the group.

The order of acidity is HI > HBr > HCl.

- The hydrogen halides are reducing agents with the reducing ability increasing down the group.

Reactions of halides with concentrated sulphuric acid

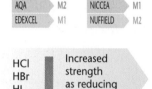

Concentrated sulphuric acid is an oxidising agent. Halide salts react with concentrated sulphuric acid producing a range of products depending on the halide used. This difference is caused by the increasing reducing power of the hydrogen halides as the group is descended.

HCl
HBr
HI
Increased strength as reducing agents

HCl does **not** reduce H_2SO_4

NaCl and H_2SO_4

Hydrogen chloride gas, HCl, is formed.

$$NaCl(s) + H_2SO_4(l) \longrightarrow NaHSO_4(s) + HCl(g)$$

The HCl formed is not a sufficiently strong reducing agent to reduce the sulphuric acid. No redox reaction takes place.

NaBr and H_2SO_4

Hydrogen bromide gas, HBr, is initially formed.

$$NaBr(s) + H_2SO_4(l) \longrightarrow NaHSO_4(s) + HBr(g)$$

Some of the hydrogen bromide reduces the sulphuric acid with the formation of sulphur dioxide and orange bromine fumes.

HBr reduces H_2SO_4

$H_2SO_4 \xrightarrow{\text{HBr}} Br_2 + SO_2$

$$2HBr(g) + H_2SO_4(l) \longrightarrow SO_2(g) + Br_2(g) + 2H_2O(l)$$

$$+6 \xrightarrow{-2} +4 \qquad \text{S reduced}$$

$$2 \times -1 \xrightarrow{+2} 0 \qquad \text{Br oxidised}$$

NaI and H_2SO_4

Hydrogen iodide gas, HI, is initially formed.

$$NaI(s) + H_2SO_4(l) \longrightarrow NaHSO_4(s) + HI(g)$$

Hydrogen iodide is a strong reducing agent and reduces the sulphuric acid to a mixture of reduced products including sulphur dioxide and hydrogen sulphide.

Reduction to SO_2 (ox no: +4)

$$2HI(g) + H_2SO_4(l) \longrightarrow SO_2(g) + I_2(s) + 2H_2O(l)$$

$$+6 \xrightarrow{-2} +4 \qquad \text{S reduced}$$

$$2 \times -1 \xrightarrow{+2} 0 \qquad \text{I oxidised}$$

HI reduces H_2SO_4 and SO_2

$H_2SO_4 \xrightarrow{\text{HI}} I_2 + SO_2$

$SO_2 \xrightarrow{\text{HI}} I_2 + H_2S$

Further reduction to H_2S (ox no: −2)

$$6HI(g) + SO_2(g) \longrightarrow H_2S(g) + 3I_2(s) + 2H_2O(l)$$

$$+4 \xrightarrow{-6} -2 \qquad \text{S reduced}$$

$$6 \times -1 \xrightarrow{+6} 0 \qquad \text{I oxidised}$$

Uses of chlorine

AQA	M2	NUFFIELD	M2
EDEXCEL	M1	OCR	M1
NICCEA	M1	WJEC	CH1

A solution of chlorine in water is green, showing the presence of chlorine.

Small amounts of chlorine are added to drinking water to kill bacteria that would make the water unsafe to drink. Stronger solutions of chlorine in water are in equilibrium (see also page 104):

$$Cl_2(aq) + H_2O(l) \rightleftharpoons HClO(aq) + HCl(aq)$$

HClO chloric(I) acid acts as a bleach and, if indicator solution is added, the initial red colour (from the acids) disappears as the bleaching action takes place.

The formation of bleach

AQA	M2	NUFFIELD	M2
EDEXCEL	M1	OCR	M1
NICCEA	M1		

Notice how the products obtained depend upon the conditions.

This is an example of how chemists are able to control the products using different conditions.

Bleach is formed when dilute aqueous sodium hydroxide and chlorine are mixed at room temperature.

This reaction is an example of *disproportionation* in which chlorine is both reduced (to chloride, Cl^-) and oxidised (to chlorate(I) ClO^-).

$$Cl_2(aq) + 2\,NaOH(aq) \longrightarrow NaCl(aq) + NaClO(aq) + H_2O(l)$$

$$0 \xrightarrow{\quad -1 \quad} -1 \qquad \textit{chlorine reduced}$$

$$0 \xrightarrow{\quad +1 \quad} 1 \qquad \textit{chlorine oxidised}$$

Further disproportionation of the chlorine takes place with hot, concentrated sodium hydroxide:

$$3\,NaClO(aq) \longrightarrow 2\,NaCl(aq) + NaClO_3(aq)$$

$$2 \times +1 \xrightarrow{\quad -4 \quad} 2 \times -1 \qquad \textit{chlorine reduced}$$

$$+1 \xrightarrow{\quad +4 \quad} +5 \qquad \textit{chlorine oxidised}$$

The industrial extraction of the chlorine

| EDEXCEL | M1 |

Chlorine

Chlorine is manufactured by electrolysis of a concentrated solution of sodium chloride (purified saturated brine). The diagram below shows the modern 'Membrane cell'.

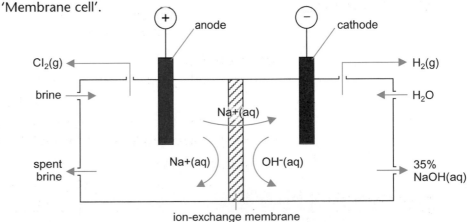

The cell reactions are:

at the cathode $\qquad 2H_2O(l) + 2e^- \longrightarrow 2OH^-(aq) + H_2(g)$

at the anode $\qquad 2Cl^-(aq) \longrightarrow Cl_2(g) + 2e^-$

> **KEY POINT**
>
> Chlorine reacts with hydroxide ions (see above). The membrane keeps these two products apart, preventing any reaction.
>
> The other products are hydrogen and the alkali, sodium hydroxide.
>
> The process consumes large amounts of electricity, which is the main expense.

Iodine–thiosulphate titrations

NICCEA ▶ M1   NUFFIELD ▶ M2

> Starch is a sensitive test for the presence of iodine. If it is added too early in this titration, the deep-blue colour perseveres and the end-point cannot be detected with accuracy.

Iodine-thiosulphate titrations can be used to determine iodine concentrations, either directly or by liberating iodine using a more powerful oxidising agent.

Iodine oxidises thiosulphate ions:

$$I_2(aq) + 2S_2O_3^{2-}(aq) \longrightarrow 2I^-(aq) + S_4O_6^{2-}(aq)$$

- Thiosulphate is added from the burette to the iodine until the deep-brown colour of iodine becomes a light-straw colour.
- Starch is now added, forming a deep-blue colour.
- Thiosulphate is added dropwise until the deep-blue colour becomes colourless. This is the end-point and shows that all the iodine has just been reacted.

The estimation of chlorine in a bleach

10.0 cm³ of a sample of bleach was diluted to 250 cm³ with water. 25.0 cm³ of this solution was pipetted into a flask. An excess of aqueous potassium iodide was then added to liberate iodine:

> *equation 1* $Cl_2(aq) + 2I^-(aq) \longrightarrow I_2(aq) + 2Cl^-(aq)$

Titration of this solution required 23.2 cm³ of 0.100 mol dm⁻³ sodium thiosulphate.

> *equation 2* $2S_2O_3^{2-}(aq) + I_2(aq) \longrightarrow 2I^-(aq) + S_4O_6^{2-}(aq)$

Calculate the concentration of chlorine in the bleach.

> Compare acid-base titrations (p.40). The essential method is the same.

From the titration results, amount (in moles) of $Na_2S_2O_3$ can be calculated:

> The concentration and volume of $Na_2S_2O_3$ are known.

$$\text{amount (in mol) of } Na_2S_2O_3 = c \times \frac{V}{1000} = 0.100 \times \frac{23.2}{1000} = 0.00232 \text{ mol}$$

From the equations, the amount (in moles) of Cl_2 can be determined:

From *equation 2*, 2 mol $S_2O_3^{2-}$ reacts with 1 mol I_2

> Using the equations, determine the number of moles of Cl_2.

∴ 0.00232 mol $S_2O_3^{2-}$ reacts with $\frac{0.00232}{2}$ mol I_2 = 0.00116 mol I_2

From *equation 1*, 1 mol Cl_2 forms 1 mol I_2

∴ amount (in mol) of Cl_2 = 0.00116 mol

The concentration in mol dm⁻³ of the diluted bleach solution can now be calculated by scaling to 1000 cm³:

> Work out the concentration of $Cl_2(aq)$, in mol dm⁻³.

25.0 cm³ of diluted bleach contains 0.00116 mol Cl_2.

1 cm³ of diluted bleach contains $\frac{0.00116}{25}$ mol Cl_2.

1 dm³ of diluted bleach contains $\frac{0.00116}{25} \times 1000 = 0.0464$ mol Cl_2.

Finally, take into account the dilution with water from 10 cm³ to 250 cm³:

> Don't forget to take into account any dilution.

The original bleach was diluted by a factor of $\frac{250}{10} = 25$

∴ concentration of Cl_2 in the bleach is 25 x 0.0464 = 1.16 mol dm⁻³.

Progress check

1 State and explain the trend in boiling points of the halogens fluorine to iodine.

2 How could you distinguish between NaCl, NaBr and NaI by a simple test?

3 Comment on the changes in oxidation number of chlorine in following reaction: $Cl_2(aq) + H_2O(l) \longrightarrow HClO(aq) + H^+(aq) + Cl^-(aq)$

3 $Cl_2 \longrightarrow HClO, Cl: 0 \longrightarrow +1$ (oxidation). $Cl_2 \longrightarrow Cl^-, Cl: 0 \longrightarrow -1$ (reduction). Chlorine has been both oxidised and reduced (disproportionation).

2 Add AgNO₃(aq). NaCl gives a white precipitate, soluble in dilute ammonia. NaBr gives a cream precipitate, soluble in concentrated ammonia. NaI gives a yellow precipitate, insoluble in concentrated ammonia.

1 Boiling points increase down the group. From F_2 to I_2, the number of electrons increase leading to greater van der Waals' forces between molecules and higher boiling points.

4.5 Extraction of metals

After studying this section you should be able to:

- *understand that carbon reduction is used to extract iron from its ore*
- *understand that electrolysis is used to extract aluminium from its ore*
- *understand that metal reduction is used to extract titanium from its ore*
- *understand the importance of economic factors and recycling*

LEARNING SUMMARY

Reduction of metals

AQA ▸ M2

The main methods for extracting metals from their ores are:

- reduction of the ore with carbon
- reduction of the molten ore by the electrolysis
- reduction of the ore with a more reactive metal.

Reduction of metal oxides with carbon

AQA ▸ M2

Production of iron

The main ore of iron, haematite, contains Fe_2O_3 and this is reduced by coke in the blast furnace. This is a continuous process in which high quality haematite, coke and limestone are fed in at the top of the furnace and hot air blown in near the bottom.

> Coke contains a very high carbon content.

> The production of iron in the blast furnace is a **continuous** process. Raw materials are added at the top of the furnace and the products removed at the base. For further production, more raw materials are added and the process continues in a continuous flow, sometimes for months on end.

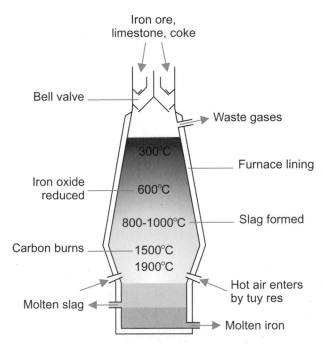

> Both C and CO are reductants in the blast furnace.

- Initially, hot coke reacts with air forming carbon monoxide.

$$2C(s) + O_2(g) \longrightarrow 2CO(g)$$

- The carbon monoxide reduces most of the iron oxide at around 1200°C:

$$Fe_2O_3(s) + 3CO(g) \longrightarrow 2Fe(l) + 3CO_2(g)$$

- In hotter parts of the furnace, coke also reacts directly with iron oxide:

$$2Fe_2O_3(s) + 3C(s) \longrightarrow 4Fe(l) + 3CO_2(g)$$

This equation represents the overall reaction for the reduction.

Limestone removes SiO_2.

The limestone is used to remove acidic impurities such as silicon dioxide (sand). This produces a slag, largely of calcium silicate, $CaSiO_3$:

$$CaCO_3 + SiO_2 \longrightarrow CaSiO_3 + CO_2$$

- Molten iron collects at the bottom of the furnace and is run off as 'pig iron.'
- Pig iron is very impure and brittle, containing about 4% of carbon (as well as Mn, Si, P and S).

Making steel

Oxygen removes impurities including C, Si, P and Mn.

Steels are alloys of iron with 0.1%–1.0% of carbon. The major process for converting impure iron from a blast furnace into steel is the Basic Oxygen Process.

- Sulphur impurities are first removed by adding magnesium powder to the molten iron. Magnesium sulphide forms in a vigorous reaction and is raked off from the surface of the molten iron.

$$Mg + S \longrightarrow MgS$$

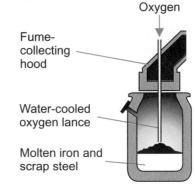

- Oxygen under pressure is blown onto the molten iron.
- Carbon is oxidised to carbon monoxide which escapes as a gas.
- The oxygen also reacts with other impurities (such as Si, P and Mn) in the iron forming a mixture of oxides which remain in the converter.
- Lime is added as a basic oxide and this reacts with any acidic oxides forming a slag which floats to the surface:

e.g. $$CaO + SiO_2 \longrightarrow CaSiO_3$$

Not all of the carbon is removed in the Basic Oxygen Converter.

- 'Mild steel' contains 0.1% carbon.
- 'High carbon steels' contain about 1.0% carbon.

Limitations of carbon reduction

Environmental pollution problems:

sulphide ores form sulphur dioxide;

carbon reduction produces greenhouse gases.

Metal ores often contain metal oxides or sulphides.

- Sulphide ores (such as galena, PbS, and sphalerite, ZnS) are first roasted in air to produce the metal oxide. The sulphur is liberated as sulphur dioxide, which contributes to acid rain.
- Oxide ores (such as haematite, Fe_2O_3) are reduced to the metal using fossil fuels such as coke from coal. Carbon dioxide, emitted as the main gaseous product, is a greenhouse gas.

Ores of some metals (such as titanium and tungsten) react with carbon forming metal carbides, so this is not a practical method for extracting these metals.

$$TiO_2 + 3C \longrightarrow TiC + 2CO$$

Reduction of metal oxides by electrolysis of the molten ore

AQA   M2
EDEXCEL   M2

For metals more reactive than zinc, reduction with carbon does not take place except at very high temperatures and these metals are usually extracted by electrolysis of the molten ore.

The extraction of aluminium

The melting point of aluminium oxide is 2045°C but this is decreased using molten cryolite. This reduces the energy requirements.

The main ore of aluminium, bauxite, contains Al_2O_3. Purified bauxite is dissolved in molten cryolite (Na_3AlF_6) at 970°C. This is a continuous process needing regular additions of aluminium oxide.

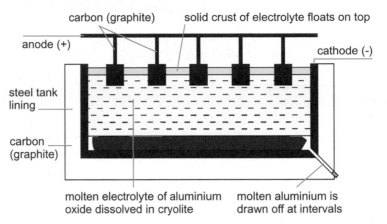

The electrodes are made of graphite and the cell reactions are:

at the cathode $\quad\quad Al^{3+} + 3e^- \longrightarrow Al$

at the anode $\quad\quad\quad 2O^{2-} \longrightarrow O_2 + 4e^-$

- Periodically the graphite anodes need replacing because, at the high temperatures used, the carbon anodes react with the liberated oxygen.

$$2C + O_2 \longrightarrow 2CO$$
$$\text{and } C + O_2 \longrightarrow CO_2$$

- The process consumes large amounts of electricity (as electricity is needed to melt the Al_2O_3 as well as to reduce it).
- The process is only economic where electricity is relatively inexpensive.

Reduction of metal halides with metals

AQA   M2

Some metals, such as titanium, become brittle if contaminated with traces of impurities such as carbon, oxygen or nitrogen.

Highly pure metals can be formed by reduction of a metal halide with a reactive metal.

Despite the high cost, pure titanium is manufactured by reduction of its molten chloride with a more reactive metal.

When pure, titanium is a light metal with a high strength and high resistance to corrosion.

When impure, it is brittle and of little use.

The production of titanium is a **batch** process. Raw materials are reacted together in a single 'batch' and the products are separated from the reaction mixture. For further production, the whole process needs to be repeated from scratch.

Production of titanium

The main ore of titanium, rutile, contains titanium(IV) oxide. Titanium is manufactured in a two-stage batch process.

- Rutile is first converted to titanium(IV) chloride using chlorine and coke at around 900°C:

$$TiO_2 + 2C + 2Cl_2 \longrightarrow TiCl_4 + 2CO$$

- The titanium(IV) chloride is purified from other chlorides (e.g. those of iron, silicon and chromium) by fractional distillation under argon or nitrogen.
- The chloride is then reduced by a more reactive metal such as sodium or magnesium:

$$TiCl_4 + 4Na \longrightarrow Ti + 4NaCl$$

- The sodium is initially held at around 550 °C but the temperature rises to nearly 1000°C during the reaction.
- An inert atmosphere of argon is used to prevent any contamination of the metal with oxygen or nitrogen.
- The sodium chloride by-product is washed out with dilute hydrochloric acid, leaving titanium as a granular powder.

Economic factors

- Chlorine and sodium have to be produced first by the electrolysis of molten sodium chloride. This process is expensive.
- High temperatures are involved in both stages of the production.
- Precautions have to be taken in the handling of the intermediate $TiCl_4$ which reacts rapidly with water.
- An argon atmosphere has to be maintained to prevent oxidation of the hot titanium as it forms.

Economic factors and recycling

AQA   M2

Costs of extracting a metal

The method used to reduce a metal on an industrial scale is dependent upon several factors:

The cost of the reducing agent

A reducing agent that is naturally available such as carbon (from coal) is cheaper than one such as sodium, which has to prepared first by a separate (and often costly) process.

The energy costs for the process

The lower the temperature of a process, the lower the energy requirement and the cheaper the process.

The purity required of the metal

It is relatively expensive to produce a metal with very high purity. There needs to be a high demand for the metal if this extra expense is to be justified.

These factors need to be weighed against each other when considering the overall cost of an extraction process.

Recycling

Metals are a valuable resource and, instead of disposal, metals are often collected and recycled.

Recycling metals conserves metal ores and saves energy.

Recycling of iron

- Iron is recycled by collection of scrap iron which is melted and reused.
- The Earth's reserves of iron ore are saved.
- The energy required to mine iron ore, transport it and smelt it is several times greater than the energy required to recycle scrap iron.

Recycling of aluminium

- Owing to its very high resistance to corrosion, used aluminium is as good as new.
- Recycling conserves the Earth's supply of aluminium.
- The electrolysis of aluminium oxide requires high temperatures and vast quantities of electricity.
- It is also cheaper and easier to refine recycled aluminium than to mine bauxite and extract the metal by electrolysis.

The cost of re-using scrap aluminium is only one-twentieth of the cost of making the pure metal.

Sample question and model answer

1

(a) State and explain the trends in atomic radius, electronegativity and boiling point of the halogens from fluorine to iodine.

> Atomic radius increases from fluorine to iodine ✓ because there are more electron shells ✓ and the extra shielding repels the outer electrons. ✓
>
> Electronegativity decreases from fluorine to iodine ✓ because the atomic radius increases ✓ and the attraction for the pair of electrons in a bond decreases. ✓
>
> Boiling point increases from F_2 to I_2 ✓ because the number of electrons in molecules increases. ✓ This results in greater van der Waals' forces between molecules. ✓

[8]

Notice that electronegativity applies to attraction for the electrons in a bond – a point often missed in exams.

Again, you must be careful with language: 'more electrons' results in 'greater van der Waals' forces'.

(b) When chlorine is bubbled into water a pale green solution is formed. If a piece of universal indicator paper is dipped into this solution it first turns red and then becomes white. When chlorine is bubbled into a cold dilute solution of sodium hydroxide a colourless solution is formed.

Explain these observations, write equations for the reactions of chlorine with water and with sodium hydroxide and state why the reaction between chlorine and sodium hydroxide is of commercial importance.

> $Cl_2 + H_2O \longrightarrow HCl + HClO$ ✓
> Acids turn universal indicator red ✓ HClO is a bleach ✓
> $Cl_2 + 2NaOH \longrightarrow NaCl + NaClO + H_2O$ ✓
> NaClO is a commercial bleach/disinfectant ✓

[5]

The reactions of chlorine with water and alkali are often tested in exams.

(c) Explain how aqueous silver nitrate and ammonia solution can be used to show the presence of iodide ions in solution.

> A yellow precipitate forms ✓ which is AgI. ✓ The precipitate is insoluble in concentrated aqueous ammonia. ✓

[3]

[Total: 16]

NEAB Equilibria and Inorganic Chemistry Q8 June 1997

More easy marks for all candidates, allowing credit for learning the chemistry. You should, of course, also know these reactions with chloride and bromide ions.

Practice examination questions

1 In the Periodic Table, describe and explain the trend in the atomic radii of the elements in Period 3 and in Group 1. [Total: 8]

2 Magnesium is in Group II of the Periodic Table.
   (a) Complete, and balance if necessary, the following equations:
      (i) $Mg(s) + H_2O(g) \longrightarrow$
      (ii) $Mg(s) + O_2(g) \longrightarrow$
      (iii) $Mg(s) + H_2O(l) \longrightarrow$ [3]

   (b) Magnesium reacts very slowly with cold water whereas barium reacts rapidly. This shows the increasing reactivity of the Group II elements on descending the group.
      (i) Suggest why the reactivity increases in this way.
      (ii) Name one other reaction of Group II elements which illustrates this trend of increasing reactivity. [3]

   (c) (i) Magnesium oxide is used as a refractory lining in some furnace linings. What property makes it suitable for its use?
      (ii) Name two other Group II compounds, and state a use for each of them. [3]

[Total: 9]

Cambridge Trends and Patterns Q2 March 1997

3

   This question concerns the halogens chlorine, bromine and iodine.
   (a) Describe and explain the trend in oxidising ability shown by the halogens. [5]

   (b) Sodium chlorate(V) can be made by reacting chlorine with hot concentrated aqueous sodium hydroxide.
$$3Cl_2(g) + 6NaOH(aq) \longrightarrow 5NaCl(aq) + NaClO_3(aq) + 3H_2O(l)$$
      (i)   Use changes in oxidation numbers to show that this is a redox reaction.
      (ii) Calculate the maximum mass of sodium chlorate(V) that could be made from the reaction of 85 dm^3 of chlorine at room temperature and pressure.
      (iii) How does chlorine react with dilute aqueous sodium hydroxide at room temperature and what use has the resulting solution? [9]

[Total: 14]

Cambridge Trends and Patterns Q7 June 1998 (modified)

4

   (a) (i) Define the term *electronegativity*.
      (ii) State and explain the trend in electronegativity of the halogens fluorine to iodine. [5]

   (b) State and explain the trend in boiling points of the halogens fluorine to iodine. [3]

   (c) Describe what is observed when aqueous bromine is added to separate aqueous solutions of sodium chloride and sodium iodide. Write equation(s) for any reaction(s) occurring. [4]

   (d) State and explain the trend in reducing properties of the halide ions shown by the experiments in part (c). [2]

[Total: 14]

NEAB Equilibria and Inorganic Chemistry Q4 June 1995

Chapter 5
Chemical energetics

The following topics are covered in this chapter:

- Enthalpy changes
- Determination of enthalpy changes
- Bond enthalpy

5.1 Enthalpy changes

After studying this section you should be able to:

- understand that reactions can be exothermic or endothermic
- construct a simple enthalpy profile diagram for a reaction
- explain and use the terms: standard conditions, enthalpy changes of reaction, formation and combustion

LEARNING SUMMARY

Energy out, energy in

AQA	M2	OCR	M3
EDEXCEL	M2	SALTERS	M1
NICCEA	M1	WJEC	CH2
NUFFIELD	M1		

(Change in heat content of chemicals)

=

– (Change in heat content of the surroundings)

The first law of thermodynamics underpins all of this section.

Enthalpy, H, is the **heat energy** that is stored in a chemical system.

Enthalpy cannot be measured experimentally. However, an **enthalpy change** can be measured from the temperature change in a chemical reaction.

> **KEY POINT**
>
> An enthalpy change ΔH is the heat energy exchange with the surroundings at constant pressure.

The conservation of energy is an important principle in science. This is often summarised by the first law of thermodynamics.

> **KEY POINT**
>
> **The first law of thermodynamics** states that energy may be exchanged between a chemical system and the surroundings but the *total* energy remains constant.

Exothermic reactions

During an exothermic reaction, heat energy is **released** to the surroundings.

Any energy **loss** from the chemicals is balanced by the same energy **gain** to the surroundings, which rise in temperature.

exothermic

chemicals lose energy: ΔH –ve

surroundings gain energy and rise in temperature: ΔT +ve

Energy pathway diagram (reaction profile)

Energy **loss** during reaction

Energy gain by the surroundings (identified by a temperature *rise*: ΔT +ve)

Chemists refer to *surroundings* as anything other than the reacting chemicals.

Surroundings often just means the water in which chemicals are dissolved.

> **KEY POINT**
>
> In an exothermic reaction, ΔH is negative:
> - heat is given out (**to** the surroundings)
> - the reacting chemicals lose energy.

87

Endothermic reactions

During an endothermic reaction, heat energy is **taken in** from the surroundings.

Any energy **gain** to the chemicals is provided by the same energy **loss** from the surroundings, which fall in temperature.

Energy pathway diagram (reaction profile)

> **endothermic**
>
> chemicals gain energy:
> ΔH **+ve**
>
> surroundings lose energy and fall in temperature:
> ΔT **–ve**

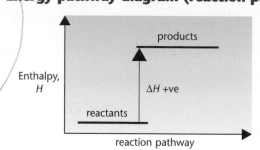

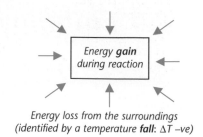

Energy loss from the surroundings (identified by a temperature **fall**: ΔT –ve)

> **KEY POINT**
>
> In an endothermic reaction, ΔH is positive
> - heat is taken in (**from** the surroundings)
> - chemicals gain energy.

Standard enthalpy changes

AQA	M2	OCR	M3
EDEXCEL	M2	SALTERS	M1
NICCEA	M1	WJEC	CH2
NUFFIELD	M1, M2		

Enthalpy changes have been measured for many reactions. Many are recorded in data books as *standard enthalpy changes* and these are discussed below.

Standard conditions

$\Delta H^{\ominus}$ refers to an enthalpy (H) change (Δ) under standard conditions ($^{\ominus}$).

> Standard pressure is 100 kPa or 1 bar.
>
> The former standard pressure of 101 kPa or 1 atmosphere is still quoted in many books.

> **KEY POINT**
>
> *Standard conditions* are:
> a pressure of 100 kPa
> a stated temperature: 298K (25°C) is usually used
> a concentration of 1 mol dm⁻³ (*for aqueous solutions*).
> A *standard state* is the physical state of a substance under standard conditions.

The standard state of water at 298K and 100 kPa is a liquid.

Standard enthalpy change of reaction

> Always use a chemical equation or an unambiguous definition with a stated enthalpy change.

> **KEY POINT**
>
> The *standard enthalpy change of reaction* $\Delta H^{\ominus}_r$ is the enthalpy change that accompanies a reaction in the molar quantities that are expressed in a chemical equation under standard conditions, all reactants and products being in their standard states.

The enthalpy change of reaction $\Delta H^{\ominus}_r$ depends upon the quantities shown in a chemical equation. $\Delta H^{\ominus}_r$ should always be quoted with an equation.

> $\Delta H^{\ominus}_r$ only has a meaning with an equation.
>
> $\Delta H^{\ominus}_r$ has units of kJ mol⁻¹
>
> mol⁻¹ means 'for the amount (in moles) shown in the equation'.

For the reaction:

$$H_2(g) \quad + \quad \tfrac{1}{2} O_2(g) \longrightarrow H_2O(l) \qquad \Delta H^{\ominus}_r = -286 \text{ kJ mol}^{-1}$$
1 mol · ½ mol · 1 mol

but with twice the quantities, there is twice the enthalpy change:

$$2H_2(g) \quad + \quad O_2(g) \longrightarrow 2H_2O(l) \qquad \Delta H^{\ominus}_r = -572 \text{ kJ mol}^{-1}$$
2 mol · 1 mol · 2 mol

Standard enthalpy change of combustion

The *standard enthalpy change of combustion* $\Delta H^{\ominus}{}_c$ is the enthalpy change that takes place when one mole of a substance reacts completely with oxygen under standard conditions, all reactants and products being in their standard states.

This means that complete combustion of 1 mole of $C_2H_4(g)$ releases 1411 kJ of heat energy to the surroundings at 298 K and 100 kPa.

e.g. $C_2H_4(g) + 3O_2(g) \longrightarrow 2CO_2(g) + 2H_2O(l)$ $\Delta H^{\ominus}{}_c = -1411 \text{ kJ mol}^{-1}$

Standard enthalpy change of formation

The *standard enthalpy change of formation* $\Delta H^{\ominus}{}_f$ is the enthalpy change that takes place when one mole of a compound in its standard state is formed from its constituent elements in their standard states under standard conditions.

This means that the formation of 1 mole of $H_2O(l)$ from 1 mole of $H_2(g)$ and ½ mole of $O_2(g)$ releases 286 kJ of heat energy to the surroundings at 298K and 100 kPa.

e.g. $H_2(g) + \frac{1}{2}O_2(g) \longrightarrow H_2O(l)$ $\Delta H^{\ominus}{}_f = -286 \text{ kJ mol}^{-1}$

For an element, the standard enthalpy change of formation is defined as zero.

The formation of $H_2(g)$ from $H_2(g)$ does not involve a chemical change so there is no enthalpy change.

Standard enthalpy change of neutralisation

Notice that
$\Delta H^{\ominus}{}_f (O_2) = 0 \text{ kJ mol}^{-1}$
$\Delta H^{\ominus}{}_r$ for any element must be zero (formation of the element from the element – no change).

The *standard enthalpy change of neutralisation* $\Delta H^{\ominus}{}_{neut}$ is the energy change that accompanies the neutralisation of an acid by a base to form 1 mole of $H_2O(l)$, under standard conditions.

$$HCl(aq) + NaOH(aq) \longrightarrow NaCl(aq) + H_2O(l) \quad \Delta H^{\ominus}{}_{neut} = -57.9 \text{ kJ mol}^{-1}$$
acid   base $\longrightarrow$ 1 mol

- In aqueous solution, strong acids and bases are completely dissociated and $\Delta H^{\ominus}{}_{neut}$ is approximately equal to the value above. This neutralisation process corresponds to the reaction: $H^+(aq) + OH^-(aq) \longrightarrow H_2O(l)$.
- For weak acids, this enthalpy change is less exothermic because some input of energy is required to dissociate the acid (see also page 109).

Progress check

1 Draw energy profile diagrams for the following reactions:
   (a) $N_2O_4(g) \longrightarrow 2NO_2(g)$ $\Delta H = +58 \text{ kJ mol}^{-1}$
   (b) $N_2(g) + 3H_2(g) \longrightarrow 2NH_3(g)$ $\Delta H = -92 \text{ kJ mol}^{-1}$

2 Write an equation to represent the enthalpy change for:
   (a) $\Delta H^{\ominus}{}_c (CH_4)$; (b) $\Delta H^{\ominus}{}_f (NO_2)$

2 (a) $CH_4(g) + 2O_2(g) \longrightarrow CO_2(g) + 2H_2O(l)$
  (b) $\frac{1}{2}N_2(g) + O_2(g) \longrightarrow NO_2(g)$

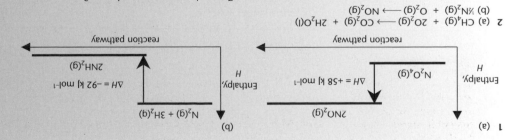

5.2 Determination of enthalpy changes

After studying this section you should be able to:

LEARNING SUMMARY

- *calculate enthalpy changes from direct experimental results, using the relationship: energy change = mcΔT*
- *use Hess's Law to construct enthalpy cycles*
- *determine enthalpy changes indirectly, using enthalpy cycles and enthalpy changes of formation and combustion*

Direct determination of enthalpy changes

AQA	M2	OCR	M3
EDEXCEL	M2	SALTERS	M1
NICCEA	M1	WJEC	CH2
NUFFIELD	M1, M2		

Calculating enthalpy changes

The **heat energy change, Q**, in the surroundings can be calculated using the relationship below.

$$Q = mc\Delta T \text{ Joules}$$

- m is the **mass** of the surroundings that experience the temperature change
- c is the **specific heat capacity** of the surroundings
- ΔT is the **temperature change** (final temperature – initial temperature)

> The **specific heat capacity** of a substance is the energy required to raise the temperature of 1 g of a substance by 1°C.

Example

Addition of an excess of magnesium to 100 cm³ of 2.00 mol dm⁻³ $CuSO_4$(aq) raised the temperature from 20.0°C to 65.0°C. Find the enthalpy change for the reaction:

$$Mg(s) + CuSO_4(aq) \longrightarrow MgSO_4(aq) + Cu(s)$$

specific heat capacity of solution $c = 4.18 \text{ J g}^{-1} \text{ K}^{-1}$;
density of solution = 1.00 g cm⁻³.

> Assume that solids, such as Mg (s) make little difference to the energy change.

Find the heat energy change

100 cm³ of solution has a mass of 100 g;

Temperature change, ΔT	= (65.0–20.0)°C
	= +45.0°C
Heat energy **gain** to surroundings, $Q = mc\Delta T$	= 100 × 4.18 × 45.0 J
	= **+18810 J**
∴ heat energy **loss** from the reacting chemicals	= **–18810 J**

> Any energy **gain** by the surroundings must have come from the same energy **loss** in the chemical reaction.

Find out the amount (in moles) that reacted

Amount (in mol) of $CuSO_4$ that reacted = $2.00 \times \dfrac{100}{1000}$ mol = 0.200 mol

Scale the quantities to those in the equation

$Mg(s)$	+	$CuSO_4(aq)$	$\longrightarrow$	$MgSO_4(aq)$	+	$Cu(s)$
1 mol		1 mol		1 mol		1 mol

∴ 0.200 mol (1/5th mol) of $CuSO_4$ ⟹ ΔH of –18810 J
∴ 1 mol of $CuSO_4$ ⟹ ΔH of 5 × –18810 = –94050 J
ΔH = –94.1 kJ mol⁻¹ (to 3 sig. figs.)

∴ enthalpy change of reaction is given by:

$$Mg(s) + CuSO_4(aq) \longrightarrow MgSO_4(aq) + Cu(s) \quad \Delta H = -94.1 \text{ kJ mol}^{-1}$$

> All values in this example are to 3 significant figures. This indicates the expected accuracy of the answer which should also be expressed to 3 significant figures.
>
> In this case,
> ΔH = –94.1 kJ mol⁻¹.

Indirect determination of enthalpy changes

AQA	M2	OCR	M3
EDEXCEL	M2	SALTERS	M1
NICCEA	M1	WJEC	CH2
NUFFIELD	M1, M2		

Hess's Law

> Hess's Law is an extension of the First Law of Thermodynamics.

Many reactions have enthalpy changes that cannot be found directly from a single experiment. Hess's Law provides a method for the *indirect* determination of enthalpy changes.

> *Hess's Law* states that, if a reaction can take place by more than one route and the initial and final conditions are the same, the total enthalpy change is the same for each route.
>
> **KEY POINT**

The diagram below shows two routes for converted reactants into products.

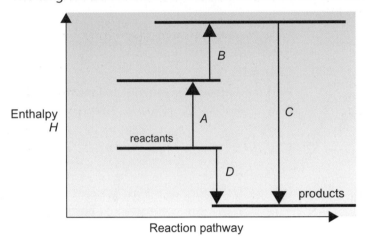

Reaction pathway

> Indirect determination of an enthalpy change uses an energy cycle based on Hess' Law.
>
> This method is used when the reaction is very difficult to carry out and related reactions can be measured more easily.

Following the arrows from reactants to products,

Route 1:   **A + B + C**
Route 2:   **D**

By Hess's Law, the total enthalpy change is the same for each route.

∴ **A + B + C = D**

If three of these enthalpy changes are known, the fourth can always be calculated.

Using $\Delta H^{\ominus}_c$ values to determine an enthalpy change indirectly

> Enthalpy changes of combustion are required for all reactants and products.

Example:

Find the enthalpy change for the reaction:

$$C(s) \;+\; 2H_2(g) \longrightarrow CH_4(g)$$

substance	C(s)	$H_2(g)$	$CH_4(g)$
$\Delta H^{\ominus}_c$ / kJ mol^{-1}	−394	−286	−890

Use the $\Delta H^{\ominus}_c$ data as a 'link' to construct an energy cycle.

- An energy cycle is constructed by linking the reactants and products to their **combustion products**.
- Note the direction of the arrows **from** the reactants and products of the reaction **to** the common combustion products.

> Always show your working. In exams, you are rewarded for a good method.
>
> If you make one small slip in a calculation, you may only lose 1 mark provided that you show clear working.

> The common combustion products here are $CO_2(g)$ and $H_2O(l)$. You can include these in your cycle but they are not required and 'combustion products' have been used here.

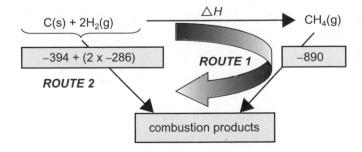

Calculate the unknown enthalpy change
By Hess's Law,

$$\text{Route 1:} \qquad \Delta H + [(-890)]$$
$$\text{Route 2:} \qquad [(-394) + (2 \times -286)]$$
$$\therefore \underbrace{\Delta H + [(-890)]}_{\text{Route 1}} = \underbrace{[(-394) + (2 \times -286)]}_{\text{Route 2}}$$

$$\Delta H = [(-394) + (2 \times -286)] - [(-890)]$$
$$\Delta H = \textbf{-76 kJ mol}^{-1}$$
$$\therefore C(s) + 2H_2(g) \longrightarrow CH_4(g) \qquad \Delta H^{\ominus} = -76 \text{ kJ mol}^{-1}$$

> **KEY POINT**
> **For enthalpy changes of combustion data only,**
> $\Delta H = \Sigma \Delta H^{\ominus}_c \text{ (reactants)} - \Sigma \Delta H^{\ominus}_c \text{ (products)}$

> Enthalpy changes of formation are required for all reactants and products that are compounds.
>
> For elements, $\Delta H^{\ominus}_f = 0$ (formation of the element from the element – no change).

Using $\Delta H^{\ominus}_f$ values to determine an enthalpy change indirectly

Example:

Find the enthalpy change for the reaction:

$$C_2H_6(g) + 3\tfrac{1}{2}O_2(g) \longrightarrow 2CO_2(g) + 3H_2O(l)$$

substance	C_2H_6 (g)	CO_2 (g)	H_2O (l)
$\Delta H^{\ominus}_f$ / kJ mol^{-1}	-85	-394	-286

Use the $\Delta H^{\ominus}_f$ data as a 'link' to construct an energy cycle
- An energy cycle is constructed linking the reactants and products with their **constituent elements**.
- Note the direction of the arrows **from** the constituent elements to the reactants and products of the reaction.

> Notice that
> $\Delta H^{\ominus}(O_2) = 0$ kJ mol^{-1}
> This can be omitted from the cycle.

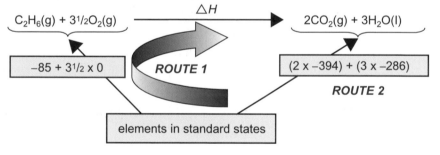

Calculate the unknown enthalpy change
By Hess's Law,

$$\text{Route 1:} \qquad [(-85) + (3\tfrac{1}{2} \times 0)] + \Delta H$$
$$\text{Route 2:} \qquad [(2 \times -394) + (3 \times -286)]$$
$$\therefore \underbrace{[(-85) + 0] + \Delta H}_{\text{Route 1}} = \underbrace{[(2 \times -394) + (3 \times -286)]}_{\text{Route 2}}$$

$$\therefore \Delta H = [(2 \times -394) + (3 \times -286)] - [(-85) + 0]$$
$$\therefore \Delta H = \textbf{-1561 kJ mol}^{-1}$$
$$C_2H_6(g) + 3\tfrac{1}{2}O_2(g) \longrightarrow 2CO_2(g) + 3H_2O(l) \qquad \Delta H^{\ominus} = -1561 \text{ kJ mol}^{-1}$$

> **KEY POINT**
> **For enthalpy changes of formation data only,**
> $\Delta H = \Sigma \Delta H^{\ominus}_f \text{ (products)} - \Sigma \Delta H^{\ominus}_f \text{ (reactants)}$

Progress check

For questions 1 and 2 assume that:

specific heat capacity of solution
$c = 4.18 \text{ J g}^{-1} \text{ K}^{-1}$;

density of solution
$= 1.00 \text{ g cm}^{-3}$.

1 Addition of zinc powder to 55.0 cm³ of aqueous copper(II) sulphate at 22.8 °C raised the temperature to 32.3 °C. 0.3175 g of copper were obtained.
   (a) Calculate the energy released in this reaction.
   (b) Write an equation, including state symbols, for this reaction.
   (c) Calculate the enthalpy change for this reaction per mole of copper formed.

2 Combustion of 1.60 g of ethanol, C_2H_5OH, raised the temperature of 150 g of water from 22.0 °C to 71.0 °C. Find the enthalpy change of combustion of ethanol.

3 Use the $\Delta H^{\ominus}_c$ data below to calculate enthalpy changes for:
   (a) $3C(s) + 4H_2(g) \longrightarrow C_3H_8(g)$
   (b) $C(s) + 2H_2(g) + \tfrac{1}{2}O_2(g) \longrightarrow CH_3OH(l)$

substance	$\Delta H^{\ominus}_c$ / kJ mol^{-1}
C(s)	−394
$H_2(g)$	−286
$C_3H_8(g)$	−2219
$CH_3OH(l)$	−726

4 Use the $\Delta H^{\ominus}_f$ data below to calculate enthalpy changes for:
   (a) $C_2H_4(g) + H_2(g) \longrightarrow C_2H_6(g)$
   (b) $SO_2(g) + 2H_2S(g) \longrightarrow 2H_2O(l) + 3S(s)$

compound	$\Delta H^{\ominus}_f$ / kJ mol^{-1}
$C_2H_4(g)$	+52
$C_2H_6(g)$	−85
$SO_2(g)$	−297
$H_2S(g)$	−21
$H_2O(l)$	−286

4 (a) −137 kJ mol⁻¹; (b) −233 kJ mol⁻¹.
3 (a) −107 kJ mol⁻¹; (b) −240 kJ mol⁻¹.
2 −883 kJ mol⁻¹.
1 (a) 2.18 kJ; (b) Zn(s) + CuSO₄(aq) ⟶ Cu(s) + ZnSO₄(aq); (c) −437 kJ mol⁻¹.

93

5.3 Bond enthalpy

After studying this section you should be able to:

- *understand and use the term bond enthalpy*
- *explain chemical reactions in terms of enthalpy changes associated with the breaking and making of chemical bonds*
- *determine enthalpy changes indirectly, using average bond enthalpies*

Bond enthalpy

AQA	M2	OCR	M3
EDEXCEL	M2	SALTERS	M1
NICCEA	M1	WJEC	CH2
NUFFIELD	M2		

Bond enthalpies are **positive** and refer to **bond breaking** – this process requires energy.

Enthalpy is stored within chemical bonds and **bond enthalpy** indicates the strength of a chemical bond in a gaseous molecule. For simple molecules, such as $H_2(g)$ and $HCl(g)$, bond enthalpy applies to the following processes:

$$H–H(g) \longrightarrow 2H(g) \qquad \Delta H = +436 \text{ kJ mol}^{-1}$$
$$H–Cl(g) \longrightarrow H(g) + Cl(g) \quad \Delta H = +432 \text{ kJ mol}^{-1}$$

> **Bond enthalpy** is the enthalpy change required to **break** and separate **1 mole of bonds** in the molecules of a gaseous element or compound so that the resulting gaseous species exert no forces upon each other.

K E Y   P O I N T

See also Activation Energy: p.99.

Average bond enthalpies

Bond enthalpies such as those above (H–H and H–Cl) apply to specific compounds. Only H_2 can have H–H bonds and the H–H bond enthalpy shown above has a definite value. However, some bonds can have different strengths in different environments.

Not all C–H bonds are created equal.

The Cl atom in CH_3Cl affects the environment of the C–H bonds.

$$H–\underset{\underset{H}{|}}{\overset{\overset{H}{|}}{C}}–H \qquad H–\underset{\underset{H}{|}}{\overset{\overset{H}{|}}{C}}–Cl$$

C—H bonds have different strengths and different bond enthalpies

Data books provide an indication of the likely bond enthalpy of a particular bond by listing **average** or **mean bond enthalpies**.

An *average* bond enthalpy indicates the strength of a *typical* bond.
The average bond enthalpy for the C–H bond is $+413 \text{ kJ mol}^{-1}$.

Bond making and bond breaking

Bond breaking requires energy: ENDOTHERMIC.
Bond making releases energy: EXOTHERMIC.

Chemical reactions involve bond breaking followed by bond making.

- Energy is first needed to break bonds in the reactants.
  Bond breaking is an endothermic process and **requires** energy.
- Energy is then released as new bonds are formed in the products.
  Bond making is an exothermic process and **releases** energy.

Bond enthalpy is an endothermic change (ΔH +ve) for bonds being broken.

When bonds are made, the enthalpy change will be the same magnitude but the opposite sign (ΔH –ve).

Using bond enthalpies to determine enthalpy changes

The enthalpy change for a reaction involving simple gaseous molecules can be determined using average bond enthalpies in an **energy cycle**:

- Enthalpy required to break bonds $= \Sigma$(bond enthalpies of reactants)
- Enthalpy released to make bonds $= -\Sigma$(bond enthalpies of products)

Notice that the relative strengths of the bonds in the reactants and the bonds in the products decide whether a reaction is exothermic or endothermic.

The enthalpy change for a reaction involving gaseous molecules can be conveniently calculated using **average** bond enthalpies.

The calculated value however is only approximate – the **actual** bond enthalpies involved may differ from the average values.

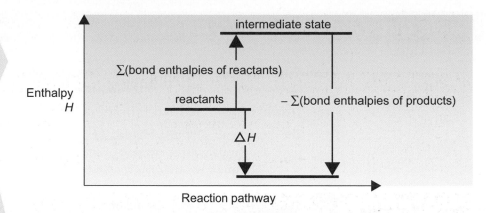

Enthalpy H

Reaction pathway

KEY POINT

From the diagram above,
$\Delta H = \Sigma(\text{bond enthalpies of reactants}) - \Sigma(\text{bond enthalpies of products})$.

For the reaction: $CH_4(g) + 2O_2(g) \longrightarrow CO_2(g) + 2H_2O(g)$,

$$4\ (C\text{–}H) + 2\ (O\text{=}O) \qquad\qquad 2\ (C\text{=}O) + 4\ (O\text{–}H)$$

$\Delta H/\text{kJ mol}^{-1}$ $(4 \times 413) + (2 \times 497)$ $(2 \times 740) + (4 \times 463)$

Bonds broken: (endothermic)   Bonds made: (exothermic)

$\Delta H = \Sigma(\text{bond enthalpies of reactants}) - \Sigma(\text{bond enthalpies of products})$
$\therefore\ \Delta H = [\ (4 \times 413) + (2 \times 497)\ ] - [\ (2 \times 740) + (4 \times 463)\ ]\ \text{kJ mol}^{-1}$
$\quad = \mathbf{-686\ kJ\ mol^{-1}}$

Progress check

1 You are provided with data for the following reaction:

$H_2(g) + I_2(g) \longrightarrow 2HI(g)$ $\Delta H = +53\ \text{kJ mol}^{-1}$
$H_2(g) + I_2(g) \longrightarrow 2H(g) + 2I(g)$ $\Delta H = +183\ \text{kJ mol}^{-1}$

(a) Draw an energy profile diagram for this reaction showing both these energy changes.

(b) Use your diagram to find out how much energy is released during bond formation in this reaction.

(c) In this reaction, what can be deduced about the relative strengths of the bonds that are broken and the bonds that are formed?

2 Use the data in the margin to find the enthalpy change for each reaction.

(a) $C_2H_4(g) + 3O_2(g) \longrightarrow 2CO_2(g) + 2H_2O(g)$
(b) $N_2(g) + 3H_2(g) \longrightarrow 2NH_3(g)$

bond	bond average enthalpy /kJmol⁻¹
C–H	+413
C=O	+740
O=O	+497
O–H	+463
C=C	+612
C–C	+347
N≡N	+945
H–H	+436
N–H	+391

2 (a) –1057 kJ mol⁻¹; (b) –93 kJ mol⁻¹.

(c) Bonds broken are stronger than bonds that are made.

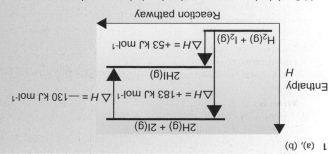

1 (a), (b)

Sample question and model answer

Enthalpy changes enable chemists to compare different fuels. A particularly useful term is the enthalpy change of combustion.

(a) Define the term standard enthalpy change of combustion.

> The standard enthalpy change of combustion is the enthalpy change that accompanies the complete combustion ✓ of one mole ✓ of a compound under standard conditions. [2]

The question uses 3 significant figures and your answer should also use this accuracy.

(b) In an experiment, the combustion of 0.00175 mol propan-1-ol C_3H_7OH produced sufficient heat to raise the temperature of 50.0 g of water by 15.0°C.

   (i) Calculate the energy released by the combustion of 0.00175 mol propan-1-ol. [Specific heat capacity, c, of water = 4.18 J g^{-1} K^{-1}.]

> energy released by combustion of propan-1-ol = energy gained by water
> energy gained by water = mass water × specific heat capacity × temperature change
>   = 50.0 × 4.18 × 15.0 J = 3135 J
>   = 3.14 kJ (to 3 sig. figs.) ✓

   (ii) Calculate the experimental enthalpy change of combustion of propan-1-ol.

Remember to include the sign! This reaction is exothermic and the negative sign is essential.

> enthalpy change from combustion of 0.00175 mol propan-1-ol = −3.14 kJ
> The definition for enthalpy change of combustion refers to 1 mole.
> ∴ $\Delta H_{combustion}$ = −3.14/.00175 ✓ = −1790 kJ mol^{-1} ✓ (to 3 sig. figs.) [3]

(c) Using the data below, show that the theoretical enthalpy change of combustion of propan-1-ol is −2022 kJ mol^{-1}.

substance	$C_3H_7OH(l)$	$CO_2(g)$	$H_2O(l)$
$\Delta H^\ominus_f$ / kJ mol^{-1}	−304	−394	−286

> The equation for this enthalpy change is:
> $C_3H_7OH(l) + 4\frac{1}{2}O_2(g) \longrightarrow 3CO_2(g) + 4H_2O(l)$ ✓
> Linking this equation to the ΔH data gives an enthalpy cycle:

The correct cycle is critical. Make sure the arrows go in the direction indicated by the definitions.

Here the data are ΔH values and the link is from the constituent elements to the reactants and products.

> $$C_3H_7OH(l) + 4\frac{1}{2}O_2(g) \xrightarrow{\Delta H} 3CO_2(g) + 4H_2O(l) \checkmark \text{(cycle)}$$
> −304 kJ mol^{-1} $(3 \times -394) + (4 \times -286)$ kJ mol^{-1}
> $3C(s) + 4H_2 + 5O_2(g)$
> From the cycle, $(-304) + \Delta H = (3 \times -394) + (4 \times -286)$ ✓
> ∴ ΔH = −2022 ✓ kJ mol^{-1} [4]

(d) This experiment was repeated several times to obtain an average 'experimental' value of −1800 kJ mol^{-1}. Compare this average 'experimental' value with the theoretical value in (c) and suggest **two** reasons for any difference.

*The word 'suggest' indicates that there may not be a 'correct answer'. Any **sensible** response would be credited.*

*DON'T blame shoddy practical techniques – these are **your** fault.*

> non-standard conditions/ heat loss/ evaporation of propanol/ incomplete combustion (up to 2 valid points) ✓ ✓ [2]

[Total: 11]

Cambridge Chemistry Foundation Q5 June 1996

Practice examination questions

1 (a) Write a chemical equation, including state symbols, for the reaction which is used to define the enthalpy of formation of one mole of aluminium chloride ($AlCl_3$). [2]

(b) State the additional condition necessary if the enthalpy change for this reaction is to be the standard enthalpy of formation, $\Delta H^{\ominus}_f$, at 298K. [1]

(c) Use the standard enthalpies of formation given below to calculate a value for the standard enthalpy change for the following reaction:

$$AlCl_3(s) + 6H_2O(l) \longrightarrow AlCl_3.6H_2O(s)$$

substance	$AlCl_3(s)$	$H_2O(l)$	$AlCl_3.6H_2O(s)$
$\Delta H^{\ominus}_f$ / kJ mol^{-1}	695	286	2680

[3]

[Total: 6]

NEAB Atomic Structure, Bonding and Periodicity Q5 Feb 1996 (modified)

2 Methanol is used as a fuel making use of combustion reaction with oxygen.

(a) (i) Define the term *standard enthalpy change of combustion*.
   (ii) State the temperature that is conventionally chosen for quoting standard enthalpy changes. [3]

(b) Calculate the standard enthalpy change of combustion of methanol using the following data.

compound	$\Delta H^{\ominus}_f$ /kJ mol^{-1}
$CH_3OH(l)$	−239
$CO_2(g)$	−394
$H_2O(l)$	−286

[3]

(c) The combustion of methanol is a highly exothermic reaction, but methanol does not spontaneously burst into flame and requires a spark for ignition. Explain why this should be so. [1]

[Total: 7]

Cambridge Chemistry Foundation Q3 June 1997 (modified)

3 (a) Define the term *standard enthalpy of formation* ($\Delta H^{\ominus}_f$). [3]

(b) Give the chemical equation for which the enthalpy change is the standard enthalpy of formation of methane. [2]

(c) Use the following data to calculate a value for the enthalpy of formation of methane.

$$C(s) + O_2(g) \longrightarrow CO_2(g) \qquad \Delta H = -394 \text{ kJ mol}^{-1}$$
$$H_2(g) + \tfrac{1}{2}O_2(g) \longrightarrow H_2O(g) \qquad \Delta H = -242 \text{ kJ mol}^{-1}$$
$$CH_4(g) + 2O_2(g) \longrightarrow CO_2(g) + 2H_2O(g) \qquad \Delta H = -802 \text{ kJ mol}^{-1} \quad [3]$$

(d) Use the following data to calculate mean bond enthalpy values for the C–H and the C–C bonds.

$$CH_4(g) \longrightarrow C(g) + 4H(g) \qquad \Delta H = 1648 \text{ kJ mol}^{-1}$$
$$C_2H_6(g) \longrightarrow 2C(g) + 6H(g) \qquad \Delta H = 2820 \text{ kJ mol}^{-1} \quad [3]$$

[Total: 11]

NEAB Atomic Structure, Bonding and Periodicity Q4 June 1995

Chapter 6

Rates and equilibrium

The following topics are covered in this chapter:

- Reaction rates
- Catalysis
- Chemical equilibrium
- Acids and bases

6.1 Reaction rates

After studying this section you should be able to:

- *recall the factors which control the rate of a chemical reaction*
- *understand, in terms of collision theory, the effect of concentration changes on the rate of a reaction*
- *explain the significance of activation energy using the Boltzmann distribution*
- *use the Boltzmann distribution to explain the effect of temperature changes on a reaction rate*
- *understand that an enthalpy change indicates the feasibility of a reaction*

What is a reaction rate?

AQA	M2	OCR	M3
EDEXCEL	M2	SALTERS	M1, M2
NICCEA	M2	WJEC	CH2
NUFFIELD	M2		

The rate of a chemical reaction is a measure of how quickly a reaction takes place.

> **The rate of a reaction is usually measured as the rate of change of concentration of a stated species in a reaction.**
>
> The units of rate are mol dm^{-3} s^{-1}.

KEY POINT

Measuring reaction rates

Although concentration can be measured, other quantities proportional to concentration, such as gas volumes, may be easier to monitor. Rates of different reactions show a very wide variation and it may be more convenient to measure a reaction rate per minute or per hour rather than per second.

> The reasons why reaction rates are affected by these factors is discussed in the rest of this chapter.

What affects a reaction rate?

The rate of a reaction may be affected by the following factors:

- the concentration of the reactants
- the surface area of solid reactants
- a temperature change
- the presence of a catalyst.

In some reactions, such as photosynthesis, the rate is affected by the presence and intensity of radiation.

How does reaction rate change during a reaction?

> **A** is being used up – its concentration decreases.
>
> **B** is being formed – its concentration increases.

For a reaction: **A** ⟶ **B**, the reaction rate = $\dfrac{\text{change of concentration}}{\text{time}}$

The reaction rate can be measured as:

- the rate of **decrease** in concentration of **A**
- the rate of **increase** in concentration of **B**.

The graphs below show how the concentrations of **A** and **B** change during the course of the reaction: **A** $\longrightarrow$ **B**.

A is used up and its concentration falls. **B** is formed and its concentration increases.

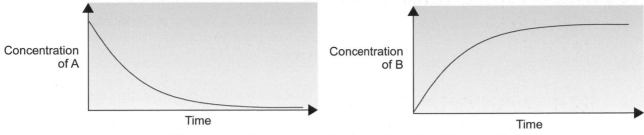

In the graph, the gradient indicates the reaction rate at any time.

- The steeper the gradient, the faster the rate of the reaction.
- The reaction is fastest at the start when the concentration of **A** is greatest.
- As the reaction proceeds, the rate slows down because the concentration of **A** decreases.
- When the reaction is complete, the graph levels off and the gradient becomes zero.

Activation energy

AQA	M2	OCR	M3
EDEXCEL	M2	SALTERS	M2
NICCEA	M2	WJEC	CH2
NUFFIELD	M2		

The activation energy of a reaction is the minimum energy required for the reaction to occur.

Only those collisions of sufficient energy to overcome the activation energy lead to a reaction.

In a gas or solution, particles are in constant motion and they collide both with each other, with any solid species and with the walls of their container. When particles collide, a reaction can only take place if the energy of the collision exceeds the activation energy of the reaction.

The *activation energy* of a reaction is the energy required to start a reaction by breaking bonds (see page 94).

- Activation energy is often supplied by a spark or by heating the reactants.
- Reactions with a small activation energy often take place very readily.
- A large activation energy may 'protect' the reactants from taking part in the reaction (at room temperature).

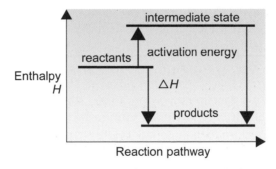

Reaction pathway

The Boltzmann distribution

AQA	M2	OCR	M3
EDEXCEL	M2	SALTERS	M2
NICCEA	M2	WJEC	CH2
NUFFIELD	M2		

The Boltzmann distribution shows the distribution of molecular energies in a gas at constant temperature.

The Boltzmann distribution

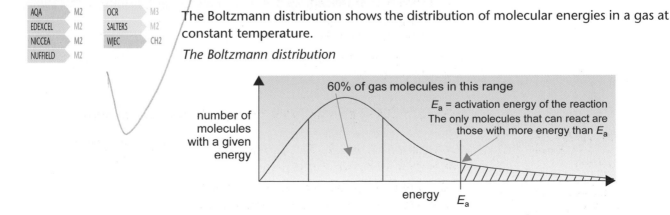

Characteristics of the Boltzmann distribution

- Most gas molecules have energies within a comparatively narrow range.
- The curve will only meet the energy axis at infinity energy. No molecules have zero energy.
- The area under the distribution curve gives the total number of gas molecules.
- Only those molecules with more energy than the activation energy of the reaction are able to react.

The effect of a concentration change on reaction rate

If the concentration of a reactant in solution or in a gas mixture is increased,

- there are more particles present per volume
- more collisions take place each second
- more collisions exceed the activation energy every second
- therefore the rate of reaction increases.

The diagram below shows distribution curves for two concentrations, C^1 and C^2. where concentration, C_2 > concentration, C_1.

- Providing the temperature is the **same**, distribution curves for different concentrations have the **same** shape.

> A reaction can only take place if the activation energy is exceeded.

> Note that the proportion of the total number of molecules exceeding the activation energy is the same. The rate increases because there are more molecules per volume and more molecules must now exceed the activation energy.

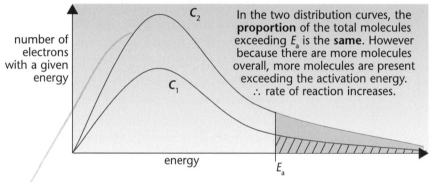

In the two distribution curves, the **proportion** of the total molecules exceeding E_a is the **same**. However because there are more molecules overall, more molecules are present exceeding the activation energy. ∴ rate of reaction increases.

For reactions involving gases, increasing the pressure also increases the concentration of any gas. This results in an increased reaction rate.

The effect of a change in surface area on reaction rate

For a reaction involving a solid, the reaction takes place at a faster rate when the solid is in a powdered form rather than as lumps.
Calcium carbonate reacts with hydrochloric acid producing carbon dioxide gas:

$$CaCO_3(s) + 2HCl(aq) \longrightarrow CaCl_2(aq) + H_2O(l) + CO_2(g)$$

By measuring the volume of carbon dioxide gas evolved with time, the rate of this reaction can be monitored. The graph below compares the reaction rates when using an excess of calcium carbonate as powder and as lumps. The same volume of the same concentration of hydrochloric acid has been reacted in both experiments.

> The gradient provides an indication of reaction rate.
>
> The steeper the curve, the faster the reaction.

> The final volume is the same in both experiments since the same amount (in mol) of hydrochloric acid has been reacted with an excess of calcium carbonate.

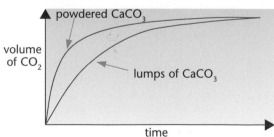

The gradient of the graph is much steeper with powdered carbonate because the surface area available for the reaction with hydrochloric acid is much greater than with lumps allowing more collisions per second.

The effect of a temperature change on reaction rate

The average kinetic energy of the particles is proportional to temperature. As temperature increases, so does the kinetic energy of gas molecules.

The diagram below shows distribution curves for a sample of gas at two temperatures, T_1 and T_2, where temperature, $T_2 >$ temperature, T_1.

Increasing the temperature **moves** the distribution curve to the right.

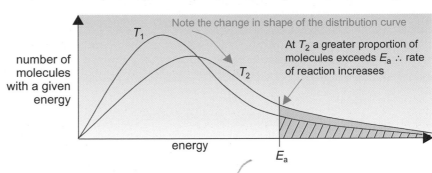

Increasing the temperature **does not change** the activation energy or the total number of molecules – the **shape of the curve changes**.

A **greater proportion** of molecules **exceeds the activation energy at higher temperature**.

The Boltzmann distribution curve is displaced to the right with the peak lower. The average energy is now increased.

The total area under each curve is a measure of the total number of molecules present, and this is the same for each curve.

An increase in temperature increases the rate of a reaction because:

- the molecules move faster and have more kinetic energy
- there are more collisions each second
- the increased kinetic energy produces more energetic collisions
- **a greater proportion of molecules exceed the activation energy.**

The feasibility of a chemical reaction

EDEXCEL	M2
NICCEA	M2
WJEC	CH2

The value of the enthalpy change for a reaction provides an indication of the *relative stability* of reactants and products and the *feasibility* of a reaction.

For example, sucrose is oxidised by oxygen in a highly exothermic reaction:

$$C_{12}H_{22}O_{11}(s) + 12O_2(g) \longrightarrow 12CO_2(l) + 11H_2O(g) \quad \Delta H = -5644 \text{ kJmol}^{-1}$$

- The reaction is *very likely* to take place because of the great amount of energy released in this reaction – the products are *thermodynamically* far more stable.
- However, the reaction does not take place spontaneously at room temperature because the reaction has a large activation energy which acts as an energy barrier – the reactants are *kinetically* stable.

Activation energy is the energy barrier of a reaction.

Progress check

1 Explain the following, in terms of collision theory and distribution graphs.
   (a) Reactions take place quicker when the reactants are more concentrated.
   (b) Reactions take place quicker at higher temperatures.

1 (a) Because there are more molecules per volume, more collisions take place each second. More collisions exceed the activation energy every second, increasing the reaction rate.
(b) The molecules move faster and have more kinetic energy. There are more collisions each second and the increased kinetic energy produces more energetic collisions. A greater proportion of molecules exceeds the activation energy.

6.2 Catalysis

After studying this section you should be able to:

- *understand what is meant by a catalyst*
- *explain how a catalyst changes the activation energy of a reaction*
- *explain what is meant by homogeneous and heterogeneous catalysis*

LEARNING SUMMARY

How do catalysts work?

AQA	M2	OCR	M1
EDEXCEL	M2	SALTERS	M1, M2
NICCEA	M2	WJEC	CH2
NUFFIELD	M2		

What is a catalyst?

A catalyst changes the rate of a chemical reaction, but is unchanged at the end of the reaction. Most catalysts speed up reaction rates although there are some (called inhibitors) that slow down reactions.

> These key points are often tested in exams.

KEY POINT

A catalyst speeds up the rate of a reaction by providing an **alternative route** for the reaction with a **lower activation energy**.
Activation energy with catalyst, E_c < Activation energy without catalyst, E_a

The effects of a catalyst on reaction route and activation energy are shown in the diagrams below.

Energy profile diagram

> Notice the way the catalyst reduces the energy barrier to the reaction.
>
> The catalyst provides an alternative route in which $E_c < E_a$

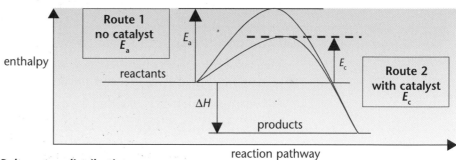

> The catalyst does **not** change the distribution curve.
>
> Notice how more molecules have an energy exceeding the new, lower activation energy.

Boltzmann distribution curve

E_c = activation energy with a catalyst. A greater proportion of molecules now exceeds the lower activation energy ∴ reaction rate increases.

number of molecules with a given energy

energy

E_c E_a

activation energy decreased

proportion of molecules exceeding the activation energy without a catalyst

Heterogeneous and homogeneous catalysis

OCR	M3	NICCEA	M2
SALTERS	M1, M2	WJEC	CH2

A **hetero**geneous catalyst is in a **different** phase from the reactants. Many reactions between **gases** are catalysed using a **solid** catalyst.

A **homo**geneous catalyst is in the **same** phase as the reactants. Many reactions between **gases** in the atmosphere are catalysed using a **gaseous** catalyst. Homogeneous catalysis also takes place in many **aqueous** reactions in which both the reactants and the catalyst are dissolved in water.

>
> hetero- = different
> homo- = same

Many heterogeneous catalysts are transition metals or their compounds.

Heterogeneous catalysis

The examples below are gaseous reactions catalysed by a solid catalyst.

Haber Process for ammonia (see also pages 107–108)

$$N_2(g) + 3H_2(g) \rightleftharpoons 2NH_3(g)$$

- The **gaseous** equilibrium process is catalysed by a **solid** iron catalyst, Fe(s).

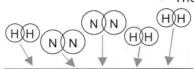

surface of iron catalyst
Reactant molecules diffuse towards the catalyst surface

Reactant molecules adsorbed at active sites on the catalyst surface

Bonds weaken allowing reaction to take place

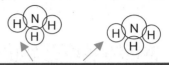

Following reaction, the product molecules diffuse away from the catalyst surface

Polluting gases temporarily bond to the catalyst's surface and it is here that reactions are able to take place. The products then leave the surface, allowing processing of more polluting gases by the catalyst.

Contact process for sulphur trioxide (and subsequently sulphuric acid)

$$2SO_2(g) + O_2(g) \rightleftharpoons 2SO_3(g)$$

- The **gaseous** equilibrium process is catalysed by **solid** vanadium(V) oxide, $V_2O_5(s)$.

Catalytic converters (see also page 120)

Many polluting gases, such as carbon monoxide, CO, and nitrogen monoxide, NO, are produced in a car engine.

These gases are formed by:

The catalytic converter has a surface area equivalent to two football pitches! The reacting gases now have a much greater chance of being adsorbed onto the catalyst so that reaction can take place.

- incomplete combustion of hydrocarbons forming CO
- reaction between N_2 and O_2 in a hot car engine forming NO.

These polluting **gases** are made to react using the catalytic converter in a car's exhaust system. The **solid** Rh/Pt/Pd catalyst is supported on a ceramic honeycomb which gives a large surface area for catalysis, a feature that greatly improves the effectiveness of many solid heterogeneous catalysts.

The catalytic converter removes much of this pollution in a series of reactions, e.g.

$$2CO(g) + 2NO(g) \longrightarrow 2CO_2(g) + N_2(g)$$

Homogeneous catalysis

A homogeneous catalyst does react but it is then regenerated, to be used again in a chain reaction – the middleman of chemistry.

During homogeneous catalysis, the catalyst **does** react forming an *intermediate state* with **lower activation energy**. Subsequently, the catalyst is regenerated, allowing further reaction to take place. Overall the catalyst is not used up, but it is actively involved in the process.

The breakdown of atmospheric ozone, catalysed by Cl· radicals

The Cl· free radicals are recycled in a chain reaction. A single Cl· free radical can break down thousands of O_3 molecules.

Chlorine free radicals, Cl· are formed in the stratosphere by the action of ultraviolet radiation on chlorofluorocarbons, CFCs. The **gaseous** Cl· free radicals catalyse the breakdown of **gaseous** ozone – the catalysis is homogeneous.

chain reaction: $O_3(g) + Cl\cdot(g) \longrightarrow O_2(g) + ClO\cdot(g)$
　　　　　　　　　$O(g) + ClO\cdot(g) \longrightarrow O_2(g) + Cl\cdot(g)$

overall:　　　　 $O_3(g) + O(g) \longrightarrow 2O_2(g)$

See also free-radical substitution of alkanes, p.121.

The O(g) free radicals are formed naturally in the stratosphere by the action of ultraviolet radiation on $O_2(g)$.

6.3 Chemical equilibrium

After studying this section you should be able to:

- explain the main features of a dynamic equilibrium
- predict the effects of changes on the position of equilibrium
- know that a catalyst does not affect the position of equilibrium
- understand why a compromise temperature and pressure may be used to obtain an economic yield in industrial processes

LEARNING SUMMARY

Reversible reactions and dynamic equilibrium

AQA	M2	OCR	M3
EDEXCEL	M2	SALTERS	M2
NICCEA	M2	WJEC	CH2
NUFFIELD	M2		

In a chemical equation,

reactants are on the left-hand side;

products are on the right-hand side.

Reversible reactions

Many chemical reactions take place until the reactants are completely used up and such reactions *'go to completion'*.

An example of a reaction that goes to completion is the oxidation of magnesium:

$$2Mg(s) + O_2(g) \longrightarrow 2MgO(s)$$

- The sign '$\longrightarrow$' indicates that the reaction takes place from left to right: this is called the *forward direction*.

Many reactions are **reversible**: they can take place in either direction.

An example of a reaction that is reversible is the formation of ammonia $NH_3(g)$ from nitrogen $N_2(g)$ and hydrogen $H_2(g)$.

$$N_2(g) + 3H_2(g) \rightleftharpoons 2NH_3(g)$$

The sign '$\rightleftharpoons$' indicates that the reaction is reversible.

The reversible reaction:
$N_2(g) + 3H_2(g) \rightleftharpoons 2NH_3(g)$
is used to illustrate dynamic equilibrium throughout this chapter.

This is a homogeneous equilibrium – one in which all reactants and products have the same phase. In this system, all are gases.

Approaching equilibrium

- If nitrogen and hydrogen are mixed together, the reaction can proceed only in the forward direction because no products are yet present:

$$N_2(g) + 3H_2(g) \longrightarrow$$

- As the reaction proceeds, ammonia is formed. The ammonia starts to react in the *reverse direction*, indicated by '$\longleftarrow$', forming nitrogen and hydrogen. At first, there is so little ammonia present that the reverse process takes place extremely slowly:

$$N_2(g) + 3H_2(g) \underset{\longleftarrow}{\longrightarrow} 2NH_3(g)$$

- As more ammonia is formed the reverse process takes place faster. Nitrogen and hydrogen are being used up and the forward reaction slows down. Eventually, the reverse process takes place at the same rate as the forward reaction and **equilibrium** is attained:

$$N_2(g) + 3H_2(g) \rightleftharpoons 2NH_3(g)$$

Dynamic equilibrium

At equilibrium, there is a balance: reactants and products are both present and the reaction *appears* to have stopped.

- Although there is no apparent change, both forward and reverse processes continue to take place – the equilibrium is *dynamic*.
- The forward reaction proceeds at the same rate as the reverse reaction.

In a chemical equation,

reactants are on the left-hand side;

products are on the right-hand side.

A reversible reaction can be approached from either direction and the terms *reactants* and *products* need to be used with caution.

- The concentrations of reactants and products are constant.

The equilibrium position:

- can be reached from either forward or reverse directions
- can only be achieved in a *closed system* – one in which no materials are being added or removed.

Equilibrium may be approached from either direction

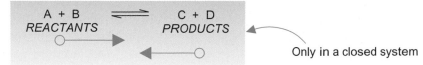

Only in a closed system

Factors affecting equilibrium

AQA	M2	OCR	M3
EDEXCEL	M2	SALTERS	M2
NICCEA	M2	WJEC	CH2
NUFFIELD	M2		

By opening up a closed equilibrium system, conditions can be changed after which the system can be allowed to reach equilibrium again.

The equilibrium position of the system may be altered by the following changes.

- Changing the concentration of a reactant or product.
- Changing the pressure of a gaseous equilibrium.
- Changing the temperature.

The likely effect on the equilibrium position can be predicted using Le Chatelier's Principle.

> *Le Chatelier's Principle* states that if a system in dynamic equilibrium is subjected to a change, processes will occur to minimise this change and to restore equilibrium.
>
> **KEY POINT**

The effect of concentration changes

A change in the concentration of a reactant will alter the rate of the forward direction. A change in the concentration of a product will alter the rate of the reverse direction. Either change results in a shift in the equilibrium position.

$$N_2(g) + 3H_2(g) \rightleftharpoons 2NH_3(g)$$

If the concentration of a reactant is increased, or the concentration of the product is decreased, e.g. by removing some of it, the equilibrium is displaced to the right and more product is obtained.

Change: **Increase** concentration of a reactant: $N_2(g)$ or $H_2(g)$

Change opposed: The equilibrium system will **decrease** the concentration of the reactant by removing it. This is achieved by a shift in equilibrium to the right, forming more $NH_3(g)$.

The effects of changes in concentration for this equilibrium are shown below. These effects will apply to any system in equilibrium.

> increase concentration of reactant OR reduce concentration of product
>
> $$N_2(g) + 3H_2(g) \rightleftharpoons 2NH_3(g)$$
>
> decrease concentration of reactant OR increase concentration of product
>
> **KEY POINT**

The effect of pressure changes

A change in total pressure may alter the equilibrium position of a system involving gases. **The direction favoured depends upon the total number of gas molecules on each side of the equilibrium.**

$$N_2(g) + 3H_2(g) \rightleftharpoons 2NH_3(g)$$

Change: **Increase** the total pressure.

Change opposed: The equilibrium system will **decrease** the pressure by reducing the total number of moles of gas molecules. This is achieved by a shift in equilibrium to the right:

$$N_2(g) \quad + \quad 3H_2(g) \quad \rightleftharpoons \quad 2NH_3(g)$$
$$\text{1 mol} \qquad\quad \text{3 mol} \qquad\qquad \text{2 mol}$$
$$\underbrace{\qquad\qquad\qquad\qquad} $$
$$\text{4 mol} \qquad\qquad\qquad\qquad \text{2 mol}$$

> Movement of gas molecules causes pressure. The more gas molecules in a volume, the greater the pressure.

> If there are more moles of gaseous reactant than there are moles of gaseous product, an increase in total pressure will displace the reaction to the right.

KEY POINT

pressure increase

more gas molecules $\xrightarrow{\hspace{3cm}}$ $N_2(g) + 3H_2(g) \rightleftharpoons 2NH_3(g)$ $\xleftarrow{\hspace{3cm}}$ fewer gas molecules

pressure decrease

Increasing the pressure also increases the concentration of any gas present and this **speeds** up the reaction.

The equilibrium position is also changed if:

- at least one of the equilibrium species is a gas
- there are different numbers of gaseous moles of reactants and of products.

The effect of temperature changes

A change in temperature alters the rates of the forward and reverse reactions by different amounts, resulting in a shift in the equilibrium position. **The direction favoured depends upon the sign of the enthalpy change.**

$$N_2(g) + 3H_2(g) \rightleftharpoons 2NH_3(g) \qquad \Delta H^\ominus = -92 \text{ kJ mol}^{-1}$$

Change: **Decrease** the temperature.

Change opposed: The equilibrium system will **increase** the temperature by releasing more heat energy. This is achieved by a shift in equilibrium in the exothermic direction, to the right.

> An exothermic process is favoured by low temperatures.
> An endothermic process is favoured by high temperatures.

> An exothermic reaction in one direction is endothermic in the opposite direction.
> The value of the enthalpy changes is the same – the sign is different.

KEY POINT

exothermic process favoured by low temperatures

$\Delta H^\ominus = +92$ kJ mol^{-1} $N_2(g) + 3H_2(g) \rightleftharpoons 2NH_3(g)$ $\Delta H^\ominus = -92$ kJ mol^{-1}

endothermic process favoured by high temperatures

Increasing the temperature *speeds* up the reaction and less time is needed to reach equilibrium. However, the equilibrium position may change to produce a lower equilibrium yield of products (see also pages 107–108).

The effect of a catalyst

There is **no change** in the equilibrium position. However, the catalyst **speeds** up both forward and reverse reactions and equilibrium is reached quicker.

Equilibria, rates and industrial processes

AQA	M2	OCR	M3
EDEXCEL	M2	WJEC	CH2
NICCEA	M2		

All courses study one or more industrial processes, usually including the Haber and Contact processes.

You should learn outline details of any process on your course.

The principles, however, are common for any similar situation.

For many industrial processes, the overall yield of a chemical product can vary considerably, depending on the conditions used.

In deciding the conditions to use, industrial chemists need to consider:

- the availability of the starting materials required for a process
- the equilibrium conditions required to ensure a good yield
- the rate of reaction which should be fast but manageable
- the cost, which should be as low as possible, taking into account energy, cost of materials and of the chemical plant
- the safety of workers from hazardous chemicals, pressure, temperature, etc.
- any effects on the environment from waste discharges, toxic fumes, etc.

Industrial chemists compare each of these factors to arrive at compromise conditions. Although the equilibrium yield might be **optimised** by using a very high temperature and pressure, this may prove **impracticable** for reasons of cost and safety.

The Haber Process for the production of ammonia

AQA	M2	OCR	M3
EDEXCEL	M2	WJEC	CH2
NICCEA	M2		

$$N_2(g) + 3H_2(g) \rightleftharpoons 2NH_3(g) \qquad \Delta H^\ominus = -92 \text{ kJ mol}^{-1}$$

The raw materials

The raw materials used should be readily available. The Haber Process uses:

- the air, as a source of $N_2(g)$
- natural gas, CH_4, and water, H_2O, from which $H_2(g)$ is extracted.

The optimum equilibrium conditions

Consider Le Chatelier's Principle

- The forward reaction produces fewer moles of gas, favoured by a **high** pressure.
- The forward reaction producing NH_3 is exothermic, favoured by a **low** temperature.

Optimising the process

These are the ideal conditions to give a maximum equilibrium yield.

The optimum equilibrium conditions for maximum equilibrium yield are:

- **high pressure and low temperature.**

The need for compromise

Compromising

Optimum conditions, reaction rate, feasibility, reality, safety and economics must be considered.

Having arrived at optimum equilibrium conditions, each must be considered to arrive at compromise conditions.

optimum condition	advantages	disadvantages
	equilibrium yield of ammonia is high	energy costs are high – it is expensive to compress gases
HIGH PRESSURE	the concentration of gases is high, increasing the rate	there are considerable safety implications of using very high pressures – vessel walls need to be very thick to withstand high pressures and weaknesses cause danger to workers and potential leakage of chemicals into the environment
LOW TEMPERATURE	equilibrium yield of ammonia is high	the reaction takes place very slowly as few molecules possess the activation energy of the reaction

The use of a catalyst

A catalyst speeds up the rates of both the forward and reverse reactions and hence less time is needed for the reaction to reach equilibrium. The increase in reaction rate allows lower temperatures to be used for a realistic reaction rate. The use of lower temperatures has the added bonus of supporting the optimum conditions for this process, increasing the equilibrium yield of ammonia.

Compromise conditions

Exam papers always test the wider aspects of chemistry to society.

Exam answers need to be sensible and related to relevant issues arising from the process itself. It is pointless to use terms such as 'dangerous', 'toxic', etc. unless substantiated with the reasons why the hazard is present.

The compromise conditions used in the Haber Process are such that:

- the temperature is sufficiently high to allow the reaction to occur at a realistic rate, but not too high to give a minimal equilibrium yield of ammonia
- a high pressure is used but not too high as to be impracticable
- the reaction rate is increased by using an iron catalyst with potassium hydroxide added to promote its activity. This enables the process to take place at lower temperatures, saving on energy costs.

The compromise conditions used (20 000 kPa and 450°C) only yield 15% $NH_3(g)$ which is removed by liquefying at −40°C. Unreacted $N_2(g)$ and $H_2(g)$ are recycled.

Economic considerations

Much of the cost of any industrial process goes on energy. Companies carry out research to reduce energy demands. Recent developments mean that the Haber Process can be carried out using:

- lower temperatures (a greater equilibrium yield and less energy for heating)
- lower pressures (less energy required to compress gases).

Progress check

1 What will be the result of an increase in pressure on the following equilibria?
(a) $N_2O_4(g) \rightleftharpoons 2NO_2(g)$
(b) $CO(g) + 2H_2(g) \rightleftharpoons CH_3OH(g)$
(c) $H_2(g) + Br_2(g) \rightleftharpoons 2HBr(g)$

2 What will be the result of an increase in temperature on the following equilibria?
(a) $N_2(g) + O_2(g) \rightleftharpoons 2NO(g)$ $\Delta H^\ominus = +180$ kJ mol^{-1}
(b) $H_2(g) + Br_2(g) \rightleftharpoons 2HBr(g)$ $\Delta H^\ominus = -9.6$ kJ mol^{-1}
(c) $2SO_2(g) + O_2(g) \rightleftharpoons 2SO_3(g)$ $\Delta H^\ominus = -197$ kJ mol^{-1}

3 What are the optimum conditions of temperature **and** pressure for a high yield of the products in the equilibria below?
(a) $CO(g) + 2H_2(g) \rightleftharpoons CH_3OH(g)$ $\Delta H^\ominus = -92$ kJ mol^{-1}
(b) $PCl_5(g) \rightleftharpoons PCl_3(g) + Cl_2(g)$ $\Delta H^\ominus = +124$ kJ mol^{-1}

3 (a) low temperature and high pressure;
(b) high temperature and low pressure.
2 (a) moves to right; (b) moves to left; (c) moves to left.
1 (a) moves to left; (b) moves to right; (c) no change.

6.4 Acids and bases

After studying this section you should be able to:

- describe and use the Brønsted–Lowry theory of acids and bases
- explain the differences between strong and weak acids
- describe the typical reactions of an acid in terms of H⁺(aq)
- describe the use of ammonia as a base in the preparation of fertilisers

Acids, bases and alkalis

| NUFFIELD | M1 | SALTERS | M2 |
| OCR | M3 | WJEC | CH2 |

An acid is a proton donor.
A base is a proton acceptor.

The Brønsted–Lowry theory treats acids and bases in terms of proton transfer.

KEY POINT

A Brønsted–Lowry acid is a proton, H⁺, donor
A Brønsted–Lowry base is a proton, H⁺, acceptor

Acids

A hydrogen ion, H⁺ is simply a hydrogen atom with an electron removed.
Because a hydrogen atom comprises one proton and one electron, removing the electron leaves just a proton.

'H⁺', 'Hydrogen ion' and proton refer to the same thing.

Common acids include:

- strong 'mineral' acids: sulphuric acid, H_2SO_4; hydrochloric acid, HCl; nitric acid, HNO_3
- weak organic acids: ethanoic acid, CH_3COOH; methanoic acid, HCOOH; benzoic acid, C_6H_5COOH; phenol, C_6H_5OH

When added to water, HCl acts as a Brønsted–Lowry acid, donating a proton to water to form the **oxonium** ion, $H_3O^+(aq)$:

$$HCl(aq) + H_2O(l) \longrightarrow H_3O^+(aq) + Cl^-(aq)$$
proton donor → H⁺ → oxonium ion

- The oxonium ion, $H_3O^+(aq)$ is the 'active' part of an aqueous acid.
- It is common practice to simply use H⁺(aq) instead of $H_3O^+(aq)$ and the dissociation of HCl can more simply be shown as:

$$HCl(aq) + aq \longrightarrow H^+(aq) + Cl^-(aq)$$

KEY POINT

All acids generate H⁺(aq) in aqueous solution.

Strong acids

H_2SO_4, HCl and HNO_3 are strong acids because dissociation in water goes virtually to completion.

A **strong** acid such as nitric acid HNO_3 is a **good** proton donor, with near to complete dissociation in water.

$$HNO_3(l) + aq \longrightarrow H^+(aq) + NO_3^-(aq)$$

A **strong** acid **completely dissociates** into ions.

Weak acids

A weak acid only partially dissociates

A **weak** acid such as ethanoic acid CH_3COOH is a **poor** proton donor. The dissociation in water is an equilibrium, with the equilibrium position well to the left-hand side of the equation.

$$CH_3COOH(aq) \rightleftharpoons H^+(aq) + CH_3COO^-(aq)$$
equilibrium position

For every 250 molecules of ethanoic acid, only 1 molecule dissociates and only a small proportion of the potential acidic power is released as H⁺.

A **weak** acid only **partially dissociates** into ions.

Bases

Common bases include:
- metal oxides: magnesium oxide, MgO; copper(II) oxide, CuO
- metal hydroxides: sodium hydroxide, NaOH; iron(III) hydroxide, $Fe(OH)_3$
- ammonia and amines: ammonia, NH_3; methylamine, CH_3NH_2

Alkalis

An alkali, such as NaOH, is a base that dissolves in water forming $OH^-(aq)$ ions:

$$NaOH(s) + aq \longrightarrow Na^+(aq) + OH^-(aq)$$

There are only a few common alkalis.
- strong alkalis: potassium hydroxide, KOH; lithium hydroxide, LiOH
- weak alkalis: ammonia, NH_3; magnesium hydroxide, $Mg(OH)_2$

Typical reactions of an acid

OCR   M3
NUFFIELD M1

Ionic equations can be written for the typical acid reactions. Note the role of $H^+(aq)$ in each of these reactions. By using $H^+(aq)$, the same reaction can be written, irrespective of the acid. (Remember, the 'active part' of an acid is $H^+(aq)$.)

Using hydrochloric acid as an example, typical reactions of an acid are shown below. The ionic equations clearly show the role of $H^+(aq)$.

Acids react with many metals, evolving hydrogen

$$2HCl(aq) + Mg(s) \longrightarrow H_2(g) + MgCl_2(aq)$$
$$2H^+(aq) + Mg(s) \longrightarrow H_2(g) + Mg^{2+}(aq)$$

Acids react with carbonates, evolving carbon dioxide

$$2HCl(aq) + CaCO_3(s) \longrightarrow CaCl_2(aq) + H_2O(l) + CO_2(g)$$
$$2H^+(aq) + CaCO_3(s) \longrightarrow H_2O(l) + CO_2(g) + Ca^{2+}(aq)$$

Acids react with bases (metal oxides), forming water

$$2HCl(aq) + CaO(s) \longrightarrow H_2O(l) + CaCl_2(aq)$$
$$2H^+(aq) + CaO(s) \longrightarrow H_2O(l) + Ca^{2+}(aq)$$

Acids react with alkalis, forming water

$$HCl(aq) + NaOH(aq) \longrightarrow NaCl(aq) + H_2O(l)$$
$$H^+(aq) + OH^-(aq) \longrightarrow H_2O(l)$$

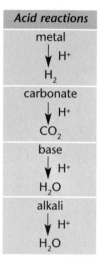

Acid reactions

metal → H^+ → H_2

carbonate → H^+ → CO_2

base → H^+ → H_2O

alkali → H^+ → H_2O

Salts

In each of these reactions a salt is formed, in which the active part of the acid, H^+, has been replaced by a metal. Examples of some common salts are shown below.

Salts are always formed in solution from these typical acid reactions. The salt can be isolated by evaporation of water.

acid		salt	example
sulphuric acid	H_2SO_4	sulphate	Na_2SO_4
hydrochloric acid	HCl	chloride	NaCl
nitric acid	HNO_3	nitrate	KNO_3

Ammonia as a base

EDEXCEL   M2
OCR   M3

Ammonia NH_3 is a colourless gas, very soluble in water forming a weakly alkaline solution:

$$NH_3(g) + H_2O(l) \rightleftharpoons NH_4^+(aq) + OH^-(aq)$$

Much of the ammonia formed in the Haber Process (see page 107) is neutralised by acids to form ammonium salts, such as ammonium nitrate, shown below.

$$NH_3(aq) + HNO_3(aq) \longrightarrow NH_4NO_3(aq)$$

Ammonium salts such as ammonium nitrate are very soluble in water and have a high nitrogen content. They are extensively used as artificial fertilisers.

Sample question and model answer

1

The curve below represents the distribution of molecular energies at a temperature T_1 for a mixture of gases which react with each other. **A** is the activation energy for the reaction.

> Notice that the distribution curve starts at the origin but it never quite reaches the x-axis, even at high energies. Remember also that the area under the curve is equal to the total number of molecules.

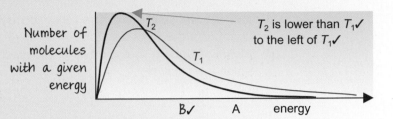

Number of molecules with a given energy

T_2 is lower than T_1 ✓
to the left of T_1 ✓

B✓   A   energy

(a) (i) Label the vertical axis.

(ii) Explain the meaning of the term *activation energy*.

This is the minimum ✓ energy required for a reaction to take place. ✓

> At different temperatures,
> the curve changes
> the activation energy stays the same.
>
> Note the difference
>
> In the presence of a catalyst,
> the activation energy changes
> the curve stays the same.

(iii) Draw a second curve on the same axes, and label it T_2, for the same mixture at a lower temperature.

(iv) By reference to the curves, state and explain in molecular terms the effect of reducing the temperature on the rate of reaction.

Reduced temperature slows down the reaction rate. ✓
The overall kinetic energy of the molecules is reduced ✓ and
fewer molecules now exceed the activation energy. ✓   [9]

(b) The reaction is repeated in the presence of a catalyst.

(i) Mark on the energy axis a possible activation energy B for the catalysed reaction.

(ii) State and explain, in molecular terms, the effect of the catalyst on the rate of reaction.

The catalyst speeds up the reaction rate. ✓ A catalyst allows the
reaction to proceed via a different route ✓ with a lower activation
energy, B. This means that a larger proportion of molecules now
exceeds the activation energy. ✓   [4]

> Always pay attention to the words that examiners use.
>
> In parts (i) and (ii), the examiner has used 'Explain'. You need to justify your answer.
>
> In part (iii), the examiner has used 'State'. You only need to 'state' your answer. You waste time here if you 'explain'.

(c) The equation for the industrial manufacture of ammonia is given below.

$$N_2(g) + 3H_2(g) \rightleftharpoons 2NH_3(g) \qquad \Delta H = -92 \text{ kJ mol}^{-1}$$

A temperature of 450°C is used in industry, together with pressures between 200 and about 250 atmospheres. A catalyst is also used.

(i) Explain why the reaction is carried out at a high pressure.

There is a greater equilibrium yield. ✓ The increase in pressure
is relieved by reducing the number of gas molecules. ✓ The
equilibrium shifts in favour to the right because there are fewer
gas molecules to the right. ✓

(ii) Explain why pressures higher than 1000 atmospheres are not used.

High pressures are very costly to generate in terms of energy. ✓

(iii) State **one** advantage and **one** disadvantage of carrying out the reaction at high temperature.

An advantage is that there is a greater reaction rate. ✓
A disadvantage is that the equilibrium yield is reduced. ✓   [6]

[Total: 19]

NEAB Kinetics and Organic Chemistry Q2 June 1995 (modified)

Practice examination questions

1 (a) Define the term *activation energy*.   [1]
   (b) The endothermic reaction between substances **P** and **Q** can be represented
        by the following equation.
        $$P(g) + Q(g) \longrightarrow R(g) + S(g)$$
        (i)   Complete the reaction profile for this reaction, shown in **Figure 1**.
              Indicate and label clearly the activation energy and the enthalpy
              change for the reaction.
        (ii) Also on **Figure 1**, draw and label the reaction profile for the reaction
              when it is catalysed.

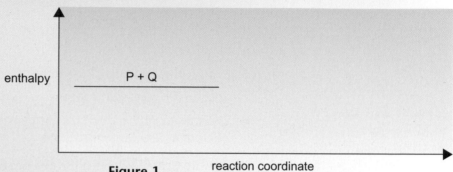

Figure 1

        (iii) Explain why a catalyst increases the rate of a chemical reaction.   [6]

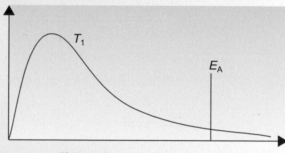

Figure 2

   (c) (i)   **Figure 2** shows the distribution of molecular energies for the mixture of
              reactants in **(b)** at a temperature T_1. Label the axes and, on the same
              axes, draw the distribution of molecular energies for the same gaseous
              mixture at a higher temperature T_2.
        (ii) In **Figure 2**, the energy corresponding to the activation energy of the
              reaction (E_A) is shown. Indicate a possible energy corresponding to the
              activation energy of the catalysed reaction and label it E_A **catalysed**.

                                                                                        [5]

                    [Total: 12] *AEB Paper 1 Q6 Jun 1997 (modified)*

2 A dynamic homogeneous equilibrium is established when hydrogen and carbon
   dioxide react as shown below:
        $$H_2(g) + CO_2(g) \rightleftharpoons H_2O(g) + CO(g)$$
   (a) (i)   Why is this reaction a *homogeneous equilibrium*?
        (ii) State **three** features of a *dynamic equilibrium*.   [5]
   (b) The equilibrium yield of steam and carbon monoxide increases as the
        temperature is increased. Determine whether the reaction as written is
        exothermic or endothermic. Explain your answer.   [2]
   (c) Explain the effect of an increase in pressure on:
        (i)   the equilibrium position; (ii) the reaction rate.   [4]
                                                                             [Total 11]

Chapter 7
Organic chemistry

The following topics are covered in this chapter:

- Basic concepts
- Hydrocarbons from oil
- Alkanes
- Alkenes
- Alcohols
- Halogenoalkanes

7.1 Basic concepts

After studying this section you should be able to:

- understand the different types of formula used for organic compounds
- understand the terms saturated and unsaturated hydrocarbon
- understand what is meant by structural isomerism
- recognise the alkanes and alkyl groups
- understand what is meant by a homologous series
- recognise common functional groups
- apply rules for naming simple organic compounds
- understand the difference between homolytic and heterolytic fission
- calculate percentage yields

LEARNING SUMMARY

Types of formula

AQA	M3	OCR	M2
EDEXCEL	M2	SALTERS	M1
NICCEA	M2	WJEC	CH2
NUFFIELD	M1		

In organic chemistry, there are many ways of representing a formula.
For the compound butane, with 4 carbon atoms and 10 hydrogen atoms:

the *empirical* formula is: C_2H_5 — The simplest, whole-number ratio of elements in a compound.

the *molecular* formula is: C_4H_{10} — The *actual* number of atoms of each element in a molecule.

the *structural* formula is: $CH_3CH_2CH_2CH_3$ — The minimal detail for an unambiguous structure.

the *displayed* formula is:

$$H-\overset{\overset{\displaystyle H}{|}}{\underset{\underset{\displaystyle H}{|}}{C}}-\overset{\overset{\displaystyle H}{|}}{\underset{\underset{\displaystyle H}{|}}{C}}-\overset{\overset{\displaystyle H}{|}}{\underset{\underset{\displaystyle H}{|}}{C}}-\overset{\overset{\displaystyle H}{|}}{\underset{\underset{\displaystyle H}{|}}{C}}-H$$

The relative placing of atoms and the bonds between them.

the *skeletal* formula — The carbon skeleton and functional groups only.

Carbon chains

AQA	M3	OCR	M2
EDEXCEL	M2	SALTERS	M1
NICCEA	M2	WJEC	CH2
NUFFIELD	M1		

Hydrocarbons

Hydrocarbons are compounds of carbon and hydrogen only.

- A saturated hydrocarbon has single bonds only.
- An unsaturated hydrocarbon contains a multiple carbon carbon bond.

saturated
single bonds only

unsaturated
contains a double bond

Alkanes

Carbon atoms can bond with other carbon atoms to form an enormous range of compounds with different carbon-chain lengths. The simplest organic compounds are a family of saturated hydrocarbons called the alkanes, shown below.

meth- C
eth- C_2
prop- C_3
but- C_4
pent- C_5
hex- C_6

Number of carbons	Name	Molecular formula	Structural formula
1	methane	CH_4	CH_4
2	ethane	C_2H_6	CH_3CH_3
3	propane	C_3H_8	$CH_3CH_2CH_3$
4	butane	C_4H_{10}	$CH_3CH_2CH_2CH_3$
5	pentane	C_5H_{12}	$CH_3CH_2CH_2CH_2CH_3$
6	hexane	C_6H_{14}	$CH_3CH_2CH_2CH_2CH_2CH_3$

Note the following points.

- The name of the alkane ends with –*ane*.
- The prefixes (*meth-*, *eth-*, …) are used to represent the number of carbon atoms. You will need to use these many times in organic chemistry and they must be learnt.

General formula

Alkanes, C_nH_{2n+2} are saturated.

- The **general formula** is the simplest algebraic representation for any member in a series of organic compounds.
- The general formula for any alkane is C_nH_{2n+2}, where n = the number of carbon atoms.

Homologous series

Members in a homologous series react similarly.

By studying the reactions of one member of the series, you know how all members in the series are likely to react.

The alkanes are an example of a **homologous series** with the following features.

- Each successive member differs by $-CH_2-$. You can see this by comparing the formula of each alkane in the table above.
- Each member in a homologous series has the same general formula. The alkanes have the general formula: C_nH_{2n+2}.
- All members in a homologous series have the same functional group and similar chemical reactions.

Note that physical properties, such as boiling point and density, do gradually change as the length of the carbon chain increases (see also page 120).

Isomerism

AQA	M3	OCR	M2
EDEXCEL	M2	SALTERS	M1
NICCEA	M2	WJEC	CH2
NUFFIELD	M1		

In addition to forming long chains, the atoms making up a molecular formula are often arranged differently, forming **isomers**.

Structural isomerism

Chain isomerism

This is a type of structural isomerism in which the carbon skeleton is different. Butane and methylpropane are chain isomers of C_4H_{10}

Structural isomers are molecules with the same molecular formula but with different structural arrangements of atoms.

The two structural isomers of C_4H_{10} are shown below:

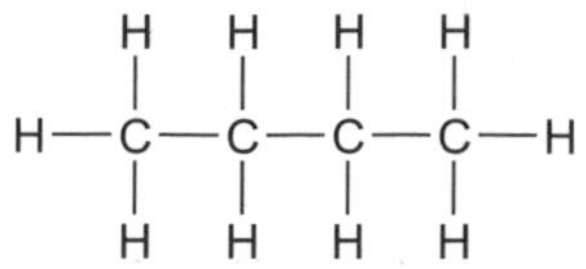

butane 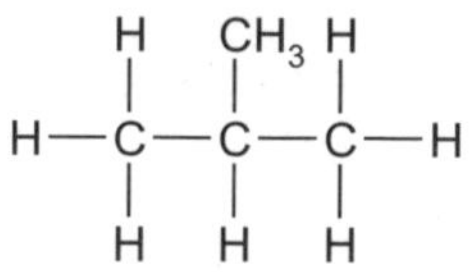   methylpropane

See also *cis–trans* isomerism, p.122.

The number of different structural isomers possible from a molecular formula increases dramatically as the carbon-chain length increases. This is shown in the table below.

molecular formula	number of isomers
C_5H_{12}	3
C_6H_{14}	5
C_7H_{16}	9
C_8H_{18}	18
C_9H_{20}	35
$C_{10}H_{22}$	75
$C_{15}H_{32}$	4,347
$C_{20}H_{42}$	366,319

Look at the different names given to the two isomers.

- The branched isomer, methylpropane, is treated as a side chain attached to a straight-chain alkane.
- Side chains, such as the methyl group CH_3-, are called *alkyl* groups with the general formula of C_nH_{2n+1}.
- An alkyl group can be regarded as an alkane with one hydrogen atom removed (to allow another group to be attached).

The naming of organic compounds is discussed in more detail on page 116.

Notice how the alkyl groups are named and how their formulae are linked to the parent alkane.

You must learn these.

Alkanes and alkyl groups

Number of carbons	Alkane C_nH_{2n+2} Formula	Name	Alkyl group C_nH_{2n+1} Formula	Name
1	CH_4	methane	CH_3-	methyl
2	C_2H_6	ethane	C_2H_5-	ethyl
3	C_3H_8	propane	C_3H_7-	propyl

Functional groups

AQA	M3	OCR	M2
EDEXCEL	M2	SALTERS	M1
NICCEA	M2	WJEC	CH2
NUFFIELD	M1		

Saturated hydrocarbon chains are comparatively unreactive. The reactivity is increased by the presence of a functional group – the reactive part of a carbon compound.

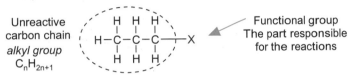

Unreactive carbon chain
alkyl group
C_nH_{2n+1}

Functional group
The part responsible for the reactions

Common functional groups

It is essential that you can instantly identify a functional group within a molecule. You can then apply the relevant chemistry.

If you continue to study Chemistry to A Level, you will meet more functional groups.

The main functional groups for AS Chemistry are shown below.

name	functional group	structural formula	general formula	prefix or suffix (for naming)
alkane	C–H	$CH_3CH_2CH_3$ *propane*	C_nH_{2n+2}	-ane
alkene	$\diagdown C = C \diagup$	$CH_3CH=CH_2$ *propene*	C_nH_{2n}	-ene
halogenoalkane	C–X (X=halogen)	CH_3CH_2Br *bromoethane*	$C_nH_{2n+1}Br$	bromo-
alcohol	C–OH	CH_3CH_2OH *ethanol*	$C_nH_{2n+1}OH$	-ol
aldehyde	$-C{\diagup\!\!\!\!\overset{O}{\diagdown}}_H$	CH_3CHO *ethanal*	$C_nH_{2n}O$	-al
ketone	$R-\overset{O}{\overset{\|}{C}}-R$	$CH_3COC_2H_5$ *butanone*	$C_nH_{2n}O$	-one
carboxylic acid	$-C{\diagup\!\!\!\!\overset{O}{\diagdown}}_{OH}$	CH_3COOH *ethanoic acid*	$C_nH_{2n+1}COOH$	-oic acid

- When studying reactions of a homologous series, the alkyl group is relatively unimportant – it is the functional group that reacts.

- The alkyl group, C_nH_{2n+1} is often represented simply as R–. This shifts the emphasis in the formula towards the reactive part of the molecule, the functional group.

The alcohols, shown below, is a homologous series with the –OH functional group.

- The alcohols have the general formula:
$$C_nH_{2n+1}OH$$
- *Any* alcohol can be represented as:
R–OH

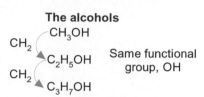

Naming of organic compounds

With so many isomers possible, it is important that each has an individual name. The examples below show the steps needed to name an organic compound.

Rule 1

- The name is based upon the longest carbon chain on an **alkane**.

Rule 2

- Any functional groups and alkyl groups are identified.
- These are then added to the name as a prefix (e.g. chloro-) or suffix (e.g. -ol) (see page 115).

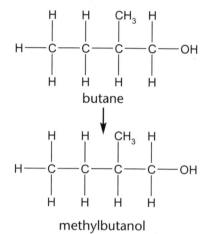

butane

methylbutanol

Rule 3

- If there is more than one possible isomer then the carbon atoms are labelled with numbers. Numbering starts from the end giving the lowest possible numbers for any functional groups and side chains. In this example, numbering from the left would give the incorrect name of 3-methylbutan-4-ol.

2-methylbutan-1-ol

Rule 4

- If there is more than one alkyl or functional group, they are placed in alphabetical order. This rule is not needed for the example above but it does need to be applied to name the compound below.

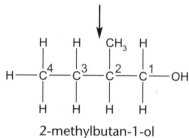

3-ethyl-2-methylpentane

Types of bond fission

AQA	M3	OCR	M2
EDEXCEL	M2	SALTERS	M2
NICCEA	M2	WJEC	CH2
NUFFIELD	M2		

The breaking of a covalent bond is called **bond fission**. Reactions of organic compounds involve bond fission followed by the formation of new bonds. Two types of bond fission are possible, **homolytic** fission and **heterolytic** fission.

Homolytic fission

These are very important principles, used throughout organic chemistry.

In homolytic fission, bond-breaking produces two species of the same (*homo-*) type:

$$A:B \longrightarrow A\bullet + \bullet B$$
free radicals

A free radical is a species with an unpaired electron (see p.121).

- In the example above, a covalent bond breaks so that one of the bonding electrons goes to each of **A** and **B**.
- Homolytic fission forms **two free-radicals**.

Heterolytic fission

Homolytic fission ⟶ free radicals

Heterolytic fission ⟶ ions

In heterolytic fission, bond-breaking produces two species of different (*hetero-*) type:

$$A{:}B \longrightarrow A{:}^- + B^+ \quad \text{or} \quad A{:}B \longrightarrow A^+ + {:}B^-$$
$$\qquad\qquad\quad ions \qquad\qquad\qquad\qquad\qquad\quad ions$$

- In the example above, a covalent bond breaks so that both the bonding electrons go to either **A** or **B**.
- Heterolytic fission forms **oppositely-charged ions**.

Percentage yields

| EDEXCEL | M2 | NUFFIELD | M1 |
| NICCEA | M2 | OCR | M2 |

Organic reactions typically produce a lower yield than expected from the balanced equation. The yield is usually expressed as a percentage yield:

> $$\text{percentage yield} = \frac{\text{actual yield}}{\text{theoretical yield}} \times 100\%$$

KEY POINT

Example

5.2 g of 1-bromobutane is formed by reacting 5.0 g of butan-1-ol with sodium bromide and concentrated sulphuric acid. Find the percentage yield of 1-bromobutane.

Factors that can contribute to a low yield:

Organic reactions often do not go to completion.

An organic compound often reacts to produce a mixture of products.

The purification stages result in loss of some of the desired product.

equation $\quad C_4H_9OH + NaBr + H_2SO_4 \longrightarrow C_4H_9Br + H_2O + NaHSO_4$

moles $\qquad$ 1 mol $\qquad\qquad\qquad\longrightarrow$ 1 mol

reacting masses $\quad$ 74 g $\qquad\qquad\qquad\longrightarrow$ 137 g

$\qquad\qquad\qquad$ 1 g $\qquad\qquad\qquad\longrightarrow \dfrac{137}{74}$ g

$\qquad\qquad\qquad$ 5.0 g $\qquad\qquad\qquad\longrightarrow 5 \times \dfrac{137}{74}$ g

∴ 5.0 g of C_4H_9OH produces a theoretical yield of 9.3 g C_4H_9Br.

Mass of C_4H_9Br that forms = 5.2 g.

$$\text{percentage yield} = \frac{\text{actual yield}}{\text{theoretical yield}} \times 100\% = \frac{5.2}{9.3} \times 100 = 56\%$$

Progress check

C_6H_{14} has 5 structural isomers. Show the structural and displayed formula for each of these isomers and name them.

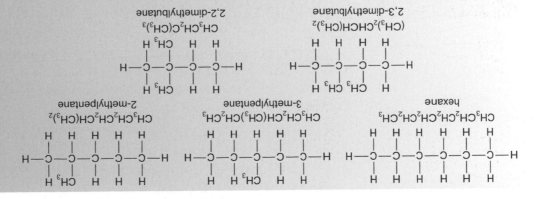

7.2 Hydrocarbons from oil

After studying this section you should be able to:

- understand how hydrocarbon fractions are separated from crude oil
- describe how low demand fractions are processed into hydrocarbons of higher demand

LEARNING
SUMMARY

Useful products from crude oil

AQA	M3	OCR	M2
NICCEA	M2	SALTERS	M1
NUFFIELD	M2	WJEC	CH2

Crude oil is a fossil fuel, formed from the decay of sea creatures over millions of years. It is a complex mixture of hydrocarbons, containing mainly alkanes. Crude oil cannot be used directly but it is the source of many useful products. The first stages in the processing of crude oil are described below.

Fractional distillation

Crude oil is heated and passed into a fractionating tower, separating the complex mixture into *fractions*. Note that the fractions are not pure and contain mixtures of hydrocarbons within a range of boiling points. The fractionating tower, shown below, shows the temperature gradient which separates the fractions according to their boiling points (see also page 120).

Fractions with the lowest boiling points are collected at the top of the tower.

On descending the tower, the temperature rises and the boiling points increase.

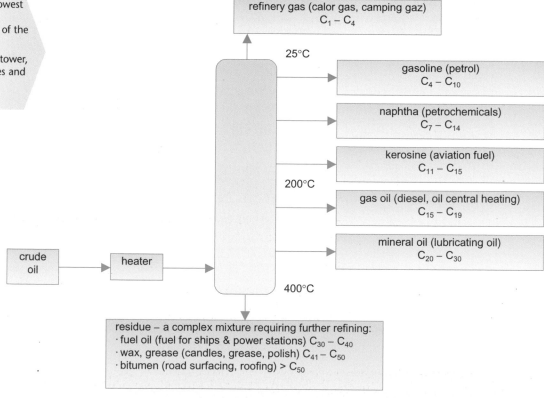

Cracking

Cracking is the starting point for the manufacture of many organics.

Cracking is the breaking down of an unsaturated hydrocarbon into smaller hydrocarbons.

The purpose of cracking is to produce high demand hydrocarbons:

- short-chain alkanes for use in petrol
- alkenes, as a feedstock for a wide range of organic chemicals, including polymers (see also page 125).

Thermal cracking heats the *naphtha fraction* with steam at a high temperature (about 800°C) and high pressure. This forms a mixture of straight-chain alkanes and alkenes (mainly ethene) with a small proportion of branched and cyclic hydrocarbons. Some hydrogen is also produced.

The following equations summarise the overall process of cracking.

$$C_{10}H_{22} \longrightarrow C_8H_{18} + C_2H_4$$
$$C_{10}H_{22} \longrightarrow C_6H_{14} + 2C_2H_4$$
$$\text{alkane} \longrightarrow \text{alkane} + \text{alkene}$$

> There are many more equations possible but all must produce both an alkane and an alkene.

> **Thermal cracking** breaks C-C bonds by homolytic fission forming free radicals.
>
> **Catalytic cracking** breaks C-C bonds by heterolytic fission forming ions.

Catalytic cracking processes heavy long-chain fractions obtained in larger quantities than required. A mixture of the vapourised fraction and a zeolite catalyst are reacted at about 450°C using a slight pressure only. The process forms a higher proportion of branched and cyclic hydrocarbons than thermal cracking (see also reforming and isomerisation below). These are used for petrol.

Reforming

> The use of leaded petrol has been phased out in the UK.
>
> Unleaded petrol and lead replacement petrol require a higher proportion of branched and cyclic hydrocarbons for effective combustion. This has increased demand for reformed alkanes.

Reforming uses heat and pressure to convert unbranched fractions into cycloalkanes (e.g. cyclohexane) and arenes (e.g. benzene). These products are used in petrol and as a feedstock for a wide range of organic chemicals including many pharmaceuticals and dyes.

Examples

cyclohexane

benzene

Isomerisation

Isomerisation converts unbranched hydrocarbons into branched hydrocarbons, needed for petrol.

Example

Progress check

1 Write down the formula of an alkane present in each of the main fractions obtained from crude oil.
2 Construct equations for two possible reactions taking place during the cracking of $C_{14}H_{30}$.

1 Any alkane with a formula C_nH_{2n+2} (see page xx) to match each fraction.
2 Any 2 equations —— alkane + alkene, e.g. $C_{14}H_{30} \longrightarrow C_8H_{18} + C_6H_{12}$

7.3 Alkanes

After studying this section you should be able to:

- *understand why different alkanes have different melting and boiling*
- *understand that combustion of alkanes provides useful energy*
- *describe the substitution reactions of alkanes with halogens*

LEARNING SUMMARY

Alkanes

General formula: C_nH_{2n+2}

AQA	M3	OCR	M2
EDEXCEL	M2	SALTERS	M1
NICCEA	M2	WJEC	CH2
NUFFIELD	M2		

Alkanes are generally unreactive. Alkanes contain only C–H and C–C bonds, which are relatively strong and difficult to break. The similar electronegativities of carbon and hydrogen give molecules which are non-polar. Alkanes are the typical 'oils' used in many non-polar solvents and they do not mix with water.

Melting points and boiling points of alkanes

Boiling points increase with increasing chain length because of the increasing strength of the intermolecular forces. This allows the alkane mixture in crude oil to be separated by fractional distillation.

> The same principle can be applied to any homologous series.

Increasing chain length

m.pt and b.pt increase

more surface of contact between molecules

greater van der Waals' forces between molecules

C_3H_8 → C_5H_{12}

For hydrocarbons of similar molecular mass, branched alkanes have lower boiling points because there are fewer interactions possible between molecules:

> The surface contact between molecules is important because it allows interaction between electrons. Remember that van der Waals' forces result from fluctuations in electron density (see p.55).

Isomers of C_4H_{10}

m.pt and b.pt increase

more surface of contact between molecules

greater van der Waals' forces between molecules

Combustion of alkanes

AQA	M3	OCR	M2
EDEXCEL	M2	SALTERS	M1
NICCEA	M2	WJEC	CH2
NUFFIELD	M2		

> Although generally unreactive, alkanes do react with oxygen and the halogens.

Combustion with oxygen is the most important reaction of alkanes giving their immediate use as fuels:

natural gas: $CH_4 + 2O_2 \longrightarrow CO_2 + 2H_2O$
petrol: $C_8H_{18} + 12\frac{1}{2}O_2 \longrightarrow 8CO_2 + 9H_2O$

The burning of fossil fuels such as alkanes provides society with many environmental problems, some of which are listed below.

> Remember that reactions with oxygen produce oxides.

- Petroleum fractions contain sulphur impurities. Unless removed before combustion, sulphur oxides are formed from which acid rain (H_2SO_3 and H_2SO_4) is produced.
- Toxic gases such as carbon monoxide, nitrogen oxides and unburnt hydrocarbons are present in car emissions. The development of catalytic converters has helped to remove these polluting gases (see also page 103).

> These environmental issues are often tested in exams.

- Excessive combustion of fossil fuels increases carbon dioxide emissions contributing to the 'Greenhouse Effect'.

Substitution of alkanes with halogens

AQA	M3	OCR	M2
EDEXCEL	M2	SALTERS	M2
NICCEA	M2	WJEC	CH2
NUFFIELD	M2		

> **KEY POINT**
>
> A *substitution reaction* involves the swapping over of one species for another.

Alkanes are substituted by halogens, such as chlorine and bromine.

$$CH_4 + Cl_2 \longrightarrow CH_3Cl + HCl$$

This reaction only takes place in the presence of ultraviolet radiation, which produces highly reactive free radicals.

> **KEY POINT**
>
> A free radical, such as $Cl\bullet$, $CH_3\bullet$,
> * is a highly reactive species with an unpaired electron
> * reacts by pairing of an unpaired electron
> * is often involved in chain reactions.

Mechanism of free-radical substitution (not EDEXCEL)

Initiation

> Make sure that you learn this mechanism – these are easy marks in exams if you do!

In this stage, the reaction starts.

Ultraviolet radiation provides energy to break Cl–Cl bonds homolytically producing chlorine free-radicals, $Cl\bullet$

$$Cl_2 \longrightarrow 2Cl\bullet$$

Propagation

In this stage, the reaction products are made.

Free radicals are recycled in a chain reaction

$$CH_4 + Cl\bullet \longrightarrow CH_3\bullet + HCl$$
$$CH_3\bullet + Cl_2 \longrightarrow CH_3Cl + Cl\bullet$$

> The chlorine free-radicals catalyse the reaction.
>
> See also the breakdown of ozone, p.103.

Termination

In this stage, free radicals react together and are removed from the reaction mixture.

$$Cl\bullet + Cl\bullet \longrightarrow Cl_2$$
$$CH_3\bullet + Cl\bullet \longrightarrow CH_3Cl$$
$$CH_3\bullet + CH_3\bullet \longrightarrow CH_3CH_3$$

Impure products

> This is answered poorly in exams.

Further substitution of the reaction products is possible, producing a mixture of products:

$$CH_3Cl \xrightarrow{Cl\bullet} CH_2Cl_2 \xrightarrow{Cl\bullet} CHCl_3 \xrightarrow{Cl\bullet} CCl_4$$

Progress check

1. Write an equation for the complete combustion of butane.
2. Butane reacts with chlorine in a substitution reaction.
   (a) What conditions are essential for this reaction? Explain your answer.
   (b) Write an equation for this reaction.
   (c) Write the structural formula of 2 possible monosubstituted products.
   (d) What is the formula for the final product obtained from complete substitution?

7.4 Alkenes

After studying this section you should be able to:

- *recognise the nature of cis-trans isomerism in alkenes*
- *understand that the double bond in alkenes takes part in addition reactions with electrophiles*
- *understand the use of alkenes in the production of organic compounds*
- *understand what is meant by addition polymerisation*

LEARNING SUMMARY

Alkenes

General formula: C_nH_{2n}

AQA	M3	OCR	M2
EDEXCEL	M2	SALTERS	M2
NICCEA	M2	WJEC	CH2
NUFFIELD	M2		

Alkenes are unsaturated compounds with a C=C double bond. The high electron density of the double bond makes alkenes more reactive than alkanes.

Naming of alkenes

Positional isomerism

A type of structural isomerism in which the functional group is in a different position on the carbon skeleton.

The position of the double bond is indicated using one number only, as shown in the example below for two isomers of C_4H_8.

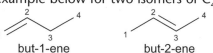

but-1-ene   but-2-ene

Only one number is needed. in but-1-ene, a double bond starting at carbon-1 must finish at carbon-2

What is a double bond?

The coverage here is greatly simplified but sufficient for AS Chemistry.

In saturated compounds, the C–H and C–C single bonds are σ-bonds. A σ-bond is on the C–H or C–C axis, formed by overlap of s- and p- atomic orbitals.

In unsaturated compounds, the C=C double bond comprises a σ- and a π-bond.

σ- and π-bonds are *molecular orbitals* – these bond together the atoms in a molecule.

A π-bond is above and below the C–C axis, formed by overlap of atomic p-orbitals. The π-bond introduces an area of high electron density in the molecule, shown in the diagram of ethene below.

overlap forms a

σ-bond

π-bond

p-orbitals

The C=C bond
planar molecule
3 σ-electron pairs around each carbon atom.
Bond-angle is approximately 120°

cis-trans (or geometric) isomerism

See also structural isomerism, p.114.

Single C–C bonds can rotate. The C=C double bond, however, restricts rotation and prevents groups from moving from one side of the double bond to the other. This gives rise to *cis-trans* (or geometric) isomerism.

No rotation possible about a C=C bond.

The *cis-trans* isomers of the but-2-ene are shown here:

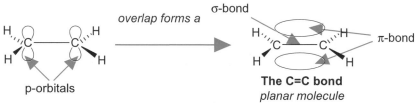

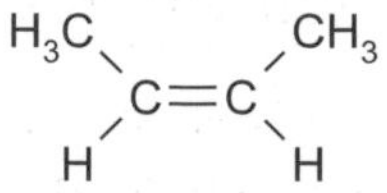

cis-isomer
(groups on same side)

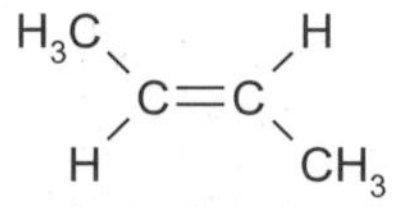

trans-isomer
(groups on opposite sides)

These are relatively easy exam marks. Make sure that you learn this thoroughly.

Note that, for *cis-trans* isomerism, there must be:

- a C=C double bond
- two **different** groups attached to **each** carbon end of the double bond.

Addition reactions of alkenes

AQA	M3	OCR	M2
EDEXCEL	M2	SALTERS	M2
NICCEA	M2	WJEC	CH2
NUFFIELD	M2		

An addition reaction of an alkene involves the opening of the double bond with the formation of a saturated addition product.

> **KEY POINT**
> An *addition reaction* involves two species adding together to make one.
> Addition usually converts an **unsaturated** compound into a **saturated** compound.

This is made possible by the high electron density of the π-bond which attracts **electrophiles**.

> **KEY POINT**
> An *electrophile*, such as Br_2, HBr, H_2SO_4, NO_2^+:
> • is an electron-deficient species
> • 'attacks' an electron-rich carbon atom or double bond by **accepting a pair of electrons**.

This type of reaction is called **electrophilic addition**.

Addition of bromine

The addition reaction between an alkene and bromine, in which the orange colour of bromine is decolourised, is used as a test for unsaturation.

unsaturated ——— addition ———▶ saturated

$$H_2C=CH_2 \ + \ Br_2 \ \xrightarrow{\text{room temperature}} \ \underset{\underset{Br \ \ Br}{|\quad|}}{H-C-C-H}$$

orange colour ——— bromine is decolorised ———▶ colourless

Mechanism of electrophilic addition (not EDEXCEL)

Be careful with the direction of the curly arrows.

Don't get confused by the charges and partial charges within this mechanism.

double bond of alkene induces a dipole on Br_2

electrophilic attack

addition product

> **KEY POINT**
> In mechanisms, *a curly arrow* is used to show the movement of **a pair of electrons**.
> The movement of the electron pair involves either:
> • the formation of a new covalent bond or
> • the heterolytic fission of a covalent bond (see page 117).
> The *curly arrow* always goes **from** an electron pair (or double bond).

Remember that a curly arrow shows the movement of an electron pair.

Addition using acidified manganate(VII)

Note that [O] is used in equations to represent an oxidising agent. Many candidates in exams get this reaction wrong by incorrectly adding Mn. $KMnO_4$ is used as a source of manganate (VII) ions.

Another useful test for unsaturation is the use of cold, dilute acidified manganate(VII). This oxidises the double bond of the alkene with the formation of a diol. The purple colour of manganate(VII) is decolourised.

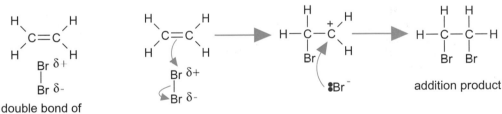

$$H_2C=CH_2 \ + \ [O] \ + \ H_2O \ \xrightarrow[\text{cold}]{H^+ / MnO_4^-} \ \underset{\underset{OH \ \ OH}{|\quad\quad|}}{H-C-C-H}$$

ethane-1,2-diol

Reduction with hydrogen

Reduction of a organic compound involves loss of oxygen OR gain of hydrogen (accompanied by a gain of electrons).

Alkenes are reduced when treated with hydrogen gas in the presence of a nickel catalyst at 150°C. The process of adding hydrogen across a double bond is sometimes referred to as **hydrogenation**.

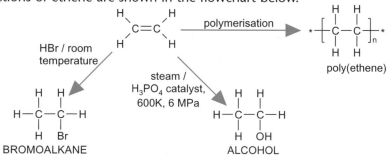

Addition of hydrogen across C=C double bond

Further addition reactions of ethene

Hydrogenation of C=C is used in the production of solid margarine from unsaturated liquid vegetable oils. Saturated fats are solid and the amount of hydrogenation can be controlled to give the correct texture of margarine.

For more details:

polymerisation, see p.125;

preparation of alcohols, see also p.128 ;

Addition reactions of alkenes are useful in organic synthesis. Some further addition reactions of ethene are shown in the flowchart below.

Although ethanol is made industrially using steam and phosphoric acid H_3PO_4, an alternative synthesis uses:

- addition with concentrated sulphuric acid at room temperature
$$C_2H_4 + H_2SO_4 \longrightarrow CH_3CH_2OSO_3H$$
- followed by hydrolysis
$$CH_3CH_2OSO_3H + H_2O \longrightarrow CH_3CH_2OH + H_2SO_4$$

Addition reactions of unsymmetrical alkenes

In the addition of HBr to propene $CH_3CH=CH_2$, two products are possible:

- the secondary bromoalkane $CH_3CHBrCH_3$ as the major product
- the primary bromoalkane $CH_3CH_2CH_2Br$ as the minor product.

Mechanism (AQA and WJEC only)

The driving force for the formation of $CH_3CHBrCH_3$ is the intermediate carbonium ion, shown in the mechanism below.

The reaction with H_2SO_4 proceeds via a similar mechanism. H_2SO_4 can be treated as $H–OSO_2OH$.

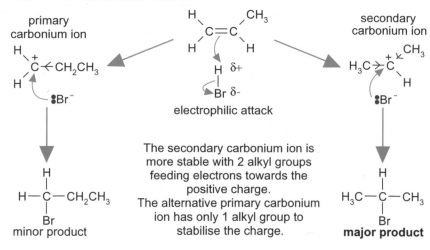

The secondary carbonium ion is more stable with 2 alkyl groups feeding electrons towards the positive charge.
The alternative primary carbonium ion has only 1 alkyl group to stabilise the charge.

Addition polymerisation of alkenes

AQA	M3	OCR	M2
EDEXCEL	M2	SALTERS	M2
NICCEA	M2	WJEC	CH2
NUFFIELD	M2		

An addition polymer is a long-chain molecule with a high molecular mass, made by joining together many small molecules called monomers:

MANY MONOMERS $\longrightarrow$ SINGLE POLYMER

n molecules $\longrightarrow$ 1 single molecule

- The **monomer** used is an **unsaturated** alkene containing a double C=C bond.
- The addition **polymer** formed is a **saturated** compound **without** a double bond.

Many different addition polymers can be formed by using different alkene monomer units, based on the ethene molecule.

The equations below shows the formation of poly(ethene), poly(chloroethene) (*pvc*) and poly(phenylethene) (*polystyrene*).

> You should be able to draw a short section of a polymer given the monomer units (and *vice versa*).

> Make sure that you can show the repeat unit of a polymer.

- Notice that the principle is the same for each addition polymerisation.

- Each of the polymer structures shows the *repeat unit* in brackets. This is repeated thousands of times in each polymer molecule.

- It is important to show the repeat unit with 'side-links' to indicate that both sides are attached also to other repeat units.

ethene — poly(ethene)

chloroethene — poly(chloroethene)

phenylethene — poly(phenylethene)

Properties of monomers and polymers

- The monomers are volatile liquids or gases.
- Polymers are solids.
- This difference in physical properties is explained by increased van der Waals' forces between the much larger polymer molecules (see page 55).

Problems with disposal

- Addition polymers are non-biodegradable and take many years to break down.
- Disposal by burning can produce toxic fumes (e.g. depolymerisation produces monomers; dioxins from combustion of chlorinated polymers).

> Rather than simply disposing of waste polymers, there is now movement towards recycling, combustion for energy production, and use as a feedstock for cracking.
>
> This eliminates problems with disposal and helps to preserve finite energy resources.

Progress check

1. Write the structural formula for the organic product for the reaction of but-2-ene with the following:
   (a) Br$_2$;   (b) H$_2$/Ni;   (c) HBr;   (d) H$_2$O/H$_3$PO$_4$.

2. Show the repeat unit for the addition polymer formed from propene.

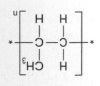

1 (a) CH$_3$CHBrCHBrCH$_3$ (b) CH$_3$CH$_2$CH$_2$CH$_3$ (c) CH$_3$CH$_2$CHBrCH$_3$
(d) CH$_3$CH$_2$CHOHCH$_3$.

Epoxyethane, $(CH_2)_2O$

AQA ▶ M3

$$\underset{H_2C-CH_2}{\overset{O}{\triangle}}$$

Epoxyethane is an important intermediate used by industry during the production of antifreeze, polyesters, solvents, plasticisers and detergents.

Production of epoxyethane

Epoxyethane is produced commercially by direct synthesis of ethene, with either oxygen or air, in the presence of a silver-based catalyst.

> Epoxyethane is prepared by partial oxidation of ethene.

$$\underset{H}{\overset{H}{>}}C=C\underset{H}{\overset{H}{<}} \quad + \quad \tfrac{1}{2}O_2 \quad \xrightarrow[\text{1–2 MPa}]{\text{Ag / }O_2\text{ 250–300°C}} \quad \underset{H_2C-CH_2}{\overset{O}{\triangle}} \quad + \quad H_2O$$

The catalyst is present as a finely-divided layer on a support such as alumina. Epoxyethane is purified by fractional distillation.

Reactions of epoxyethane

Hazards from preparing and handling epoxyethane

Epoxyethane is a colourless gas, b.p. 10 °C, which is both flammable and explosive.

Epoxyethane is also toxic and may cause irritation to the respiratory system and neurological effects.

The three-membered ring in epoxyethane is strained making the molecule highly reactive towards nucleophiles (see page 130) such as water and alcohols. Reaction brings about ring opening.

Hydrolysis of epoxyethane

Much of the epoxyethane produced is converted into ethane-1,2-diol which is used directly as antifreeze and indirectly to make polyesters such as *Terylene*. In industry, this is done by treating epoxyethane with an excess of water at 60°C in the presence of sulphuric acid.

$$\underset{H_2C-CH_2}{\overset{O}{\triangle}} \quad + \quad H_2O \quad \xrightarrow[\text{H}_2\text{SO}_4\text{ catalyst}]{\text{excess }H_2O\text{ / 60°C}} \quad \underset{CH_2-CH_2}{\overset{OH\ \ OH}{|\ \ \ \ |}}$$

The resulting aqueous solution is concentrated to 70% by evaporation and then purified by fractional distillation.

The properties of ethane-1,2-diol make it ideal for antifreeze:

- low melting point (–12°C)
- high boiling point (198°C)
- complete miscibility with water (with which it readily forms hydrogen bonds).

Ethane-1,2-diol is also used in the manufacture of polyesters, such as *Terylene*.

7.5 Alcohols

After studying this section you should be able to:

- understand the polarity and physical properties of alcohols
- describe how ethanol is prepared
- describe oxidation reactions of alcohols
- describe nucleophilic substitution reactions of alcohols
- describe esterification and dehydration of alcohols
- describe tests for the presence of the hydroxyl group

LEARNING SUMMARY

Alcohols

General formula: $C_nH_{2n+1}OH$

AQA	M3	OCR	M2
EDEXCEL	M2	SALTERS	M2
NICCEA	M2	WJEC	CH2
NUFFIELD	M1		

Types and naming of alcohols

The functional group in alcohols is the **hydroxyl** group, C–OH.

Alcohols can be classified as primary, secondary or tertiary, depending on how many alkyl groups are bonded to **C**–OH. You can see how these types of alcohols are named in the diagrams below.

primary alcohol	**secondary alcohol**	**tertiary alcohol**
1 alkyl group attached to C–OH	*2 alkyl groups attached to C–OH*	*3 alkyl groups attached to C–OH*

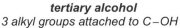

propan-1-ol — propan-2-ol — methylpropan-2-ol

Polarity of alcohols

The properties of alcohols are dominated by the hydroxyl group, C–OH.

Carbon, oxygen and hydrogen have different electronegativities, and alcohols have polar molecules:

Oxygen is more electronegative than both carbon and hydrogen – both carbon and hydrogen are electron deficient.

The polarity produces electron-deficient carbon and hydrogen atoms, indicated in the diagram above. Depending upon the reagents used, alcohols can react by breaking either the C–O or O–H bond.

Physical properties of alcohols

The –OH group dominates the physical properties of short-chain alcohols.

Hydrogen bonding takes place between alcohol molecules, resulting in:

- higher melting and boiling points than alkanes of comparable M_r
- solubility in water.

hydrogen bonding between ethanol molecules

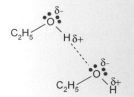

The solubility of alcohols in water decreases with increasing carbon chain length as the non-polar contribution to the molecule becomes more important.

miscible with water	**insoluble in water**
large polar contribution	*most of the molecule is a non-polar carbon chain*

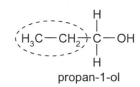 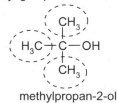

polar head — **polar head**

non-polar tail — **non-polar tail**

Preparation of ethanol

Ethanol finds widespread uses: in alcoholic drinks, as a solvent in the form of methylated spirits, and as a fuel. There are two main methods for its production.

Fermentation of sugars (for alcoholic drinks)

Yeast is added to an aqueous solution containing sugars.

E.g. $$C_6H_{12}O_6(aq) \longrightarrow 2C_2H_5OH(aq) + 2CO_2(g)$$

Hydration of ethene (for industrial alcohol)

Ethene and steam are passed over a phosphoric acid catalyst at 330°C under high pressure (see page 124).

$$C_2H_4(g) + H_2O(g) \longrightarrow C_2H_5OH(l)$$

The preparation method used for ethanol may depend on the raw materials available.

An oil-rich country may use oil to produce ethene, then ethanol. However, this method uses up finite oil reserves.

In a country without oil reserves (especially in hot climates), sugar may provide a renewable source for ethanol production.

Combustion of alcohols

Alcohols such as ethanol and methanol are used as fuels, making use of combustion.

$$C_2H_5OH + 3O_2 \longrightarrow 2CO_2 + 3H_2O$$

Ethanol is used as a petrol substitute in countries with limited oil reserves.

Methanol is used as a petrol additive in the UK to improve combustion of petrol. It also has increasing importance as a feedstock in the production of organic chemicals.

Oxidation of alcohols

The oxidation of different alcohols is an important reaction in organic chemistry. It links alcohols with aldehydes, ketones and carboxylic acids, shown below.

 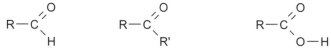

aldehyde, RCHO   ketone, RCOR'   carboxylic acid, RCOOH

- Aldehydes and ketones both contain the **carbonyl group** C=O.
- Carboxylic acids contain the **carboxyl** group COOH.

The stages of oxidation are shown below:

hydroxyl group C–OH $\longrightarrow$ carbonyl group C=O $\longrightarrow$ carboxyl group COOH

Oxidation of primary alcohols

A primary alcohol can be oxidised to an aldehyde and then to a carboxylic acid.

> Oxidation of an organic compound involves gain of oxygen OR loss of hydrogen (accompanied by loss of electrons from the organic compound).

KEY POINT

The stages of oxidation are shown below:

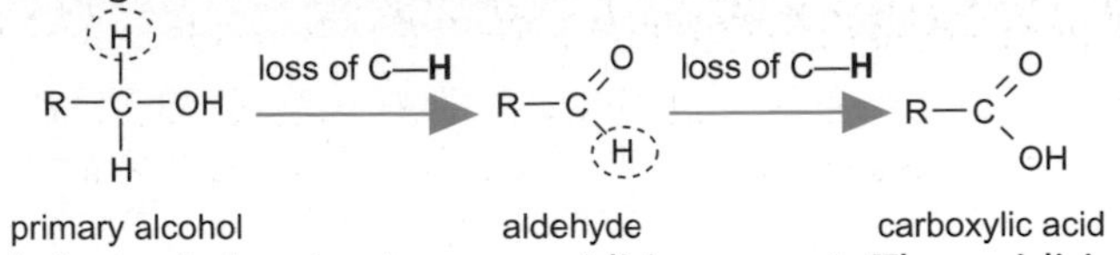

primary alcohol   aldehyde   carboxylic acid

This is carried out using an oxidising agent. The oxidising agent used is a mixture of concentrated sulphuric acid, H_2SO_4 (source of H^+) and potassium dichromate, $K_2Cr_2O_7$ (source of $Cr_2O_7^{2-}$).

> For balanced equations, the oxidising agent can be shown simply as [O].

- By heating and distilling the product immediately, oxidation can be stopped at the aldehyde stage.

$$CH_3CH_2OH + [O] \longrightarrow CH_3CHO + H_2O$$

- By refluxing with an excess of the oxidising agent, further oxidation takes place to form the carboxylic acid.

$$CH_3CH_2OH + 2[O] \longrightarrow CH_3COOH + H_2O$$

Oxidation of secondary alcohols

> The orange dichromate ions, $Cr_2O_7^{2-}$, are reduced to green Cr^{3+} ions.

By heating with $H^+/Cr_2O_7^{2-}$ a secondary alcohol can be oxidised to form a ketone. No further oxidation normally takes place.

secondary alcohol $\longrightarrow$ ketone

secondary alcohol   ketone

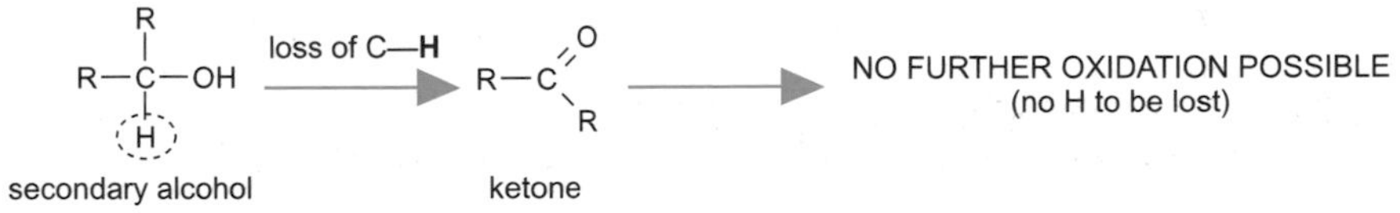

The equation for the oxidation of propan-2-ol is shown below.

$$CH_3(CHOH)CH_3 + [O] \longrightarrow CH_3COCH_3 + H_2O$$

Oxidation of tertiary alcohols

Tertiary alcohols are not oxidised under normal conditions.

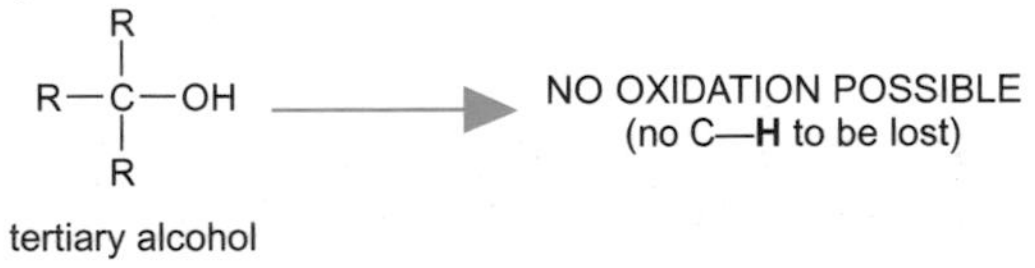

tertiary alcohol

Distinguishing between aldehydes and ketones (AQA and Nuff only)

> Tollens' reagent is a solution of silver nitrate (as a source of Ag^+ ions) in ammonia.
> $Ag^+ \longrightarrow Ag$
>
> Fehling's (or Benedict's) solution is an alkaline solution of Cu^{2+} ions.
> $Cu^{2+} \longrightarrow Cu^+$

The presence of an aldehyde can easily be detected using two tests.

Heat aldehyde with Tollens' reagent

- Tollens' reagent $\longrightarrow$ silver mirror

Heat aldehyde with Fehling's (or Benedict's) solution

- Benedict's / Fehling's solution $\longrightarrow$ brick-red precipitate of Cu_2O

Ketones cannot be oxidised and **do not** react with the above solutions and this provides an easy method for distinguishing between an aldehyde and a ketone.

Reduction of carbonyl compounds

WJEC   CH2

This is the reverse of the oxidation of alcohols shown above. Alcohols can be prepared by reduction of carbonyl compounds using a suitable reducing agent such as sodium borohydride, $NaBH_4$, in dry ether (at room temperature).

> For balanced equations, the reducing agent can be shown simply as [H].

$$CH_3CHO \quad + \quad 2[H] \quad \longrightarrow \quad CH_3CH_2OH$$
aldehyde   primary alcohol

$$C_2H_5COCH_3 \quad + \quad 2[H] \quad \longrightarrow \quad C_2H_5CH(OH)CH_3$$
ketone   secondary alcohol

Substitution reactions of alcohols

EDEXCEL	M2	OCR	M2
NICCEA	M2	SALTERS	M2

The electron-deficient carbon atom of the polar $C^{\delta+}$–$O^{\delta-}$ bond of alcohols attracts *nucleophiles*, allowing a **substitution** reaction to take place in which the nucleophile replaces the OH group. This is called **nucleophilic substitution** (see also halogenoalkanes, page 132).

> **KEY POINT**
>
> A nucleophile, such as OH^-, Br^-, H_2O, NH_3, CH_3OH, CN^-
> - is an electron-rich species
> - 'attacks' an electron-deficient carbon atom by **donating a pair of electrons**.

An alternative method is to use phosphorus halides.

For chloroalkanes, PCl_5 is used (see p.131).

For iodoalkanes, phosphorus and iodine is used.

Reaction with bromide ions, Br⁻ in acid, H⁺

$$C_2H_5OH + Br^- + H^+ \longrightarrow C_2H_5Br + H_2O$$

The polarity of the C–OH bond allows a reaction with nucleophiles such as Br⁻, which attack the electron-deficient carbon atom. The presence of an acid, H⁺, is essential as it initially reacts with the alcohol forming a better leaving group:

Use an acid with a nucleophile in substitution reactions of alcohols.

'H_2O' is a better 'leaving group' than OH

- Most reactions of alcohols require the presence of an acid.

Dehydration of alcohols

AQA	M3	OCR	M2
EDEXCEL	M2	SALTERS	M2
NUFFIELD	M1		

When an alcohol is refluxed with a concentrated acid catalyst such as sulphuric acid H_2SO_4, or phosphoric acid H_3PO_4, an **elimination** reaction takes place: water is eliminated with formation of an alkene.

$$C_2H_5OH \longrightarrow C_2H_4 + H_2O$$

Use hot concentrated acid for elimination of water from an alcohol.

> **KEY POINT**
>
> An *elimination* reaction involves one species breaking up into two – the opposite to addition.
>
> Elimination usually converts a **saturated** compound into an **unsaturated** compound.

Elimination (1 $\longrightarrow$ 2) is the opposite process to addition (2 $\longrightarrow$ 1).

Formation of esters

NICCEA M2 SALTERS M2
OCR M2

Esters are formed by reaction of a carboxylic acid with an alcohol.

Esters are formed by reaction of an alcohol with a carboxylic acid in the presence of an acid catalyst (e.g. concentrated sulphuric acid).

$$\text{carboxylic acid} + \text{alcohol} \longrightarrow \text{ester} + \text{water}$$

The esterification of ethanol by ethanoic acid is shown below:

$$CH_3COOH + CH_3OH \longrightarrow CH_3COOCH_3 + H_2O$$

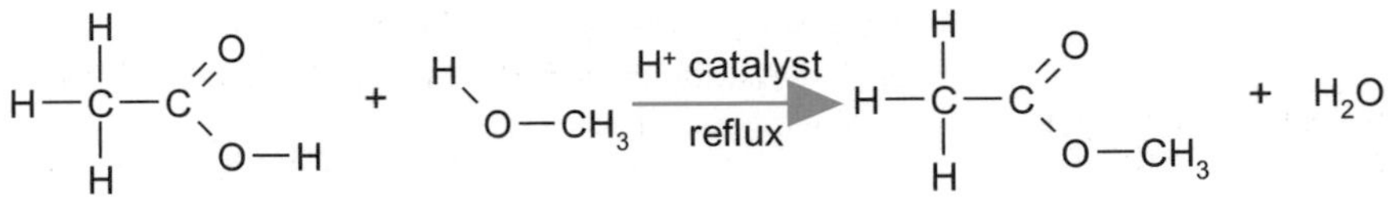

Tests for alcohols

EDEXCEL M2 NUFFIELD M1
NICCEA M2 OCR M2

Sodium releases H_2 from any compound with a hydroxyl group: alcohols, carboxylic acids (and water).

Reaction with sodium

Alcohols react with sodium, releasing hydrogen gas. The reaction involves breaking of the O–H bond of the alcohol.

$$2C_2H_5OH + 2Na \longrightarrow 2C_2H_5O^-Na^+ + H_2$$
$$\text{sodium ethoxide}$$

Reaction with phosphorus pentachloride PCl$_5$

(EDEXCEL and NICCEA only)

Alcohols react with PCl$_5$, releasing hydrogen chloride gas which forms misty fumes in air.

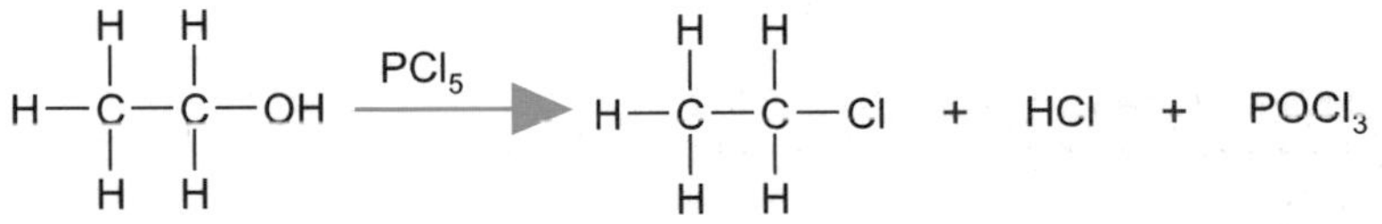

$$C_2H_5OH + PCl_5 \longrightarrow C_2H_5Cl + HCl + POCl_3$$

In the presence of ammonia, HCl fumes form dense white fumes of ammonium chloride:

$$NH_3 + HCl \longrightarrow NH_4Cl$$

Progress check

1 Why does ethanol dissolve in water?

2 Write equations to show possible oxidations for the following:
   (a) butan-1-ol;   (b) butan-2-ol.

3 What are the 2 possible organic products of the reaction between butan-2-ol and concentrated sulphuric acid at 170°C?

3 $CH_3CH=CHCH_3$ and $CH_3CH_2CH=CH_2$
(b) $CH_3CH_2CHOHCH_3 + [O] \longrightarrow CH_3CH_2COCH_3 + H_2O$
$CH_3CH_2CH_2CH_2OH + 2[O] \longrightarrow CH_3CH_2CH_2COOH + H_2O$
2 (a) $CH_3CH_2CH_2CH_2OH + [O] \longrightarrow CH_3CH_2CH_2CHO + H_2O$
1 Alcohol –OH group forms H-bonds with water molecules.

7.6 Halogenoalkanes

After studying this section you should be able to:

- understand the nature of the polarity in halogenoalkanes
- describe nucleophilic substitution reactions of halogenoalkanes
- explain the relative rates of hydrolysis of different halogenoalkanes
- describe elimination reactions of halogenoalkanes
- understand that reaction conditions can be used to control the products

LEARNING SUMMARY

Halogenoalkanes General formula: $C_nH_{2n+1}X$, where X = F, Cl, Br or I

AQA	M3	OCR	M2
EDEXCEL	M2	SALTERS	M2
NICCEA	M2	WJEC	M2
NUFFIELD	M2		

Like alcohols, halogenoalkanes can be primary, secondary or tertiary.

Types and naming of halogenoalkanes

1-bromopropane
primary

2-bromopropane
secondary

2-bromo-2-methylpropane
tertiary

Polarity of halogenoalkanes

Carbon and halogens have different electronegativities and halogenoalkanes have polar molecules with a polar C–X bond.

$C^{\delta+}–F^{\delta-}$
$C^{\delta+}–Cl^{\delta-}$
$C^{\delta+}–Br^{\delta-}$
$C^{\delta+}–I^{\delta-}$

polarity **decreases**

chlorine is more electronegative than carbon – electron flow from carbon to chlorine – dipole produced

The polarity produces an electron-deficient carbon atom, $C^{\delta+}$ which is important in the reactions of halogenoalkanes. The polarity decreases from fluorine to iodine, reflecting the decrease in electronegativity down the halogen group.

Nucleophilic substitution reactions of halogenoalkanes

AQA	M3	OCR	M2
EDEXCEL	M2	SALTERS	M2
NICCEA	M2	WJEC	M2
NUFFIELD	M2		

The electron-deficient carbon atom of the polar $C^{\delta+}–X^{\delta-}$ bond attracts **nucleophiles**, allowing a **nucleophilic substitution** reaction to take place in which the nucleophile replaces the halogen atom (see also alcohols, page 130).

The hydrolysis of halogenoalkanes

> A **hydrolysis** reaction is a reaction in which water breaks a bond.
>
> **KEY POINT**

Reaction with aqueous hydroxide ions, OH^-(aq) produces an alcohol.

$$C_2H_5Br + OH^-(aq) \xrightarrow[reflux]{OH^-/H_2O} C_2H_5OH + Br^-$$

The hydroxide ion, OH^-, behaves as a nucleophile.

Mechanism of nucleophilic substitution (not EDEXCEL or WJEC)

Primary halogenoalkanes react in a one-step mechanism.

NaOH(aq) is used as a source of OH⁻ (aq).

The OH⁻ ion behaves as a nucleophile by donating an electron pair.

Note the usefulness of the curly arrow in showing movement of an electron pair:

- the $:OH^-$ nucleophile donates its electron pair to the electron-deficient $C^{\delta+}$ of the bromoalkane, forming a covalent bond;
- the C–Br bond is broken by heterolytic fission with loss of its electron pair to form a $:Br^-$ ion.

Comparing the hydrolysis reactions of halogenoalkanes

The rates of hydrolysis of different halogenoalkanes can be investigated as follows:

- Alkaline hydrolysis with $NaOH(aq)/H_2O$, reflux:

$$R–X + OH^- \longrightarrow R–OH + X^-$$

An alternative method uses water itself for the hydrolysis. The halogenoalkane is heated with a mixture of water, ethanol and $AgNO_3$(aq). The rate of hydrolysis is monitored directly by observing precipitation of replaced halide ions as silver halides.

- Acidification with dilute nitric acid. This remove excess NaOH which would otherwise form a precipitate with the silver nitrate in the next stage, preventing detection of any silver halides.
- Addition of $AgNO_3$(aq) shows the presence of aqueous halide ions (see also Halogens, page 77).

$$Ag^+(aq) + X^-(aq) \longrightarrow AgX(s)$$

The intensity of any precipitate indicates the extent and rate of the hydrolysis.

$C^{\delta+}–F^{\delta-}$

$C^{\delta+}–Cl^{\delta-}$ polarity

$C^{\delta+}–Br^{\delta-}$ **increases**

$C^{\delta+}–I^{\delta-}$

bond	bond enthalpy /kJmol⁻¹
C–F	467
C–Cl	364
C–Br	290
C–I	228

> **KEY POINT**
>
> **Two factors – which is more important?**
> - **Polarity** predicts that the more polar C–F bond would attract water most easily and would give the fastest reaction.
> - **Bond enthalpy** predicts that the C–I bond would be broken most easily and would give the fastest reaction.
>
> For this reaction, bond enthalpy is more important than polarity.
> **The rate of hydrolysis increases as the C–X bond weakens.**

Further examples of nucleophilic substitution

Nucleophilic substitution of halogenoalkanes is used in organic synthesis to introduce different functional groups onto the carbon skeleton. The reaction scheme below shows how halogenoalkanes react with ammonia and cyanide ions.

For hydrolysis, water is used as the solvent.

For other nucleophilic substitutions, it is important that water is absent or hydrolysis will take place.

Ethanol is used as the alternative solvent.

Elimination reactions of halogenoalkanes

AQA	M3	NUFFIELD	M2
EDEXCEL	M2	OCR	M2
NICCEA	M2	WJEC	CH2

When a halogenoalkane is refluxed with hydroxide ions in **anhydrous** conditions (using NaOH in **ethanol** as a solvent at 78°C), an **elimination** reaction takes place.

$$H-\underset{\underset{H}{|}}{\overset{\overset{H}{|}}{C}}-\underset{\underset{H}{|}}{\overset{\overset{H}{|}}{C}}-Br \ + \ OH^- \quad \xrightarrow[\text{reflux}]{OH^-/\text{ethanol}} \quad \overset{H}{\underset{H}{}}C=C\overset{H}{\underset{H}{}} \ + \ Br^- \ + \ H_2O$$

Mechanism (AQA only)

Water is eliminated from the primary halogenoalkane in a **one-step** mechanism:

$$H-\underset{\underset{H}{|}}{\overset{\overset{H}{|}}{C}}-\underset{\underset{H}{|}}{\overset{\overset{H}{|}}{C}}\overset{\delta-}{\underset{\delta+}{}}Br \quad \xrightarrow{\text{single step}} \quad \overset{H}{\underset{H}{}}C=C\overset{H}{\underset{H}{}} \ + \ H_2O \ + \ \ddot{}Br^-$$

$$\overset{\bullet\bullet}{O}H^-$$

The hydroxide OH⁻ ion behaves as a base, accepting a proton H⁺ to form H_2O.

Elimination versus hydrolysis

AQA	M3	NUFFIELD	M2
EDEXCEL	M2	OCR	M2
NICCEA	M2	WJEC	CH2

Use different reaction conditions to control the type of reaction.

Commonly tested in exams.

The reaction of hydroxide ions with halogenoalkanes forms different organic products, depending upon the reaction conditions used.

> **aqueous** conditions ⟶ **alcohol**: OH⁻ acts as a **nucleophile**
> **anhydrous** conditions ⟶ **alkene**: OH⁻ acts as a **base**
>
> **KEY POINT**

The ability to control *which* reaction takes place solely by changing the conditions is an important tool for the synthetic chemist.

Uses of halogenoalkanes

Chlorofluorocarbons, CFCs (refrigerants, propellants, blowing polystyrene, dry cleaning, degreasing agents).

Chloroethene and tetrafluroethene to produce the plastics *pvc* and *ptfe*.

Halogenoalkanes as synthetic intermediates in chemistry.

Polychlorinated compounds as herbicides.

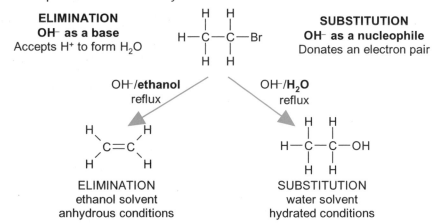

Progress check

1 What is meant by the term nucleophilic substitution?

2 Why does 1-bromopropane react with nucleophiles but propane does not?

3 Why do bromoalkanes react more readily than chloroalkanes?

4 What are the organic products for the reactions of 1-bromopropane with:
(a) NaOH/H_2O;   (b) NH_3/ethanol;   (c) NaOH/ethanol?

4 (a) $CH_3CH_2CH_2OH$ (b) $CH_3CH_2CH_2NH_2$ (c) $CH_3CH=CH_2$.
3 C–Br bond is weaker than C–Cl bond.
2 C–Br bond is polar, C–H bonds are not polar.
1 Replacement of an atom or group in an organic molecule by a nucleophile.

Sample question and model answer

1

(a) Explain, using examples, the following terms used in organic chemistry.

(i) general formula

This is the simplest algebraic representation for any member of a series of organic compounds, e.g. alkanes: C_nH_{2n+2} ✓

(ii) homologous series

This is a series containing compounds with similar chemical properties ✓ *with each successive member differing by CH_2* ✓*, e.g. 'alkanes'* ✓

(iii) homolytic fission and heterolytic fission

Fission means the breaking of a bond. ✓

In homolytic fission, one electron from the bond is released to each atom with free radicals forming. ✓ *e.g.: $Cl_2 \longrightarrow 2Cl\bullet$* ✓

In heterolytic fission, both electrons in the bonding pair are released to one of the atoms with ions forming ✓ *e.g.: $Br_2 \longrightarrow Br^+ + Br^-$* ✓

9 marks $\longrightarrow$ [7]

(b) Cracking of the unbranched compound **E**, C_6H_{14}, produced the saturated compound **F** and an unsaturated hydrocarbon **G** (M_r, 42). Compound **F** reacted with bromine in u.v. light to form a monobrominated compound **H** and an acidic gas **I**. Compound **G** reacted with hydrogen bromide to form a mixture of two compounds **J** and **K**.

(i) Use this evidence to suggest the identity of each of compounds **E** to **K**. Include equations for the reactions in your answer.

***E** is unbranched. $\therefore$ **E** = $CH_3CH_2CH_2CH_2CH_2CH_3$* ✓

***G** has M_r = 42; must have a double bond: must contain 3 carbons $\therefore$ **G** = C_3H_6.* ✓

*Cracking produces an alkane and an alkene, **F** is an alkane: **F** = C_3H_8* ✓

The equation is: $C_6H_{14} \longrightarrow C_3H_6 + C_3H_8$ ✓

*Alkanes react with bromine in u.v. by substitution. $\therefore$ **H** = C_3H_7Br* ✓*, **I** = HBr* ✓

The equation is: $C_3H_8 + Br_2 \longrightarrow C_3H_7Br + HBr$ ✓

*Alkenes react with bromine by addition. HBr can be added across a double bond in two ways to give: **J** and **K** as $CH_3CH_2CH_2Br$* ✓ *and $CH_3CHBrCH_3$.* ✓

The equation is: $C_3H_6 + HBr \longrightarrow C_3H_7Br$ ✓

(ii) Oil companies often 'reform' compounds such as **E**. Explain why this is done and suggest two organic products of the reforming of **E**.

Reforming converts low-demand fractions into more useful hydrocarbons ✓ *e.g. compound **E**, hexane could be converted into benzene* ✓ *or 2-methylpentane.*

(iii) Predict the structure of the polymer that could be formed from compound **G**.

$$\begin{array}{c} \left[\begin{array}{cc} H & CH_3 \\ | & | \\ -C-C- \\ | & | \\ H & H \end{array} \right]_n \end{array}$$

✓✓(1 for no double bond, 1 for correct skeleton) [14]

Practice examination questions

1 Petrol and camping gas are examples of fuels that contain hydrocarbons.

(a) Petrol is a mixture of alkanes containing 6 to 10 carbon atoms. Some of these alkanes are isomers of one another.

   (i) Explain the term *structural isomers*.

   (ii) State the molecular formula of an alkane present in petrol.

   [2]

(b) The major hydrocarbon in camping gas is butane. Some camping gas was reacted with chlorine to form a mixture of isomers.

   (i) What conditions are required for this reaction to take place?

   (ii) Two isomers, **A** and **B**, were separated from the mixture. These isomers had a molar mass of 92.5 g mol^{-1}.
   Deduce the molecular formula of these two isomers.

   (iii) Draw the displayed formulae of **A** and **B** and name each compound.

   [6]
   [Total: 8]
   Cambridge Chemistry Foundation Q3 June 1998 (modified)

2 Two types of isomerism in organic compounds are structural isomerism and *cis-trans* isomerism.

(a) Draw displayed formulae for:

   (i) two structural isomers of $C_2H_4Cl_2$;

   (ii) the *cis* and the *trans* isomers of $C_2H_2Cl_2$.

   [4]

(b) Suggest how the *cis-trans* isomers of $C_2H_2Cl_2$ could be converted into one of the structural isomers of $C_2H_4Cl_2$.

   [2]

(c) Under certain conditions, the *cis-trans* isomers of $C_2H_2Cl_2$ can polymerise.

   (i) Draw a short section of the polymer formed showing **two** repeat units.

   (ii) Polymers are often disposed of by combustion. In this process, in addition to carbon dioxide and steam, toxic gases may be formed. Suggest **two** toxic gases that could be formed during the combustion of the polymer in **c(i)**.

   [3]
   [Total: 9]
   Cambridge Chemistry Foundation Q3 Feb 1997

Practice examination questions *(continued)*

3 (a) Identify compounds **A–C** in the flowchart below.

$$CH_3CH=CHCH_3 \xrightarrow[\text{heat}]{HBr} A \xrightarrow[\text{heat}]{OH^-(aq)} B \xrightarrow[\text{heat}]{Cr_2O_7^{2-}/H^+} C$$

[3]

(b) Name the type of reaction taking place in each reaction above and state the role of the reagents.

(i) $CH_3CH=CHCH_3 \longrightarrow A$

(ii) $A \longrightarrow B$

(iii) $B \longrightarrow C$

[6]

(c) What reagents and conditions would be needed to convert but-2-ene directly into compound **B**?

[2]

(d) Draw a short section of the addition polymer formed by but-2-ene showing two repeat units.

[2]

[Total: 13]

4 Reduction of a branched alkene **A** with hydrogen formed a compound **B**. On analysis, **B** had the composition by mass of C, 82.8 %; H, 17.2 % (M_r = 58). Reaction of **A** with steam in the presence of a catalyst produced a mixture of two isomers **C** and **D**.

Identify possible structures for compounds **A–D**. Your answer should show all working and include structural formulae for each structure and equations for each reaction.

[9]

Cambridge Chemistry Foundation Q7 Feb 1997 (modified)

5 (a) There are four structural isomers of molecular formula C_4H_9Br. The structural formula of two of these isomers are compound **A**: $CH_3CH_2CHBrCH_3$ and compound **B**: $(CH_3)_3CBr$.

(i) Draw the remaining two structural isomers.

(ii) Give the name of **B**.

[3]

(b) All four structural isomers of C_4H_9Br undergo similar reactions with ammonia.

(i) What type of reaction is involved in these reactions?

(ii) Draw the structural formula of the product formed by the reaction of **A** with ammonia.

[2]

(c) Ethene, C_2H_4, reacts with bromine to give 1,2-dibromoethane.

(i) Give the name of the mechanism involved.

(ii) Show the mechanism for this reaction.

[4]

[Total: 9]

AEB Module Paper 6 Q4 Jun 1998 (modified)

Practice examination answers

Chapter 1 Atomic structure

1 (a)

particle	relative mass	relative charge	
proton	1	1+	✓
neutron	1	0	✓
electron	1/1840	1–	✓

[3]

   (b) (i)   neutron ✓; ^{7}Li has 1 more neutron ✓

        (ii) electron ✓; ^{32}S^{2-} has 2 more electrons ✓

        (iii) proton ✓; ^{40}Ca^{2+} has 1 more proton ✓   [6]

   [Total: 9]

2 (a) Difference: different numbers of neutrons. ✓

        Similarities: same numbers of protons ✓ and electrons. ✓   [3]

   (b) ^{7}Li ✓   [1]

   (c) (i) $1s^22s^22p^63s^23p^6$ ✓;   (ii) p block ✓   [2]

   (d) (i)   Al(g) $\longrightarrow$ Al$^+$(g) + e$^-$; ✓✓   (ii) Electron lost is in a different shell ✓, closer to the nucleus ✓, less shielding ✓, experiencing a greater nuclear attraction ✓.   [6]

   [Total: 12]

3 (a) Energy required to remove 1 electron ✓ from each atom in 1 mole ✓ of gaseous atoms ✓ to form 1 mole of gaseous 1+ ions.   [3]

   (b) Nucleus of beryllium has 1 more proton ✓. The extra electron in beryllium is in the same sub-shell (2s)/The attraction of the 2s electrons from the nucleus is greater in Be than in Li ✓.   [2]

   (c) $1s^22s^22p^1$ ✓   [1]

   (d) The extra electron in boron is in the 2p sub-shell ✓. Less energy is required to remove an electron from the higher energy 2p sub-shell ✓.   [2]

   (e) 1200 kJ mol^{-1}. The value will be greater than that of C ✓. The extra electron in nitrogen is in the same sub-shell (2p)/The attraction of the 2p electrons from the nucleus is greater in N than in C ✓.   [2]

   [Total: 10]

Chapter 2 Atoms, moles and equations

1 (a) (i)   0.011 ✓;   (ii) 0.055 ✓✓;

        (iii) 132 cm^3 using 24 dm^3; 136 cm^3 using $pV=nRT$ (AQA and WJEC) ✓✓✓   [6]

   (b) 0.167 M ✓✓✓   [3]

   [Total: 9]

2 (a) (i)   510 ✓✓;   (ii) 1020 ✓;   (iii) 204 dm^3 ✓   [4]

   (b) 51 kg ✓✓✓   [3]

   (c) NaOH is a strong alkali ✓; products of NaOH are soluble ✓   [2]

   [Total: 9]

3 (a) (i)   0.075 ✓;   (ii) 3.25 g ✓✓   [3]

   (b) 2 mol dm^{-3} ✓✓   [2]

   (c) (i)   good conductor of heat/low melting point ✓

        (ii) high reactivity of sodium/molten metals are hot/burn ✓   [2]

   [Total: 7]

Chapter 3 Bonding and structure

1 (a) A shared ✓ pair ✓ of electrons [2]

(b) The lone pair of electrons from N of an ammonia molecule ✓ bonds covalently to H+ to form an ammonium ion ✓ [2]

(c) 109.5° ✓. 107° ✓. Bonding angle is reduced by repelling effect of a lone pair which is greater than that of a bonded pair ✓ [3]

(d) Hydrogen bonding ✓. This arises from dipole-dipole attraction ✓ between $H^{\delta+}$ of an ammonia molecule and $N^{\delta-}$ on a different ammonia molecule ✓. [3]
[Total: 10]

2 Electron pairs repel one another ✓ and assume a position as **far away** from one another as possible ✓. Lone pairs repel more than bonded pairs ✓. Diagrams are shown below:

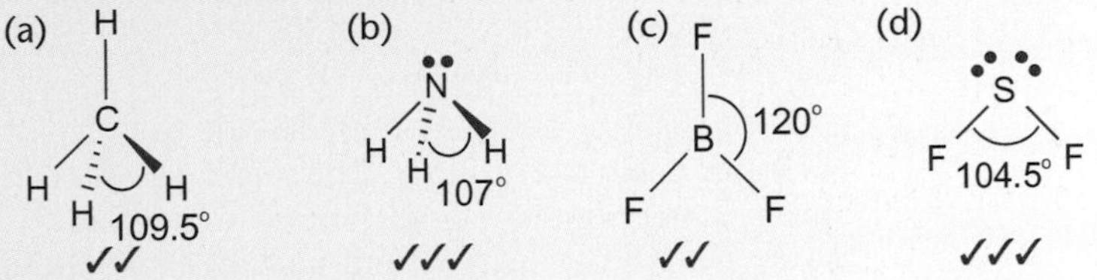

(a) H ... C ... H H H 109.5° ✓✓   tetrahedral ✓

(b) N ... H H H 107° ✓✓✓   pyramidal ✓

(c) F B 120° F F ✓✓   trigonal planar ✓

(d) S F 104.5° F ✓✓✓   non-linear ✓

1 mark for each diagram
1 mark for each bond angle
1 mark for lone pairs on NH_3 and SF_2 [Total: 17]

3 (a) (i) covalent ✓; (ii) van der Waals' forces ✓. [2]

(b) $I_2(g)$ ✓. [1]

(c) Weak van der Waals' forces break on vapourising ✓. [1]

(d) Strong ✓ ionic bonds ✓ hold the lattice together. [2]

(e) They vibrate faster ✓. [1]

(f) On melting, the ionic bonds remain close together and are just loosened from their rigid lattice positions ✓. On vapourising, the ions must be separated, breaking the ionic attraction ✓. [2]
[Total: 9]

4 (a) Water has hydrogen bonding ✓. This is dipole-dipole attraction ✓ between $H^{\delta+}$ of a water molecule and a lone pair ✓ on $O^{\delta-}$ on a different water molecule ✓. [4]

(b) $CHCl_3$ has dipole-dipole interactions ✓ between $H^{\delta+}$ of a $CHCl_3$ molecule ✓ and $Cl^{\delta-}$ on a different $CHCl_3$ molecule ✓. any 2 ⟶ [2]

(c) Ar has van der Waals' forces only ✓. These are caused by oscillating dipoles ✓ which cause induced dipoles on another Ar atom ✓. H-bonding > dipole-dipole > van der Waals' ✓. The intermolecular forces are broken on boiling ✓. The stronger the intermolecular forces, the higher the boiling point ✓. any 4 ⟶ [4]
[Total:10]

Chapter 4 Periodic Table

1 Across Period 3, atomic radius decreases ✓ because there are more protons in the nucleus across the period ✓ increasing the attraction ✓ on the electrons in the same outer shell ✓.

Down Group 1, atomic radius increases ✓ as more shells are being added ✓. The shielding effect increases as the number of inner shells increases ✓. The attraction experienced by the outer electrons from the nucleus decreases ✓. [Total: 8]

2 (a) (i) $Mg(s) + H_2O(g) \longrightarrow MgO(s) + H_2(g)$ ✓

(ii) $2Mg(s) + O_2(g) \longrightarrow 2MgO$ ✓

(iii) $Mg(s) + 2H_2O(l) \longrightarrow Mg(OH)_2 + H_2(g)$ ✓ [3]

(b) (i) On descending outer electrons are further from the nucleus ✓ with greater shielding ✓. Therefore, outer electrons are removed more easily.

(ii) Reaction with O_2/Cl_2/named acid ✓ [3]

(c) (i) very high melting point/strong forces holding lattice together ✓

(ii) CaO: cement ✓; $Mg(OH)_2$: indigestion remedy ✓ *many others possible*. [3]
[Total: 9]

3 (a) On descending the halogens, the atomic radii increase ✓ resulting in less nuclear attraction at the edge of the atom ✓. There are more electron shells between the nucleus and the edge of the atom to shield the nuclear charge ✓. The nuclear attraction at the edge of the atom decreases down the group ✓. Therefore the oxidising power of the halogens decreases down the group ✓. [5]

   (b) (i) $Cl_2 \longrightarrow NaCl$ (ox no: Cl 0 → −1) ✓ reduction ✓;
             $Cl_2 \longrightarrow NaClO_3$ (ox no: Cl 0 → +5) ✓ oxidation ✓

        (ii) 126 g ✓✓✓

        (iii) $Cl_2 + 2NaOH \longrightarrow NaOCl + NaCl + H_2O$ ✓; bleach ✓ [9]

[Total: 14]

4 (a) (i) Electronegativity is a measure of the attraction of an atom in a molecule for the pair of electrons in a covalent bond ✓.

        (ii) Electronegativity decreases from fluorine to iodine ✓. The atomic radius increases down the group ✓. The shielding effect also increases down the group as the number of electron shells increases ✓. The smaller the halogen atom, the greater the nuclear attraction experienced by the bonding electrons and the greater the electronegativity ✓. [5]

   (b) The boiling points increase from fluorine to iodine ✓. The number of electrons increase ✓, increasing the van der Waals' forces between molecules ✓. [3]

   (c) With aqueous sodium chloride, there is no change ✓. With aqueous sodium iodide, the solution turns dark brown ✓ as iodine is displaced ✓.
        $Br_2(aq) + 2I^-(aq) \longrightarrow 2Br^-(aq) + I_2(aq)$ ✓ [4]

   (d) I^- is a stronger reducing agent than Cl^- ✓ because I^- reduces Br_2, whereas Cl^- does not reduce Br_2 ✓. [2]

[Total: 14]

Chapter 5 Chemical energetics

1 (a) $Al(s) + 1\frac{1}{2}Cl_2(g) \longrightarrow AlCl_3(s)$ ✓✓ [1, equation; 1, state symbols] [2]

   (b) 100 kPa ✓ [1]

   (c) +269 kJ mol⁻¹ ✓✓✓ [3]

[Total: 6]

2 (a) (i) The enthalpy change that takes place when one mole of a substance ✓ reacts completely with oxygen ✓ under standard conditions, all reactants and products being in their standard states; (ii) 298K ✓ [3]

   (b) −727 kJ mol⁻¹ ✓✓✓ [3]

   (c) Activation energy must be overcome/energy is needed to break bonds ✓. [1]

[Total: 7]

3 (a) The enthalpy change that takes place when one mole of a compound ✓ in its standard states is formed from its constituent elements ✓ in their standard states under standard conditions. ✓. [3]

   (b) $C(s) + 2H_2(g) \longrightarrow CH_4(g)$ ✓✓ [1, equation; 1, state symbols] [2]

   (c) −76 kJ mol⁻¹ ✓✓✓ [3]

   (d) C–H: +412 kJ mol⁻¹ ✓; C–C: +348 kJ mol⁻¹ ✓✓ [3]

[Total: 11]

Chapter 6 Rates and equilibrium

1 (a) Activation energy is the minimum energy required for a reaction to take place ✓. [1]

(b) (i) ✓✓✓; (ii) ✓

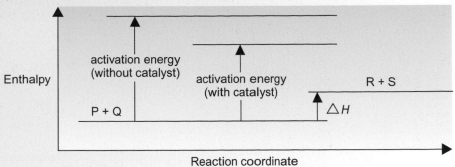

activation energy (without catalyst)

activation energy (with catalyst)

R + S

Enthalpy

P + Q

$\triangle H$

Reaction coordinate

(iii) A catalyst provides an alternative reaction route with a lower activation energy. ✓✓ [6]

(c) (i) ✓✓✓; (ii) ✓✓

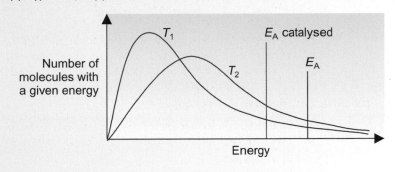

T_1

E_A catalysed

E_A

Number of molecules with a given energy

T_2

Energy

[5]
[Total: 12]

2 (a) (i) A homogeneous equilibrium has all equilibrium species in the same phase ✓. All species in this equilibrium are gaseous ✓.

(ii) The forward reaction proceeds at the same rate as the reverse reaction ✓. The concentrations of reactants and products are constant ✓. Equilibrium can only be achieved in a closed system ✓. [5]

(b) Endothermic ✓. With increased temperature an equilibrium shifts in the endothermic direction ✓. [2]

(c) (i) Equilibrium position is unaffected ✓. There are equal numbers of gaseous moles on either side of the equilibrium ✓.

(ii) Rate increases ✓. Increasing the pressure also increases the concentration ✓. [4]
[Total: 11]

Chapter 7 Organic chemistry

1 (a) (i) same **molecular** formula with a different arrangement of atoms/structural formula ✓.

(ii) allow any alkane formula from C_6 to C_{10}, e.g. C_8H_{18} ✓ [2]

(b) (i) u.v. / sunlight ✓; (ii) C_4H_9Cl ✓

(iii) any correct structures ✓✓ with names ✓✓, e.g.

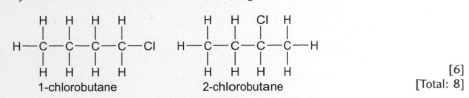

1-chlorobutane   2-chlorobutane

[6]
[Total: 8]

Organic chemistry (continued)

2 (a)(i) correct structures

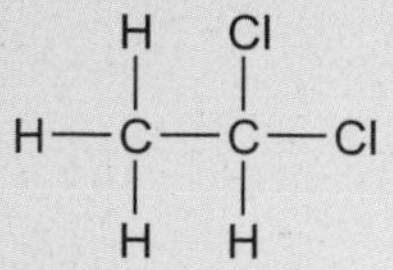

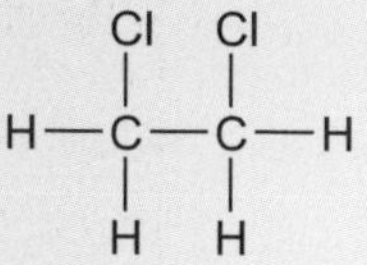

H—C—C—Cl with H, Cl top and H, H bottom ✓   H—C—C—H with Cl, Cl top and H, H bottom ✓

(ii) correct structures

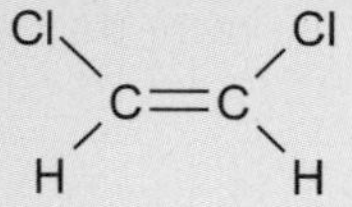

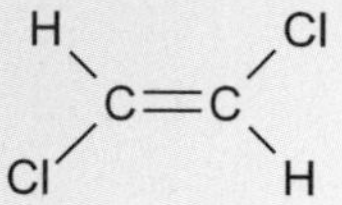

Cl,H / C=C / H,Cl ✓   H,Cl / C=C / Cl,H ✓   [4]

(b) H_2 ✓ and a named catalyst/Ni/Pt/Pd ✓.   [2]

(c) (i)

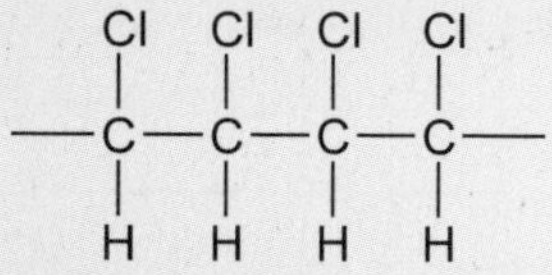

—C—C—C—C— with Cl, Cl, Cl, Cl top and H, H, H, H bottom ✓

(ii) any two from CO, HCl, Cl_2, $COCl_2$, CCl_4, $CHCl_3$ ✓✓   [3]

[Total: 9]

3 (a) A = $CH_3CH_2CHBrCH_3$ ✓;   B = $CH_3CH_2CHOHCH_3$ ✓;   C = $CH_3CH_2COCH_3$ ✓   [3]

(b) (i) addition ✓, electrophile ✓;

(ii) substitution ✓, nucleophile ✓;

(iii) oxidation ✓, oxidising agent ✓.   [6]

(c) steam ✓ and H_3PO_4 ✓   [2]

(d)

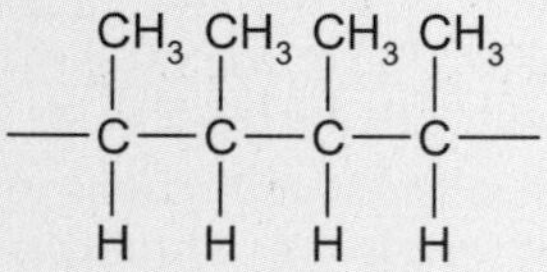

—C—C—C—C— with CH_3, CH_3, CH_3, CH_3 top and H, H, H, H bottom   no double bond ✓, correct skeleton ✓   [2]

[Total: 13]

4 Empirical formula of **B** = C_2H_5 ✓✓; Molecular formula of **B** = C_4H_{10} ✓
A = $(CH_3)_2C=CH_2$ ✓; B = $(CH_3)_2CHCH_3$ ✓;
C and D = $(CH_3)_2CHCH_2OH$ ✓ and $(CH_3)_2COHCH_3$ ✓;
$C_4H_8 + H_2 \longrightarrow C_4H_{10}$ ✓
$C_4H_8 + H_2O \longrightarrow C_4H_9OH$ ✓

[Total: 9]

5 (a) (i) $CH_3CH_2CH_2CH_2Br$ ✓ and $(CH_3)_2CHCH_2Br$ ✓;

(ii) 2-bromo-2-methylpropane ✓   [3]

(b) (i) nucleophilic substitution ✓;   (ii) $CH_3CH_2CHNH_2CH_3$ ✓   [2]

(c) (i) electrophilic addition ✓

(ii) arrows stage 1 ✓;   arrow stage 2 ✓;   carbonium ion ✓

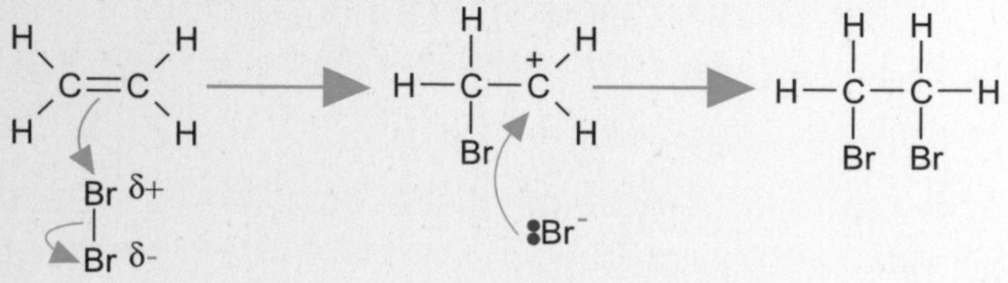

H,H / C=C / H,H → H—C—C with Br, H, H → H—C—C—H with Br, Br
Br $\delta+$ / Br $\delta-$; Br^-

[4]

[Total: 9]

The Periodic Table

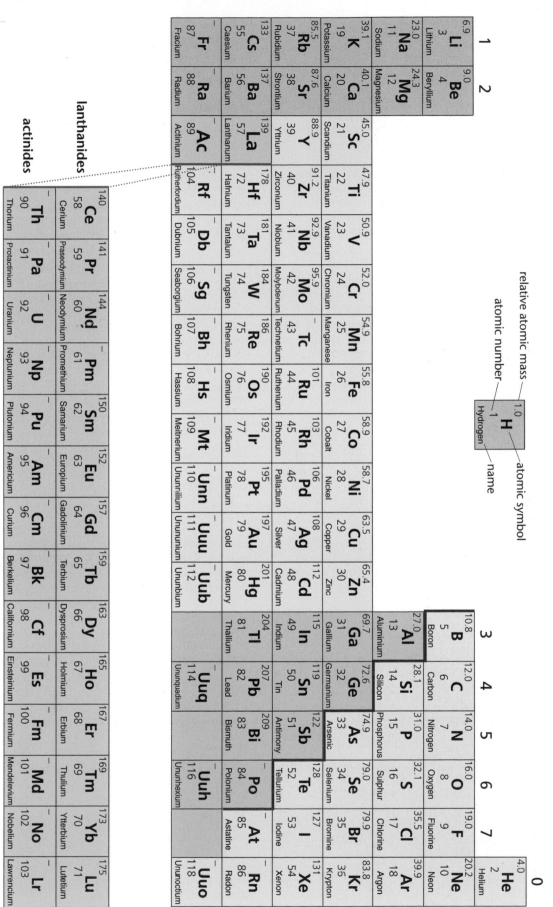

Key:

relative atomic mass — 1.0
atomic number — 1
atomic symbol — H
name — Hydrogen

Group 1	Group 2
6.9 **Li** 3 Lithium	9.0 **Be** 4 Beryllium
23.0 **Na** 11 Sodium	24.3 **Mg** 12 Magnesium
39.1 **K** 19 Potassium	40.1 **Ca** 20 Calcium
85.5 **Rb** 37 Rubidium	87.6 **Sr** 38 Strontium
133 **Cs** 55 Caesium	137 **Ba** 56 Barium
— **Fr** 87 Francium	— **Ra** 88 Radium

lanthanides

actinides

45.0 **Sc** 21 Scandium	47.9 **Ti** 22 Titanium	50.9 **V** 23 Vanadium	52.0 **Cr** 24 Chromium	54.9 **Mn** 25 Manganese	55.8 **Fe** 26 Iron	58.9 **Co** 27 Cobalt	58.7 **Ni** 28 Nickel	63.5 **Cu** 29 Copper	65.4 **Zn** 30 Zinc
88.9 **Y** 39 Yttrium	91.2 **Zr** 40 Zirconium	92.9 **Nb** 41 Niobium	95.9 **Mo** 42 Molybdenum	— **Tc** 43 Technetium	101 **Ru** 44 Ruthenium	103 **Rh** 45 Rhodium	106 **Pd** 46 Palladium	108 **Ag** 47 Silver	112 **Cd** 48 Cadmium
139 **La** 57 Lanthanum	178 **Hf** 72 Hafnium	181 **Ta** 73 Tantalum	184 **W** 74 Tungsten	186 **Re** 75 Rhenium	190 **Os** 76 Osmium	192 **Ir** 77 Iridium	195 **Pt** 78 Platinum	197 **Au** 79 Gold	201 **Hg** 80 Mercury
89 **Ac** Actinium	— **Rf** 104 Rutherfordium	— **Db** 105 Dubnium	— **Sg** 106 Seaborgium	— **Bh** 107 Bohrium	— **Hs** 108 Hassium	— **Mt** 109 Meitnerium	— **Uun** 110 Ununnilium	— **Uuu** 111 Unununium	— **Uub** 112 Ununbium

Group 3	Group 4	Group 5	Group 6	Group 7	Group 0
					4.0 **He** 2 Helium
10.8 **B** 5 Boron	12.0 **C** 6 Carbon	14.0 **N** 7 Nitrogen	16.0 **O** 8 Oxygen	19.0 **F** 9 Fluorine	20.2 **Ne** 10 Neon
27.0 **Al** 13 Aluminium	28.1 **Si** 14 Silicon	31.0 **P** 15 Phosphorus	32.1 **S** 16 Sulphur	35.5 **Cl** 17 Chlorine	39.9 **Ar** 18 Argon
69.7 **Ga** 31 Gallium	72.6 **Ge** 32 Germanium	74.9 **As** 33 Arsenic	79.0 **Se** 34 Selenium	79.9 **Br** 35 Bromine	83.8 **Kr** 36 Krypton
115 **In** 49 Indium	119 **Sn** 50 Tin	122 **Sb** 51 Antimony	128 **Te** 52 Tellurium	127 **I** 53 Iodine	131 **Xe** 54 Xenon
204 **Tl** 81 Thallium	207 **Pb** 82 Lead	209 **Bi** 83 Bismuth	— **Po** 84 Polonium	— **At** 85 Astatine	— **Rn** 86 Radon
— **Uut** 113 Ununtrium	— **Uuq** 114 Ununquadium	— **Uup** 115 Ununpentium	— **Uuh** 116 Ununhexium		— **Uuo** 118 Ununoctium

lanthanides

| 140 **Ce** 58 Cerium | 141 **Pr** 59 Praseodymium | 144 **Nd** 60 Neodymium | — **Pm** 61 Promethium | 150 **Sm** 62 Samarium | 152 **Eu** 63 Europium | 157 **Gd** 64 Gadolinium | 159 **Tb** 65 Terbium | 163 **Dy** 66 Dysprosium | 165 **Ho** 67 Holmium | 167 **Er** 68 Erbium | 169 **Tm** 69 Thulium | 173 **Yb** 70 Ytterbium | 175 **Lu** 71 Lutetium |

actinides

| 140 **Th** 90 Thorium | — **Pa** 91 Protactinium | — **U** 92 Uranium | — **Np** 93 Neptunium | — **Pu** 94 Plutonium | — **Am** 95 Americium | — **Cm** 96 Curium | — **Bk** 97 Berkelium | — **Cf** 98 Californium | — **Es** 99 Einsteinium | — **Fm** 100 Fermium | — **Md** 101 Mendelevium | — **No** 102 Nobelium | — **Lr** 103 Lawrencium |

Index

acid-base titrations 40
acids 40, 78, 89, 109, 110
activation energy 99, 102
addition reactions 123
alcohols 116, 127–131
aldehyde 129
alkanes 114–5, 120–1, 132–4
alkenes 115, 122–5
alkyl group 115–16
aluminium extraction 83
ammonia 103, 107–8, 110
amount of substance 33–4
atomic mass 30–2
atomic number 19, 64–5
atomic radius 25, 65, 76
atomic structure 19–27
atoms 19–20, 33
Avogadro constant 33

balancing equations 38–9
bases 40, 109, 110
Basic Oxygen Process 82
beryllium 27, 75
blast furnace 81
bleach 79, 80
boiling points 66–7
Boltzmann distribution 99–100, 101, 102
bond enthalpy 94–5
bond fission 116–17
bonded pair electrons 50–1
bonds 58–61
boron 27
Brønsted-Lowry theory 109

carbon chain extension 133–4
carbon isotopes 20
carbon-12 30, 33, 34
carboxylic acid 129
catalysis 102–3
catalyst 102, 106, 108
catalytic converters 103
chemical bonding 44–9
chlorine 79, 80, 103
cis-trans isomerism 122
combustion 120, 128
covalent bonding 47–8, 52–3, 61
covalent compounds 59–60, 61
cracking 118–19
crude oil products 118–19

dative covalent bonds 48
dehydration 130

delocalised electrons 49
diamond 60
dipole-dipole interactions 55, 56
double bonds 51, 122
electrolysis 82–3
electron shielding 25, 26
electronegativity 52–3, 76
electronic configurations 23, 24, 27, 71
electron-pair repulsion 50–1
electrons 19–24, 25
electrophilic addition 123
elements 64
elimination 134
empirical formula 37
endothermic reactions 88
energy shells 21–4, 26
enthalpy calculation 90–2, 94–5
enthalpy changes 87–92, 94–5, 101
epoxyethane 126
equations 38–9, 69–70
equilibria 104–8
esters 131
ethanol 128
ethene hydration 128
exothermic reactions 87

fermentation 128
first ionisation energy 66
first law of thermodynamics 87
fission 116–17
flame tests 72
formula type 37
fractional distillation 118
free-radical substitution 121
functional groups 115

gas volumes 36
giant ionic lattices 46, 58, 61
giant metallic lattice 60, 61
giant molecular structure 60, 61
graphite 60
Group 1 and 2 elements 71–5
Group 7 elements 76–80

Haber Process 103, 107–8
half-equations 70
halide reactions 78

halide testing 77
halogen reactivity 77
halogenoalkanes 132–4
Hess's law 91–2
homologous series 114
hydrocarbons 113–14
hydrogen bonds 56–7, 61
hydrogen halides 78
hydrolysis 132–4

ideal gas equation 36
intermolecular forces 55–7, 58
iodine 59
iodine-thiosulphate titration 80
ionic bonding 45–7, 52, 53, 54
ionic charge 46
ionic compounds 58–9
ionic formulae 46–7
ionisation energy 25–7, 66
ions 25
iron production 81–2
isomerisation 119
isomerism 114–15, 122
isotopes 20

ketone 129

Le Chatelier's Principle 105, 107
lone pairs 50–1

mass number 20
mass spectrometer 31
metal extraction 81–4
metal halide reduction 83–4
metal recycling 84
metals 49, 60, 61, 67, 68
molar mass 34
molarity 35
mole 33–6
molecular formula 37
molecule 32, 47
multiple covalent bonding 48

neutrons 19, 20
Noble Gases 44, 55
non-polar bonds 53
nuclear charge 25, 26, 65, 76
nucleophilic substitution 130, 132–3

Octet Rule 44, 48
orbitals 21–4, 27

organic formulae 113
organic nomenclature 113, 116, 122, 127, 132
oxidation 68–70, 128
oxidation number 69–70
ozone 103

Pauling scale 52
Periodic Table 24, 26, 46, 52, 64–84, 143
periodicity 65
permanent dipole 53, 56
polar bonds 53
polarisation 54, 127
product quantities 39
protons 19–20

reacting quantities 39
reaction rates 98–101, 102
redox reactions 68–70
reduction 68–70, 78, 124, 129
reduction of metal oxides 81
relative atomic mass 30–1, 34
relative formula mass 32
relative mass 33
relative molecular mass 32, 36
reversible reactions 104–8

s-block elements 71–5
second ionisation energy 26
shapes of molecules 50–1
solutions 35
standard enthalpy changes 88–9
state symbols 38
steel 82
sub-atomic particles 19
sub-shells 21–4, 26–7
substitution 121, 130, 132–3
symmetrical molecules 53

titanium 83–4
titrations 40

unsymmetrical molecules 53

van der Waals' forces 55, 61
water 56–7, 59

Chemistry

Rob Ritchie

Contents

Specification lists   4

AS/A2 Level Chemistry courses   11

Different types of questions in A2 examinations   13

Exam technique   15

What grade do you want?   17

Four steps to successful revision   18

Chapter 1 Reaction rates

1.1 Orders and the rate equation   19

1.2 Determination of reaction mechanisms   25

Sample question and model answer   27

Practice examination questions   28

Chapter 2 Chemical equilibrium

2.1 The equilibrium constant, K_c 29

2.2 The equilibrium constant, K_p 33

2.3 Acids and bases   35

2.4 The pH scale   38

2.5 pH changes   42

Sample question and model answer   46

Practice examination questions   47

Chapter 3 Energy changes in chemistry

3.1 Enthalpy changes   49

3.2 Electrochemical cells   55

3.3 Predicting redox reactions   59

Sample question and model answer   61

Practice examination questions   62

Chapter 4 The Periodic Table

4.1 Oxides of the Period 3 elements   64

4.2 Chlorides of the Period 3 elements   67

4.3 The transition elements   70

4.4 Transition element complexes   73

4.5 Ligand substitution of complex ions   77

4.6 Redox reactions of transition metal ions   79

4.7 Catalysis   82

4.8 Reactions of metal aqua-ions   85

Sample question and model answer   88
Practice examination questions   89

Chapter 5   Isomerism, aldehydes, ketones and carboxylic acids

5.1   Isomerism and functional groups   91
5.2   Aldehydes and ketones   94
5.3   Carboxylic acids   99
5.4   Esters   102
5.5   Acylation   104
    Sample question and model answer   107
    Practice examination questions   108

Chapter 6   Aromatics, amines, amino acids and polymers

6.1   Arenes   109
6.2   Reactions of arenes   113
6.3   Phenols   117
6.4   Amines   119
6.5   Amino acids   123
6.6   Polymers   126
    Sample question and model answer   130
    Practice examination questions   131

Chapter 7   Analysis and synthesis

7.1   Infra-red spectroscopy   132
7.2   Mass spectrometry   134
7.3   N.m.r. spectroscopy   136
7.4   Organic synthetic routes   140
    Sample question and model answer   142
    Practice examination questions   143

Chapter 8   Synoptic assessment

What is synoptic assessment?   144
Worked synoptic sample questions and model answers   145
Practice examination questions   147

Practice examination answers   149
Periodic table   158
Index   159

Specification lists

AQA Chemistry

MODULE	SPECIFICATION TOPIC	CHAPTER REFERENCE	STUDIED IN CLASS	REVISED	PRACTICE QUESTIONS
Module 4 (M4) *Further physical and organic chemistry*	Kinetics	1.1			
	Equilibria	2.1, 2.2			
	Acids and bases	2.3–2.5			
	Nomenclature and isomerism in organic chemistry	5.1			
	Compounds containing the carbonyl group	5.2–5.5			
	Aromatic chemistry	6.1, 6.2			
	Amines	6.4			
	Amino acids	6.5			
	Polymers	6.6			
	Organic synthesis and analysis	7.4			
	Structure determination	7.7–7.3			
Module 5 (M5) *Thermodynamics and further inorganic chemistry*	Thermodynamics	3.1			
	Periodicity	4.1, 4.2			
	Redox equilibria	3.2, 3.3			
	Transition metals	4.3–4.7			
	Reactions of inorganic compounds in aqueous solution	4.8			

Examination analysis

Units 1, 2 and 3 comprise AS Chemistry.		50%

A2 Chemistry comprises three unit tests. All questions are compulsory.

Unit 4	Structured questions: short and extended answers	1 hr 30 min test	15%
Unit 5	Structured questions: short and extended answers (includes synoptic assessment)	2 hr test	20%

Unit 6 comprises two components:

6(a)	Objective questions: multiple-choice and either matching pairs **or** multiple completion items (synoptic assessment)	1 hr test	10%
6(b)	Centre-assessed coursework **or** practical examination	2 hr test	5%

OCR Chemistry

MODULE	SPECIFICATION TOPIC	CHAPTER REFERENCE	STUDIED IN CLASS	REVISED	
Module 2814 (M4) *Chains, rings and spectroscopy*	Arenes	6.1–6.3			
	Carbonyl compounds	5.2			
	Carboxylic acids and esters	5.3, 5.4			
	Nitrogen compounds	6.4, 6.5			
	Stereoisomerism and organic synthesis	5.1, 7.4			
	Polymerisation	6.6			
	Spectroscopy	7.1–7.3			
Module 2815/1 (M5.1) *Trends and patterns*	Lattice enthalpy	3.1			
	The Periodic Table: Period 3	4.1–4.2			
	The Periodic Table: Transition elements	4.3–4.8			
Module 2815/2 (M5.2) *Biochemistry*					
Module 2815/2 (M5.3) *Environmental chemistry*					
Module 2815/4 (M5.4) *Methods of analysis and detection*		7.1–7.3			
Module 2815/5 (M5.5) *Gases, liquids and solids*					
Module 2815/6 (M5.6) *Transition elements*	Electrode potentials	3.2, 3.3			
	Ligands and complexes	4.4			
	Colour	4.4			
	Chemistry of Transition metals	4.6, 4.7			
Module 2816/1 (M6) *Unifying concepts in chemistry*	How far?	1.1, 1.2			
	How fast?	2.1, 2.2			
	Acids, bases and buffers	2.3–2.5			

Examination analysis

Units 2811, 2812 and 2813 comprise AS Chemistry		50%

A2 Chemistry comprises three unit tests. All questions are compulsory.

Unit 2814 Structured questions: short and extended answers	1 hr 30 min test	15%

Unit 2815 comprises two components:		
2815/1 Structured questions: short and extended answers (includes synoptic assessment)	1 hr test	7.5%
2815/2–6 One option only Structured questions: short and extended answers	50 min test	7.5%

Unit 2816 comprises two components:		
2816/1 Structured questions: short and extended answers (synoptic assessment)	1 hr 15 min test	10%
2816/2 Centre-assessed coursework (includes synoptic assessment) **or** 2816/3 Practical examination (includes synoptic assessment)	1 hr 30 min test	10%

OCR Chemistry, Salters

MODULE	SPECIFICATION TOPIC	CHAPTER REFERENCE	STUDIED IN CLASS	REVISED	PRACTICE QUESTIONS
Module 2849 (M4) *Chemistry of Materials*	What's in a Medicine?				
	Designer polymers	5.3–5.5, 6.4, 6.6			
	Engineering proteins	1.1, 2.1, 5.1, 6.5, 6.6, 7.3			
	The steel story	4.3–4.8, 3.2, 3.3			
Module 2854 (M5) *Chemistry by design*	Aspects of agriculture	1.1, 2.2, 3.2, 4.1, 4.2			
	Colour by design	5.4, 6.1, 6.2, 6.4			
	The oceans	2.3–2.5, 3.1			
	Medicines by design	5.2, 7.1–7.4			

Note that the Salters approach 'drip-feeds' concepts throughout the course. You may need to revisit the chapters and practice questions in this Study Guide throughout your A2 course.

Examination analysis

Units 2850, 2848 and 2852 comprise AS Chemistry.		*50%*

A2 Chemistry comprises two unit tests. All questions are compulsory.

Unit 2853	Structured questions: short and extended answers	*1 hr 30 min test*	*15%*
Unit 2854	Structured questions: short and extended answers (includes synoptic assessment)	*2 hr test*	*20%*
Unit 2855	Individual investigation: salternally assessed coursework (includes synoptic assessment)		*15%*

Edexcel Chemistry

MODULE	SPECIFICATION TOPIC	CHAPTER REFERENCE	STUDIED IN CLASS	REVISED	PRACTICE QUESTIONS
Module 4 (M4) *Periodicity, quantitative equilibria and functional group chemistry*	Energetics II	3.1			
	Periodic Table II (Period 3 and Group 4)	4.1, 4.2			
	Chemical equilibria II	2.1, 2.2			
	Acid–base equilibria	2.3–2.5			
	Organic Chemistry II (acids, esters, carbonyl compounds, acid chlorides, nitrogen compounds and further halogeno compounds)	5.1–5.5, 6.4, 6.5			
Module 5 (M5) *Transition metals, quantitative kinetics and applied organic chemistry*	Redox equilibria (application)	3.2, 3.3			
	Transition metal chemistry	4.3–4.8			
	Organic chemistry III (reaction mechanisms and aromatic compounds)	6.1–6.4			
	Chemical kinetics II	1.1, 1.2			
	Organic chemistry IV (analysis, synthesis and applications)	7.1–7.4 6.6			

Examination analysis

Units 1, 2 and 3 comprise AS Chemistry   *50%*

A2 Chemistry comprises three unit tests.

Unit 4   Structured questions: short answers. All questions are compulsory.   *1 hr 30 min test 15%*

Unit 5   Section A: structured questions
Section B: structured and extended questions with a choice of questions.
Parts of Section A and all of Section B will form part of the synoptic assessment.   *1 hr 30 min test 15%*

Unit 6 comprises two components:

6A   Centre-assessed coursework or
Practical examination   *1 hr 45 min test 10%*

6B   Synoptic paper
Section A: Interpretation of data from laboratory situations
Section B: Structured and extended questions with a choice of questions   *1 hr 30 min test 10%*

Edexcel Chemistry (Nuffield)

MODULE	SPECIFICATION TOPIC	CHAPTER REFERENCE	STUDIED IN CLASS	REVISED	PRACTICE QUESTIONS
Module 4 (M4) Energy and reactions	How fast? Rates of reaction	1.1, 1.2			
	Arenes: benzene and phenol	6.1–6.3			
	Entropy				
	How far? Reversible reactions	2.1–2.5			
	Oxidation products of alcohols	5.2–5.5, 7.1			
Module 5 (M5) Special studies	Biochemistry				
	Chemical engineering				
	Food science				
	Materials science				
	Mineral process chemistry				
Module 6 M6) Applying chemistry	The Born–Haber cycle, structure and bonding	3.1, 4.1, 4.2			
	Redox equilibria	3.2, 3.3			
	Natural and synthetic polymers	5.1, 6.4–6.6			
	The transition elements	4.3–4.8			
	Organic synthesis	7.4			
	Instrumental methods	7.1–7.3			

Examination analysis

Units 1, 2 and 3 comprise AS Chemistry.	50%

A2 Chemistry comprises three unit tests. All questions are compulsory.

Unit 4 Structured questions: short and extended answers	1 hr 30 min test	15%
Unit 5 Investigations and applications		
Unit 5A Investigation, general practical competence		7.5%
Unit 5B One option only		
Structured questions: short and extended answers	45 min test	7.5%
Unit 6 Synoptic (Open Book)	2 hr test	20%

Unit 6 will contain structured questions, with a prerequisite of understanding of topics 1 to 15.

WJEC Chemistry

MODULE	SPECIFICATION TOPIC	CHAPTER REFERENCE	STUDIED IN CLASS	REVISED	PRACTICE QUESTIONS
Module CH4 (M4) *Spectroscopy and further organic chemistry*	Spectroscopy	4.4, 7.1, 7.3			
	Isomerism and aromaticity	5.1, 6.1, 6.2			
	Organic compounds containing halogens				
	Organic compounds containing oxygen	5.2–5.5, 6.3			
	Organic compounds containing nitrogen	6.4, 6.5			
	Organic synthesis and analysis	6.6, 7.1–7.4			
Module CH5 (M5) *Further physical and inorganic chemistry*	Redox	3.2, 3.3, 4.6			
	Chemistry of the s block				
	Chemistry of the p block				
	Transition elements	4.3–4.8			
	Periodicity	4.1, 4.2			
	Chemical kinetics	1.1, 1.2			
	Energy changes and equilibria	2.1–2.5, 3.1			

Examination analysis

Units 1, 2 and 3 comprise AS Chemistry.		50%
A2 Chemistry comprises three unit tests. All questions are compulsory.		
Unit CH4 Structured questions and objective questions	1 hr 40 min test	15%
Unit CH5 Structured questions and objective questions (includes synoptic assessment)	1 hr 40 min test	15%
Unit CH6 comprises two components:		
CH6a Structured questions: short and extended answers and comprehension (includes synoptic assessment)	1 hr 10 min test	10%
CH6b Centre-assessed coursework (includes synoptic assessment)		10%

NICCEA Chemistry

MODULE	SPECIFICATION TOPIC	CHAPTER REFERENCE	STUDIED IN CLASS	REVISED	PRACTICE QUESTIONS
Module 4 (M4) *Further organic, physical and inorganic chemistry*	Lattice energy	3.1			
	Kinetics	1.1, 1.2			
	Equilibrium	2.1–2.5			
	Electrode potentials	3.2, 3.3			
	Isomerism	5.1			
	Aldehydes and ketones	5.2			
	Carbohydrates				
	Carboxylic acids	5.3, 5.5			
	Esters, fats and oils	5.4, 5.5, 6.6			
	Periodic trends	4.1, 4.2			
	Oxy-acids of non-metals and their salts				
Module 5 (M5) *Analytical, transition metals and further organic chemistry*	Analytical techniques	7.1–7.3			
	Transition elements	4.3–4.8			
	Arenes	6.1, 6.2			
	Amines	6.4			
	Amino acids	6.5, 6.6			
	Nitriles	7.4			

Examination analysis

Units 1, 2 and 3 comprise AS Chemistry.	50%

A2 Chemistry comprises three unit tests. All questions are compulsory.

Unit 4   Objective questions and structured questions: short and extended answers (including synoptic assessment)   1 hr 15 min test   15%

Unit 5   Objective questions and structured questions: short and extended answers (including synoptic assessment)   1 hr 15 min test   15%

Unit 6 comprises two components:

6A   Synoptic paper: structured questions: short and extended answers   1hr 30 min test   13.3%

6B   Centre-assessed coursework   6.7%

AS/A2 Level Chemistry courses

AS and A2

All Chemistry A Level courses being studied from September 2000 are in two parts, with three separate modules in each part. Students first study the AS (Advanced Subsidiary) course. Some will then go on to study the second part of the A Level course, called A2. Advanced Subsidiary is assessed at the standard expected halfway through an A Level course: i.e., between GCSE and Advanced GCE. This means that new AS and A2 courses are designed so that difficulty steadily increases:

- AS Chemistry builds from GCSE Science
- A2 Chemistry builds from AS Chemistry.

How will you be tested?

Assessment units

For AS Chemistry, you will be tested by three assessment units. For the full A Level in Chemistry, you will take a further three units. AS Chemistry forms 50% of the assessment weighting for the full A Level.

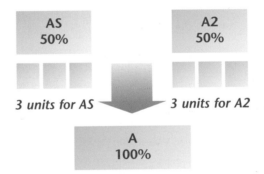

Each unit can normally be taken in either January or June. Alternatively, you can study the whole course before taking any of the unit tests.

If you are disappointed with a module result, you can resit each module once. You will need to be very careful about when you take up a resit opportunity because you will have only one chance to improve your mark. The higher mark counts.

Synoptic assessment

Synoptic assessment involves the explicit drawing together of knowledge, understanding and skills learned in different parts of the Advanced GCE course. The A2 units which draw together different parts of the course are assessed at the end of the course. More details are provided in Chapter 8 (pages 144–148).

Coursework

Coursework may form part of your A Level Chemistry course, depending on which specification you study. Where students have to undertake coursework, it is usually for the assessment of practical skills but this is not always the case. More details are provided on page 12.

Key skills

It is important that you develop your key skills of Communication, Application of Number and Information Technology throughout your AS and A2 courses. These are important skills that you need whatever you do beyond AS and A Levels. To gain the key skills qualification, you will need to demonstrate your ability to put your ideas across to other people, collect data and use up-to-date technology in your work. You will have many opportunities during A2 Chemistry to develop your key skills.

What skills will I need?

For A2 Chemistry, you will be tested by *assessment objectives*: these are the skills and abilities that you should have acquired by studying the course. The assessment objectives for A2 Chemistry are shown below.

Knowledge with understanding

- recall of facts, terminology and relationships
- understanding of principles and concepts
- drawing on existing knowledge to show understanding of the responsible use of chemistry in society
- selecting, organising and presenting information clearly and logically.

Application of knowledge and understanding, analysis and evaluation

- explaining and interpreting principles and concepts
- interpreting and translating, from one form into another, data presented as continuous prose or in tables, diagrams and graphs
- carrying out relevant calculations
- applying knowledge and understanding to familiar and unfamiliar situations
- assessing the validity of chemical information, experiments, inferences and statements.

You must also present arguments and ideas clearly and logically, using specialist vocabulary where appropriate.

Experimental and investigative skills

Chemistry is a practical subject and part of the assessment of A2 Chemistry will test your practical skills. This may be done during your lessons or you may be tested in a more formal practical examination. You will be assessed on four main skills:

- planning
- implementing
- analysing evidence and drawing conclusions
- evaluating evidence and procedures.

The skills may be assessed in separate practical exercises. They may also be assessed all together in the context of a single 'whole investigation'.

You will receive guidance about how your practical skills will be assessed from your teacher. This Study Guide concentrates on preparing you for the written examinations testing the subject content of A2 Chemistry.

AO4 Synthesis of knowledge, understanding and skills (Synoptic assessment)

- bringing together knowledge, principles and concepts from different areas of chemistry and apply them in a particular context
- using chemical skills in contexts which bring together different areas of the subject.

More details of synoptic assessment in Chemistry are discussed in Chapter 8 of this Study Guide (pages 144–148).

Different types of questions in A2 examinations

In A2 Chemistry examinations, different types of question are used to assess your abilities and skills. Unit tests mainly use structured questions requiring both short answers and more extended answers.

Short-answer questions

A short-answer question may test recall or it may test understanding. Short answer questions normally have space for the answers printed on the question paper.

Here are some examples (the answers are shown in *italics*):

What is meant by isotopes?

Atoms of the same element with different masses.

Calculate the amount (in mol) of H_2O in 4.5 g of H_2O.

1 mol H_2O has a mass of 18 g. $\therefore$ 4.5 g of H_2O contains 4.5/18 = 0.25 mol H_2O.

Structured questions

Structured questions are in several parts. The parts usually have a common context and they often become progressively more difficult and more demanding as you work your way through the question. A structured question may start with simple recall, then test understanding of a familiar or an unfamiliar situation.

Most of the practice questions in this book are structured questions, as this is the main type of question used in the assessment of A2 Level Chemistry.

When answering structured questions, do not feel that you have to complete one question before starting the next. The further you are into a question, the more difficult the marks are to obtain. If you run out of ideas, go on to the next question. You need to respond to as many parts to questions on an exam paper as possible. You will not score well if you spend so long trying to perfect the first questions that you do not reach later questions at all.

Here is an example of a structured question that becomes progressively more demanding.

(a) Write down the atomic structure of the two isotopes of potassium: ^{39}K and ^{41}K.

    (i) ^{39}K ...19... protons; ...20... neutrons; ...19... electrons. ✓

    (ii) ^{41}K ...19... protons; ...22... neutrons; ...19... electrons. ✓   [2]

(b) A sample of potassium has the following percentage composition by mass:
^{39}K: 92%; ^{41}K: 8% Calculate the relative atomic mass of the potassium sample.

    92 × 39/100 + 8 × 41/100 = 39.16 ✓   [1]

(c) What is the electronic configuration of a potassium atom?

    $1s^2 2s^2 2p^6 3s^2 3p^6 4s^1$ ✓   [1]

(d) The second ionisation energy of potassium is much larger than its first ionisation energy.

    (i) Explain what is meant by the *first ionisation energy* of potassium.

        The energy required to remove an electron ✓ *from each atom in 1 mole* ✓ *of gaseous atoms* ✓

    (ii) Why is there a large difference between the values for the first and the second ionisation energies of potassium?

The 2nd electron removed is from a different shell ✔ which is closer to the nucleus and experiences more attraction from the nucleus. ✔ This outermost electron experiences less shielding from the nucleus because there are fewer inner electron shells than for the 1st ionisation energy. ✔

[6]

Extended answers

In A2 Level Chemistry, questions requiring more extended answers may form part of structured questions or may form separate questions. They may appear anywhere on the paper and will typically have between 5 and 15 marks allocated to the answers as well as several lines of answer space. These questions are also often used to assess your abilities to communicate ideas and put together a logical argument.

The correct answers to extended questions are often less well-defined than to those requiring short answers. Examiners may have a list of points for which credit is awarded up to the maximum for the question.

An example of a question requiring an extended answer is shown below.

The table below gives data on the oxides of Period 3 in the Periodic Table.

oxide	Na_2O	MgO	Al_2O_3	SiO_2	P_4O_{10}	SO_3
melting point /K	1548	3100	2290	1880	853	306
conductivity of molten compound	good	good	medium	poor	poor	poor

Describe and explain the trends in formula, structure and bonding shown by these data.

[11 marks]

Points that the examiners might look for include:

In the formula, the number of oxygen atoms bonded increases across the period. The oxidation number increases steadily by one for each successive element. ✔ This is because the number of electrons in the outer shell increases by one for each element in the period. ✔

The trend in structure is from giant to simple molecular ✔ with a changeover between silicon and phosphorus. ✔ Structure is related to the magnitude of the melting points. ✔

The high melting points (Na_2O ✔ SiO_2) result from strong forces between ions (for Na_2O – Al_2O_3) ✔ and between atoms (for SiO_2). ✔ The low melting points (P_4O_{10} ✔ SO_3) result from weak intermolecular forces or van der Waals' forces. ✔

The trend in bonding is from ionic to covalent ✔ with a changeover between Al_2O_3 and SiO_2. ✔ This is related to the conductivity of the molten oxide. ✔
Good conductivity results from mobile ions (Na_2O ✔ $Al2O_3$. ✔)
Poor conductivity shows that no mobile ions are present (SiO_2 ✔ SO_3. ✔)

13 marking points → 11 marks

In this type of response, there may be an additional mark for a clear, well-organised answer, using specialist terms. In addition, marks may be allocated for legible text with accurate spelling, punctuation and grammar.

Other types of questions

Free-response and open-ended questions allow you to choose the context and to develop your own ideas.

Multiple-choice or objective questions require you to select the correct response to the question from a number of given alternatives.

Exam technique

A2 Chemistry builds from the knowledge and understanding acquired after studying AS Chemistry. This Study Guide has been written so that you will be able to tackle A2 Chemistry from an AS Chemistry background.

You should not need to search for important chemistry from AS Chemistry because cross-references have been included as 'Key points from AS' to the AS Chemistry Study Guide: Revise AS.

What are examiners looking for?

Examiners use instructions to help you to decide the length and depth of your answer.

If a question does not seem to make sense, you may have misread it – read it again!

State, define or list

This requires a short, concise answer, often recall of material that can be learnt by rote.

Explain, describe or discuss

Some reasoning or some reference to theory is required, depending on the context.

Outline

This implies a short response, almost a list of sentences or bullet points.

Predict or deduce

You are not expected to answer by recall but by making a connection between pieces of information.

Suggest

You are expected to apply your general knowledge to a 'novel' situation, one which you have not directly studied during the AS Chemistry course.

Calculate

This is used when a numerical answer is required. You should always use units in quantities and significant figures should be used with care.

Look to see how many significant figures have been used for quantities in the question and give your answer to this degree of accuracy.

If the question uses 3 sig figs, then give your answer to 3 sig figs also.

Some dos and don'ts

Dos

Do answer the question.

- No credit can be given for good Chemistry that is irrelevant to the question.

Do use the mark allocation to guide how much you write.

- Two marks are awarded for two valid points - writing more will rarely gain more credit and could mean wasted time or even contradicting earlier valid points.

Do use diagrams, equations and tables in your responses.

- Even in 'essay-type' questions, these offer an excellent way of communicating chemistry.

Do write legibly.

- An examiner cannot give marks if the answer cannot be read.

Do write using correct spelling and grammar. Structure longer essays carefully.

- Marks are now awarded for the quality of your language in exams.

Don'ts

Don't fill up any blank space on a paper.

- In structured questions, the number of dotted lines should guide the length of your answer.
- If you write too much, you waste time and may not finish the exam paper. You also risk contradicting yourself.

Don't write out the question again.

- This wastes time. The marks are for the answer!

Don't contradict yourself.

- The examiner cannot be expected to choose which answer is intended.

Don't spend too much time on a part that you find difficult.

- You may not have enough time to complete the exam. You can always return to a difficult calculation if you have time at the end of the exam.

What grade do you want?

Everyone would like to improve their grades but you will only manage this with a lot of hard work and determination. You should have a fair idea of your natural ability and likely grade in chemistry and the hints below offer advice on improving that grade.

For a Grade A

You will need to be a very good all-rounder.

- You must go into every exam knowing the work extremely well.
- You must be able to apply your knowledge to new, unfamiliar situations.
- You need to have practised many, many exam questions so that you are ready for the type of question that will appear.

The exams test all areas of the specification and any weaknesses in your chemistry will be found out. There must be no holes in your knowledge and understanding. For a Grade A, you must be competent in all areas.

For a Grade C

You must have a reasonable grasp of chemistry but you may have weaknesses in several areas and you will be unsure of some of the reasons for the chemistry.

- Many Grade C candidates are just as good at answering questions as Grade A students but holes and weaknesses often show up in just some topics.
- To improve, you will need to master your weaknesses and you must prepare thoroughly for the exam. You must become a better all-rounder.

For a Grade E

You cannot afford to miss the easy marks. Even if you find chemistry difficult to understand and would be happy with a Grade E, there are plenty of questions in which you can gain marks.

- You must memorise all definitions.
- You must practise exam questions to give yourself confidence that you do know some chemistry. In exams, answer the parts of questions that you know first. You must not waste time on the difficult parts. You can always go back to these later.
- The areas of chemistry that you find most difficult are going to be hard to score on in exams. Even in the difficult questions, there are still marks to be gained. Show your working in calculations because credit is given for a sound method. You can always gain some marks if you get part of the way towards the solution.

What marks do you need?

As a rough guide, you will need to score an average of 40% for a Grade E, 60% for a Grade C and 80% for a Grade A.

average	80%	70%	60%	50%	40%
grade	A	B	C	D	E

- When you get your results, identify a realistic 'target grade' for the A Level from your average mark.

Four steps to successful revision

Step 1: Understand

- Study the topic to be learned slowly. Make sure you understand the logic or important concepts.
- Mark up the text if necessary – underline, highlight and make notes.
- Re-read each paragraph slowly.

GO TO STEP 2

Step 2: Summarise

- Now make your own revision note summary:
  What is the main idea, theme or concept to be learned?
  What are the main points? How does the logic develop?
  Ask questions: Why? How? What next?
- Use bullet points, mind maps, patterned notes.
- Link ideas with mnemonics, mind maps, crazy stories.
- Note the title and date of the revision notes
  (e.g. Chemistry: Reaction rates, 3rd March).
- Organise your notes carefully and keep them in a file.

This is now in **short-term memory**. You will forget 80% of it if you do not go to Step 3.
GO TO STEP 3, but first take a 10 minute break.

Step 3: Memorise

- Take 25 minute learning 'bites' with 5 minute breaks.
- After each 5 minute break test yourself:
  Cover the original revision note summary
  Write down the main points
  Speak out loud (record on tape)
  Tell someone else
  Repeat many times.

The material is well on its way to **long-term memory**.
You will forget 40% if you do not do step 4. **GO TO STEP 4**

Step 4: Track/Review

- Create a Revision Diary (one A4 page per day).
- Make a revision plan for the topic, e.g. 1 day later, 1 week later, 1 month later.
- Record your revision in your Revision Diary, e.g.
  Chemistry: Reaction rates, 3rd March 25 minutes
  Chemistry: Reaction rates, 5th March 15 minutes
  Chemistry: Reaction rates, 3rd April 15 minutes
  ... and then at monthly intervals.

Chapter 1

Reaction rates

The following topics are covered in this chapter:

- *Orders and the rate equation*
- *Determination of reaction mechanisms*

1.1 Orders and the rate equation

Key points from AS

- **Reaction rates**
  Revise AS pages 98–101

During the study of reaction rates in AS Chemistry, reaction rates were described in terms of activation energy and the Boltzmann distribution. For A2 Chemistry, you will build upon this knowledge and understanding by measuring and calculating reaction rates using rate equations.

Orders and the rate equation

AQA	M4	SALTERS	M4, M5
EDEXCEL	M5	WJEC	CH5
OCR	M6	NICCEA	M4
NUFFIELD	M4		

The rate of a reaction is usually measured as the **change in concentration** of a reaction species **with time**.

- Units of rate = $\underbrace{mol\ dm^{-3}}_{\text{concentration}}\ \underbrace{s^{-1}}_{\text{per time}}$

Key points from AS

- **What is a reaction rate?**
  Revise AS pages 98–99

Orders of reaction

For a reaction: $A + B + C \longrightarrow$ products,

> [A] means the concentration of reactant A in mol dm^{-3}.

- the **order** of the reaction shows how the reaction rate is affected by the concentrations of **A**, **B** and **C**.

If the order is 0 (zero order) with respect to reactant **A**,
- the rate is unaffected by changes in concentration of **A**:
  $rate \propto [A]^0$

If the order is 1 (first order) with respect to a reactant **B**,
- the rate is doubled by doubling of the concentration of reactant **B**:
  $rate \propto [B]^1$

If the order is 2 (second order) with respect to a reactant **C**,
- the rate is quadrupled by doubling of the concentration of reactant **C**:
  $rate \propto [C]^2$

Combining the information above,

> A number raised to the power of zero = 1 and zero order species can be omitted from the rate equation.

$$rate \propto [A]^0[B]^1[C]^2$$
$$\therefore\ rate = k[B]^1[C]^2$$

- This expression is called the **rate equation** for the reaction.
- k is the rate constant of the reaction.

The rate equation

> The rate of a reaction is determined from experimental results.
>
> It **cannot** be determined from the chemical equation.

The rate equation of a reaction shows how the rate is affected by the concentration of reactants. A rate equation can only be determined from experiments.

In general, for a reaction: $A + B \longrightarrow C$,

the reaction rate is given by: $rate = k[A]^m[B]^n$

- m and n are the **orders of reaction** with respect to **A** and **B** respectively.
- The **overall order** of reaction is m + n.
- The reaction rate is measured as the change in concentration of a reaction species with time.
- The units of rate are $mol\ dm^{-3}\ s^{-1}$.

The rate constant k indicates the rate of the reaction:

$\qquad$ a large value of $k \longrightarrow$ fast rate of reaction

$\qquad$ a small value of $k \longrightarrow$ slow rate of reaction.

The effect of temperature on rate constants

> **Key points from AS**
>
> - **Activation energy**
>   *Revise AS page 99*

An increase in temperature speeds up the rate of most reactions by **increasing** the rate constant k.

The table below shows the increase in the rate constant for the decomposition of hydrogen iodide with increasing temperature.

$$2HI(g) \longrightarrow H_2(g) + I_2(g)$$

temperature/°C	283	356	427	508
rate constant, k / $dm^3\ mol^{-1}\ s^{-1}$	7.04×10^{-7}	6.04×10^{-5}	2.32×10^{-3}	7.90×10^{-2}

> For many reactions, the rate doubles for each 10°C increase in temperature.

A reaction with a large activation energy has a **small rate constant**. Such a reaction may need a considerable temperature rise to increase the value of the rate constant sufficiently for a reaction to take place at all.

Measuring rates using the initial rates method

AQA	M4	SALTERS	M4
EDEXCEL	M5	WJEC	CH5
OCR	M6	NICCEA	M4
NUFFIELD	M4		

For a reaction, $X + Y \longrightarrow Z$,

a series of experiments can be carried out using different **initial** concentrations of the reactants **X** and **Y**.

It is important to change only one variable at a time, so two series of experiments will be required.

- **Series 1** – The concentration of **X is changed** whilst the concentration of **Y is kept constant**.
- **Series 2** – The concentration of **Y is changed** whilst the concentration of **X is kept constant**.

For each experiment, we need to:

- plot a **concentration/time** graph

> The initial rate of the reaction is the tangent of a concentration/time graph at time = 0.

- measure the **initial rate** from the graph as the **tangent** drawn at **time = 0** (see page 23).

The results below show the initial rates using different concentrations of **X** and **Y**.

Experiment	[X(aq)] /mol dm^{-3}	[Y(aq)] /mol dm^{-3}	initial rate /mol dm^{-3} s^{-1}
1	1.0×10^{-2}	1.0×10^{-2}	0.5×10^{-3}
2	2.0×10^{-2}	1.0×10^{-2}	2.0×10^{-3}
3	2.0×10^{-2}	2.0×10^{-2}	4.0×10^{-3}

Determine the orders

Comparing experiments 1 and 2: [Y(aq)] has been kept constant

- [X(aq)] has been doubled,   rate × 4
  ∴ order with respect to X(aq) = 2.

Comparing experiments 2 and 3: [X(aq)] has been kept constant

- [Y(aq)] has been doubled,   rate doubles
  ∴ order with respect to Y(aq) = 1.

Use the orders to write the Rate Equation

- $rate = k[X(aq)]^2[Y(aq)]$

- The overall order of this reaction is (2 + 1) = 3rd order

Calculate the rate constant for the reaction

Rearrange the rate equation:

$$\text{The rate constant, } k = \frac{rate}{[X(aq)]^2[Y(aq)]}$$

Calculate k using values from one of the experiments.

$$k = \frac{(2.0 \times 10^{-3})}{(2.0 \times 10^{-2})^2 (1.0 \times 10^{-2})} = 500 \text{ dm}^6 \text{ mol}^{-2} \text{ s}^{-1}$$

In this example, the results from Experiment 2 have been used.

You will get the same value of k from the results of **any** of the experimental runs.

Try calculating k from Experiment 1 and from Experiment 3. All give the same value.

Units of rate constants

The units of a rate constant depend upon the rate equation for the reaction. We can determine units of k by substituting units for rate and concentration into the rearranged rate equation. The table below shows how units can be determined.

Notice how units need to be worked out afresh for reactions with different overall orders.

You do not need to memorise these but it is important that you are able to work out these when needed.

See also p. 30 in which the units of the equilibrium constant K_c are discussed.

For a zero order reaction:	For a **first order** reaction:
$rate = k[A]^0 = k$ units of k = **mol dm^{-3} s^{-1}**	$rate = k[A]$ ∴ $k = \dfrac{rate}{[A]}$ Units of $k = \dfrac{(\text{mol dm}^{-3} \text{ s}^{-1})}{(\text{mol dm}^{-3})} = $ **s^{-1}**
For a second order reaction:	For a **third order** reaction:
$rate = k[A]^2$ ∴ $k = \dfrac{rate}{[A]^2}$ Units of $k = \dfrac{(\text{mol dm}^{-3} \text{ s}^{-1})}{(\text{mol dm}^{-3})^2}$ = **dm^3 mol^{-1} s^{-1}**	$rate = k[A]^2[B]$ ∴ $k = \dfrac{rate}{[A]^2[B]}$ Units of $k = \dfrac{(\text{mol dm}^{-3} \text{ s}^{-1})}{(\text{mol dm}^{-3})^2 (\text{mol dm}^{-3})}$ = **dm^6 mol^{-2} s^{-1}**

Progress check

1 A chemical reaction is first order with respect to compound **P** and second order with respect to compound **Q**.
   (a) Write the rate equation for this reaction.
   (b) What is the overall order of this reaction?
   (c) By what factor will the rate increase if:
      (i) the concentration of **P only** is doubled
      (ii) the concentration of **Q only** is doubled
      (iii) the concentrations of **P** and **Q** are **both** doubled?
   (d) What are the units of the rate constant of this reaction?

2 The reaction of ethanoic anhydride, $(CH_3CO)_2O$, with ethanol, C_2H_5OH, can be represented by the equation:

$$(CH_3CO)_2O + C_2H_5OH \longrightarrow CH_3CO_2C_2H_5 + CH_3CO_2H$$

The table below shows the initial concentrations of the two reactants and the initial rates of reaction.

Experiment	*$[(CH_3CO)_2O]$ /mol dm^{-3}*	*$[C_2H_5OH]$ /mol dm^{-3}*	*initial rate /mol dm^{-3} s^{-1}*
1	0.400	0.200	6.60×10^{-4}
2	0.400	0.400	1.32×10^{-3}
3	0.800	0.400	2.64×10^{-3}

(a) State and explain the order of reaction with respect to:
  (i) ethanoic anhydride (ii) ethanol.

(b) (i) Write an expression for the overall rate equation.
  (ii) What is the overall order of reaction?

(c) Calculate the value, with units, for the rate constant, *k*.

1 (a) *rate* = $k[P][Q]^2$
(b) 3rd order
(c) (i) rate doubles (ii) rate quadruples (iii) rate × 8
(d) dm^6 mol^{-2} s^{-1}

2 (a) (i) 1st order; concentration doubles; rate doubles
(ii) 1st order; concentration doubles; rate doubles.
(b) (i) *rate* = $k[(CH_3CO)_2O][C_2H_5OH]$ (ii) 2nd order.
(c) 8.25×10^{-3} dm^3 mol^{-1} s^{-1}.

Measuring rates using graphs

EDEXCEL	M5	SALTERS	M4
OCR	M6	WJEC	CH5
NUFFIELD	M4	NICCEA	M4

Concentration/time graphs

It is often possible to measure the concentration of a reactant or product at various times during the course of an experiment.

- From the results, a concentration/time graph can be plotted.
- The shape of this graph can indicate the order of the reaction by measuring the **half-life** of a reactant.

Key points from AS

- **What is a reaction rate?**
  Revise AS pages 98–99

KEY POINT

The half-life of a reactant is the time taken for its concentration to reduce by half.

- **A first-order reaction has a constant half-life.**

Example of a first order graph

The reaction between bromine and methanoic acid is shown below.

$Br_2(aq) + HCOOH(aq) \longrightarrow 2Br^-(aq) + 2H^+(aq) + CO_2(g)$

- During the course of the reaction, the orange bromine colour disappears as Br_2 reacts to form colourless Br^- ions. The order with respect to bromine can be determined by monitoring the rate of disappearance of bromine using a colorimeter.
- The concentration of the other reactant, methanoic acid, is kept virtually constant by using an excess of methanoic acid.

The concentration/time graph from this reaction is shown on the next page. Notice that the **half-life is constant** at 200 s, showing that this reaction is first order with respect to $Br_2(aq)$.

The half-life curve of a 1st order reaction is concentration independent.

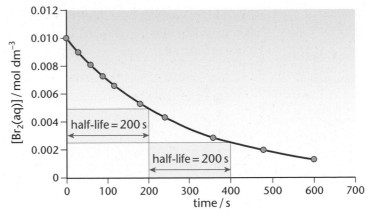

The shapes of concentration/time graphs for zero, first and second order reactions are shown below.

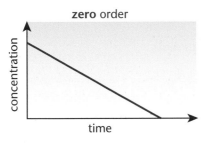

zero order

The concentration falls at a steady rate with time
• half-life **decreases** with time.

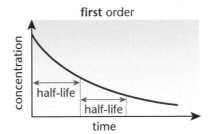

first order

The concentration halves in equal time intervals
• **constant** half-life.

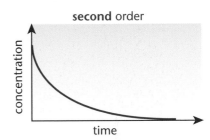

second order

The half-life gets progressively longer as the reaction proceeds
• half-life **increases** with time.

• The 1st order relationship can be confirmed by plotting a graph of log[**X**] against time, which gives a straight line.

Measuring rates from tangents

The gradient of the concentration/time graph is a measure of the rate of a reaction.

For the reaction: **A** ⟶ **B**, the graph below shows how the concentration of **A** changes during the course of the reaction.

> At any instant of time during the reaction, the rate can be measured by drawing a tangent to the curve.

KEY POINT

At the start of the reaction (*t* = 0):
• the tangent is steepest
• the concentration of **A** is greatest
• and the reaction rate is fastest.

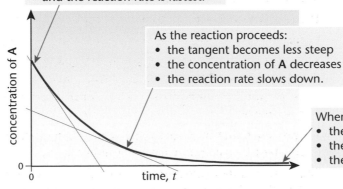

As the reaction proceeds:
• the tangent becomes less steep
• the concentration of **A** decreases
• the reaction rate slows down.

When the reaction is complete:
• the concentration of **A** is very small
• the gradient becomes zero
• the reaction stops.

At an instant of time, *t*, during the reaction:

• the rate of **decrease** in concentration of **A** = $-\dfrac{d[A]}{dt}$ *(the tangent)*

A reaction rate is expressed as a positive value.

For the reactant **A**, the **negative** sign shows a **decreasing** concentration with time.

For the product **B**, the **positive** sign shows **increasing** concentration.

The negative sign shows that the concentration of **A** decreases during the reaction. The rate could also be followed by measuring the rate of **increase** in concentration of the product **B**.

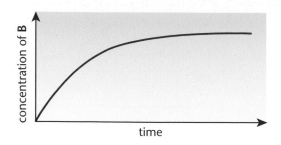

B is formed and its concentration increases during the reaction.

$$\text{rate of } \textbf{increase} \text{ in concentration of } \textbf{B} = + \frac{d[\textbf{B}]}{dt}$$

Plotting a rate/concentration graph

- A concentration/time graph is first plotted.
- **Tangents** are drawn at several time values on the concentration/time graph, giving values of reaction rates.
- A second graph is plotted of **rate** against **concentration**.
- The shape of this graph confirms the order of the reaction.

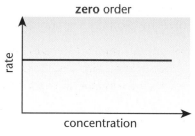

zero order

Rate ∝ $[\textbf{A}]^0$ ∴ rate = constant
- Rate unaffected by changes in concentration.

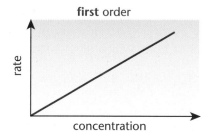

first order

Rate ∝ $[\textbf{A}]^1$
- Rate doubles as concentration doubles.

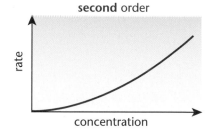

second order

Rate ∝ $[\textbf{A}]^2$
- Rate quadruples as concentration doubles.

- The 2nd order relationship can be confirmed by plotting a graph of rate against $[\textbf{A}]^2$, which gives a straight line.

Progress check

1 The data below shows how the concentration of a reactant **A** changes during the course of a reaction.

Time / 10^4 s	0	0.36	0.72	1.08	1.44
[A] /mol dm^{-3}	0.240	0.156	0.104	0.068	0.045

(a) Plot a concentration/time graph and, by drawing tangents, estimate:
   (i) the initial rate
   (ii) the rate 1×10^4 s after the reaction has started.
(b) Determine the half-life of the reaction and show that the reaction is first order with respect to **A**.

1 (a) (i) 2.9×10^{-5} mol dm^{-3} s^{-1} (ii) 8.7×10^{-6} mol dm^{-3} s^{-1}
(b) Half-life = 5.8×10^3 s. Successive half-lives are constant.

1.2 Determination of reaction mechanisms

After studying this section you should be able to:

- understand what is meant by a rate-determining step
- predict a rate equation from a rate-determining step
- predict a rate-determining step from a rate equation
- use a rate equation and the overall equation for a reaction to predict a possible reaction mechanism

Predicting reaction mechanisms

EDEXCEL	M5	WJEC	CH5
OCR	M6	NICCEA	M4
NUFFIELD	M4		

The rate-determining step

Chemical reactions often take place in a series of steps. The detail of these steps is the **reaction mechanism**.

> The rate equation can provide clues about a likely reaction mechanism by identifying the **slowest** stage of a reaction sequence, called the **rate-determining step**.

KEY POINT

Key points from AS

- **The hydrolysis of halogenoalkanes**
  Revise AS page 132–133

Predicting reaction mechanisms from rate equations

Halogenoalkanes are hydrolysed by hot aqueous alkali.

$$RBr + OH^- \longrightarrow ROH + Br^-$$

Rates for the hydrolysis reactions of primary and tertiary halogenoalkanes can be measured experimentally to determine rate equations. We can predict the rate-determining step for each of these reactions from the rate equations obtained.

The hydrolysis of primary halogenoalkanes

The primary halogenoalkane, $CH_3CH_2CH_2CH_2Br$, is hydrolysed by aqueous alkali:

$$CH_3CH_2CH_2CH_2Br + OH^- \longrightarrow CH_3CH_2CH_2CH_2OH + Br^-$$

A balanced equation represents the **overall** reaction.

It does not reveal the mechanism that achieves it.

Experiments show that the rate equation for this reaction is:

$$rate = k[CH_3CH_2CH_2CH_2Br]\,[OH^-]$$

This rate equation shows that:

- the rate is determined by a **slow reaction step** which must involve **both $CH_3CH_2CH_2CH_2Br$ and OH^-**.

This supports the **single step** mechanism below, matching the rate equation.

Evidence for the mechanism comes from the rate equation.

$$CH_3CH_2CH_2CH_2Br + OH^- \longrightarrow CH_3CH_2CH_2CH_2OH + Br^-$$

$$rate = k[CH_3CH_2CH_2CH_2Br]\,[OH^-]$$

The hydrolysis of tertiary halogenoalkanes

The tertiary halogenoalkane, $(CH_3)_3CBr$, is hydrolysed by aqueous alkali:

$$(CH_3)_3CBr + OH^- \longrightarrow (CH_3)_3COH + Br^-$$

Experiments show that the rate equation for this reaction is:

$$rate = k[(CH_3)_3CBr]$$

This rate equation shows that:

- the rate is determined by a slow reaction step which involves **only $(CH_3)_3CBr$**
- the concentration of OH^- has **no effect** on the reaction rate.

A possible step in the reaction mechanism including only $(CH_3)_3CBr$ is:

$$(CH_3)_3CBr \longrightarrow (CH_3)_3C^+ + Br^-$$

$$rate = k[(CH_3)_3CBr]$$

You are **not** expected to remember these examples.

You **are** expected to interpret data to identify a rate-determining step and to suggest a possible reaction mechanism for a reaction.

This supports the **two step** mechanism below, matching the rate equation.

Step 1: the rate-determining step

$$(CH_3)_3CBr \longrightarrow (CH_3)_3C^+ + Br^- \qquad \textbf{SLOW}$$

Step 2

$$(CH_3)_3C^+ + OH^- \longrightarrow (CH_3)_3COH \qquad \textbf{FAST}$$

In this reaction sequence, the rate is determined by the **slow first step**.

The rate equation only includes the reacting species for this rate-determining step.

$$rate = k[(CH_3)_3CBr]$$

- The rate of the reaction is controlled mainly by the slowest step of the mechanism – the **rate-determining step**.

- The rate equation only includes reacting species involved in the slow rate-determining step.

- The orders in the rate equation match the number of species involved in the rate-determining step.

Progress check

1 CH_3CH_2Br and cyanide ions, CN^- react as follows:

$CH_3CH_2Br + CN^- \longrightarrow CH_3CH_2CN + Br^-$

Show two different possible reaction routes for this reaction and write down the expected rate equation for each route.

1 Single step: $CH_3CH_2Br + CN^- \longrightarrow CH_3CH_2CN + Br^-$

$rate = k[CH_3CH_2Br][CN^-]$

Two steps: $CH_3CH_2Br \longrightarrow CH_3CH_2^+ + Br^-$ slow

$CH_3CH_2^+ + CN^- \longrightarrow CH_3CH_2CN$ fast

$rate = k[CH_3CH_2Br]$

Sample question and model answer

The oxidation of nitrogen monoxide to nitrogen dioxide in car exhausts may involve carbon monoxide and oxygen.

$$NO(g) + CO(g) + O_2(g) \longrightarrow NO_2(g) + CO_2(g)$$

The rate of this reaction can be followed colorimetrically because $NO_2(g)$ is coloured.

(a) (i) Use the axes below to show how the concentration of $NO_2(g)$ produced varies with time.

Note the key points for the marks:

1 mark for line going through the origin

1 mark for the graph levelling off

1 mark for showing how the initial rate is measured.

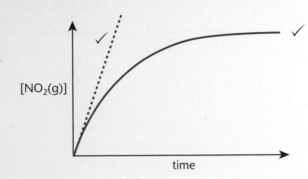

(ii) Show, with reference to your sketch, how the initial rate of reaction could be deduced.

Draw a tangent to the curve at time = 0 s (see above in diagram). ✓ [3]

(b) The results from a series of experiments are shown below:

Experiment	[NO(g)] /mol dm^{-3}	[CO(g)] /mol dm^{-3}	[O$_2$(g)] /mol dm^{-3}	Initial rate /mol dm^{-3} s^{-1}
1	1.00×10^{-3}	1.00×10^{-3}	1.00×10^{-1}	4.40×10^{-4}
2	2.00×10^{-3}	1.00×10^{-3}	1.00×10^{-1}	1.76×10^{-3}
3	2.00×10^{-3}	2.00×10^{-3}	1.00×10^{-1}	1.76×10^{-3}
4	4.00×10^{-3}	1.00×10^{-3}	2.00×10^{-1}	7.04×10^{-3}

(i) Deduce the order of reaction with respect to NO(g). Show your reasoning.

Straightforward marks. Show your logic.

double concentration of NO only, rate quadruples. ✓
∴ second order with respect to NO. ✓

(ii) State the order of reaction with respect to CO(g) and to $O_2(g)$.

CO: double concentration of CO only, rate does not change:
∴ zero order with respect to CO. ✓
O_2: double concentration of NO AND O_2, rate quadruples.
The effect of doubling [NO] quadruples the rate.
∴ changing [O_2] has no effect and zero order with respect to O_2. ✓

(iii) Write an expression for the rate equation.

rate = $k[NO(g)]^2$ ✓

You can use any of the results – they all give the same value for k.

(iv) Calculate a value for the rate constant, k. State the units of k.

Using results from the 1st experiment: $4.40 \times 10^{-4} = k \times (1.00 \times 10^{-3})^2$ ✓

Substitute units into the rate expression to find the units of k.

$$\therefore k = \frac{4.40 \times 10^{-4}}{(1.00 \times 10^{-3})^2} = 440 ✓$$

units: $\dfrac{mol\ dm^{-3}\ s^{-1}}{(mol\ dm^{-3})^2} = mol^{-1}\ dm^3\ s^{-1}$ ✓ $k = 440\ dm^3\ mol^{-1}\ s^{-1}$ [8]

[Total: 11]

Cambridge How Far, How Fast? Q1 June 1996

Practice examination questions

1

The reaction between hydrogen peroxide and iodide ions in an acid solution can be written as follows.

$$H_2O_2(aq) + 2I^-(aq) + 2H^+(aq) \longrightarrow I_2(aq) + 2H_2O(l)$$

A student wanted to investigate the rate of this reaction.

(a) Suggest how the student could follow the rate of this reaction in the laboratory. [2]

(b) The student obtained the following results.

concentration of $H_2O_2(aq)$/mol dm^{-3}	concentration of $I^-(aq)$/mol dm^{-3}	concentration of $H^+(aq)$/mol dm^{-3}	initial rate /10^{-6} mol dm^{-3} s^{-1}
0.0010	0.10	0.10	2.8
0.0020	0.10	0.10	5.6
0.0020	0.10	0.20	5.6
0.0010	0.40	0.10	11.2

(i) From these results, deduce the order of reaction with respect to each reactant. Explain your reasoning. [6]

(ii) Using your answers to (b) (i), write a rate equation for this reaction. [1]

(iii) Calculate the value of the rate constant, k, for this reaction. [2]

(c) (i) What is meant by the term rate-determining step in a reaction sequence? [1]

(ii) Using your answer to (b) (ii), suggest an equation for the rate-determining step for the reaction between H_2O_2 and I^- in the presence of acid. [2]

[Total: 14]

Cambridge How Far, How Fast? Q2 March 2000

2

(a) A chemical reaction is first order with respect to compound **X** and second order with respect to compound **Y**.

(i) Write the rate equation for this reaction. [2]

(ii) What is the overall order of this reaction? [1]

(iii) By what factor will the rate increase if the concentrations of **X** and **Y** are both doubled? [1]

(b) The table below shows the initial concentrations of two compounds, **A** and **B**, and also the initial rate of the reaction that takes place between them at constant temperature.

Experiment	[A]/mol dm^{-3}	[B]/mol dm^{-3}	Initial rate/mol dm^{-3} s^{-1}
1	0.2	0.2	3.5×10^{-4}
2	0.4	0.4	1.4×10^{-3}
3	0.8	0.4	5.6×10^{-3}

(i) Determine the overall order of the reaction between **A** and **B**. Explain how you reached your conclusion. [2]

(ii) Determine the order of reaction with respect to compound **B**. Explain how you reached your conclusion. [2]

(iii) Write the rate equation for the overall reaction. [1]

(iv) Calculate the value of the rate constant, stating its units. [2]

[Total: 11]

Assessment and Qualifications Alliance Unit 4 Specimen Test Q2 2000

Chapter 2

Chemical equilibrium

The following topics are covered in this chapter:

- The equilibrium constant K_c
- The equilibrium constant K_p
- Acids and bases
- The pH scale
- pH changes

Key points from AS

- **Chemical equilibrium**
  Revise AS pages 104–108

During the study of equilibrium in AS Chemistry, the concept of dynamic equilibrium is introduced and le Chatelier's principle is used to predict how a change in conditions may alter the equilibrium position. For A2 Chemistry, you will build upon this knowledge and understanding to find out the exact position of equilibrium using the Equilibrium Law.

2.1 The equilibrium constant, K_c

After studying this section you should be able to:

- deduce, for homogeneous reactions, K_c in terms of concentrations
- calculate values of K_c given appropriate data
- understand that K_c is changed only by changes in temperature
- understand how the magnitude of K_c relates to the equilibrium position
- calculate, from data, the concentrations present at equilibrium

LEARNING SUMMARY

The equilibrium law

AQA	M4	SALTERS	M4	
EDEXCEL	M4	WJEC	CH5	
OCR	M6	NICCEA	M4	
NUFFIELD	M4			

The equilibrium law states that, for an equation:

$$a\,\mathbf{A} + b\,\mathbf{B} \rightleftharpoons c\,\mathbf{C} + d\,\mathbf{D},$$

$$K_c = \frac{[\mathbf{C}]^c\,[\mathbf{D}]^d}{[\mathbf{A}]^a\,[\mathbf{B}]^b}$$

- $[\mathbf{A}]^a$, etc., are the **equilibrium** concentrations of the reactants and products of the reaction.

The overall equation for the reaction is used to write K_c.

- Each product and reactant has its equilibrium concentration raised to the **power** of its **balancing number** in the equation.

- The equilibrium concentrations of the **products** are multiplied together on **top** of the fraction.

- The equilibrium concentrations of the **reactants** are multiplied together on the **bottom** of the fraction.

Working out K_c

For the equilibrium: $H_2(g) + I_2(g) \rightleftharpoons 2HI(g)$

*For A2 Chemistry, most courses only use **homogeneous equilibria** – all species are in the same phase:*

all gaseous (g)
or
all aqueous (aq)
or
all liquid (l).

applying the equilibrium law above: $K_c = \dfrac{[HI(g)]^2}{[H_2(g)]\,[I_2(g)]}$

Equilibrium concentrations of $H_2(g)$, $I_2(g)$ and $HI(g)$ are shown below:

$[H_2(g)]$ /mol dm^{-3}	$[I_2(g)]$ /mol dm^{-3}	$[HI(g)]$ /mol dm^{-3}
0.0114	0.00120	0.0252

$$K_c = \frac{[HI(g)]^2}{[H_2(g)]\,[I_2(g)]} = \frac{0.0252^2}{0.0114 \times 0.00120} = 46.4$$

The units must be worked out afresh for each equilibrium.

See also p. 21 in which the units of the rate constant are discussed.

Units of K_c

- In the K_c expression, each concentration value is replaced by its units:

$$K_c = \frac{[HI(g)]^2}{[H_2(g)][I_2(g)]} = \frac{(mol\ dm^{-3})^2}{(mol\ dm^{-3})(mol\ dm^{-3})}$$

- For this equilibrium, the units cancel.
- ∴ in the equilibrium: $H_2(g) + I_2(g) \rightleftharpoons 2HI(g)$, K_c has no units.

Properties of K_c

The magnitude of K_c indicates the extent of a chemical equilibrium.

- $K_c = 1$ indicates an equilibrium halfway between reactants and products.
- $K_c = 100$ indicates an equilibrium well in favour of the products.
- $K_c = 1 \times 10^{-2}$ indicates an equilibrium well in favour of the reactants.

K_c indicates how FAR a reaction proceeds; **not** how FAST.
K_c is unaffected by changes in concentration or pressure.
K_c can **only** be changed by altering the temperature.

Although changes in K_c affect the equilibrium yield, the actual conditions used may be different.

In an exothermic reaction, K_c increases and the equilibrium yield increases as temperature decreases.

However, a decrease in temperature may slow down the reaction so much that the reaction is stopped.

In practice, a compromise will be needed where equilibrium and rate are considered together – a reasonable equilibrium yield must be obtained in a reasonable length of time.

How does K_c vary with temperature?

It cannot be stressed too strongly that K_c can be changed only by altering the temperature. How K_c changes depends upon whether the reaction gives out or takes in heat energy.

In an **exothermic** reaction, K_c **decreases** with increasing temperature. Raising the temperature reduces the equilibrium yield of products.

$$H_2(g) + I_2(g) \rightleftharpoons 2HI(g): \qquad \Delta H^\ominus_{298} = -9.6\ kJ\ mol^{-1}$$

temperature /K	500	700	1100
K_c	160	54	25

In an **endothermic** reaction, K_c **increases** with increasing temperature. Raising the temperature increases the equilibrium yield of products.

$$N_2(g) + O_2(g) \rightleftharpoons 2NO(g): \qquad \Delta H^\ominus_{298} = +180\ kJ\ mol^{-1}$$

temperature /K	500	700	1100
K_c	5×10^{-13}	4×10^{-8}	1×10^{-5}

Progress check

1 For each of the following equilibria, write down the expression for K_c. State the units for K_c for each reaction.
    (a) $N_2O_4(g) \rightleftharpoons 2NO_2(g)$ (c) $H_2(g) + Br_2(g) \rightleftharpoons 2HBr(g)$
    (b) $CO(g) + 2H_2(g) \rightleftharpoons CH_3OH(g)$ (d) $2SO_2(g) + O_2(g) \rightleftharpoons 2SO_3(g)$

2 Explain whether the two reactions, **A** and **B**, are exothermic or endothermic.

| temperature /K | numerical value of K_c | |
	reaction **A**	reaction **B**
200	5.51×10^{-8}	4.39×10^4
400	1.46	4.03
600	3.62×10^2	3.00×10^{-2}

2 Reaction A: endothermic; K_c increases with increasing temperature.
Reaction B: exothermic; K_c decreases with increasing temperature.

1 (a) $K_c = \frac{[NO_2(g)]^2}{[N_2O_4(g)]}$ (b) $K_c = \frac{[CH_3OH(g)]}{[CO(g)][H_2(g)]^2}$ (c) $K_c = \frac{[HBr(g)]^2}{[H_2(g)][Br_2(g)]}$ (d) $K_c = \frac{[SO_3(g)]^2}{[SO_2(g)]^2[O_2(g)]}$

Determination of K$_c$ from experiment

AQA	M4	SALTERS	M4
EDEXCEL	M4	WJEC	CH5
OCR	M6	NICCEA	M4
NUFFIELD	M4		

The equilibrium constant K_c can be calculated using experimental results. The example below shows how to:

- determine the equilibrium concentrations of the components in an equilibrium mixture
- calculate K_c.

The ethyl ethanoate esterification equilibrium

> The most important stage in this calculation is to find the **change** in the number of moles of each species in the equilibrium.

0.200 mol CH_3COOH and 0.100 mol C_2H_5OH were mixed together with a trace of acid catalyst in a total volume of 250 cm^3. The mixture was allowed to reach equilibrium:

$$CH_3COOH + C_2H_5OH \rightleftharpoons CH_3COOC_2H_5 + H_2O.$$

Analysis of the mixture showed that 0.115 mol of CH_3COOH were present at equilibrium.

First summarise the results

It is useful to summarise the results as an 'I.C.E.' table: 'Initial/Change/Equilibrium'.

	CH$_3$COOH	C$_2$H$_5$OH	CH$_3$COOC$_2$H$_5$	H$_2$O
Initial no. of moles	0.200	0.100	0	0
Change in moles				
Equilibrium no. of moles	0.115			

From the CH_3COOH values:

- moles of CH_3COOH that reacted = 0.200 – 0.115 = **0.085** mol

Find the change in moles of each component in the equilibrium

The amount of each component can be determined from the balanced equation. The change in moles of CH_3COOH is already known.

> In this experiment, 0.085 mol CH_3COOH has **reacted** with 0.085 mol C_2H_5OH to form 0.085 mol $CH_3COOC_2H_5$ and 0.085 mol H_2O.

equation:	CH_3COOH +	C_2H_5OH $\rightleftharpoons$	$CH_3COOC_2H_5$ +	H_2O
molar quantities:	1 mol	1 mol $\longrightarrow$	1 mol	1 mol
change/mol:	**–0.085**	–0.085	+0.085	+0.085

Equilibrium concentrations are determined for each component

> Note that the total volume is 250 cm^3 (0.250 dm^3). The concentration must be expressed as mol dm^{-3}.

	CH$_3$COOH +	C$_2$H$_5$OH $\rightleftharpoons$	CH$_3$COOC$_2$H$_5$ +	H$_2$O
Initial amount/mol	0.200	0.100	0	0
Change in moles	–0.085	–0.085	+0.085	+0.085
Equilibrium amount/mol	0.115	0.015	0.085	0.085
Equilibrium conc. /mol dm^{-3}	$\dfrac{0.115}{0.250}$	$\dfrac{0.015}{0.250}$	$\dfrac{0.085}{0.250}$	$\dfrac{0.085}{0.250}$

Write the expression for K$_c$ and substitute values

$$K_c = \frac{[CH_3COOC_2H_5]\,[H_2O]}{[CH_3COOH]\,[C_2H_5OH]} = \frac{\dfrac{0.085}{0.250} \times \dfrac{0.085}{0.250}}{\dfrac{0.115}{0.250} \times \dfrac{0.015}{0.250}}$$

Calculate K$_c$

$\therefore K_c = 4.19$ (no units: all units cancel)

Progress check

1 Several experiments were set up for the $H_2(g)$, $I_2(g)$ and $HI(g)$ equilibrium. The equilibrium concentrations are shown below.

$[H_2(g)]$ /mol dm^{-3}	$[I_2(g)]$ /mol dm^{-3}	$[HI(g)]$ /mol dm^{-3}
0.0092	0.0020	0.0296
0.0077	0.0031	0.0334
0.0092	0.0022	0.0308
0.0035	0.0035	0.0235

Calculate the value for K_c for each experiment. Hence show that each experiment has the same value for K_c (allowing for experimental error). Work out an average value for K_c.

2 2 moles of ethanoic acid, CH_3COOH were mixed with 3 moles of ethanol and the mixture allowed to reach equilibrium.

$$CH_3COOH + C_2H_5OH \rightleftharpoons CH_3COOC_2H_5 + H_2O$$

At equilibrium, 0.5 moles of ethanoic acid remained.
(a) Work out the equilibrium concentrations of each component in the mixture. (You will need to use V to represent the volume but this will cancel out in your calculation.)
(b) Use these values to calculate K_c.

3 When 0.50 moles of $H_2(g)$ and 0.18 moles of $I_2(g)$ were heated at 500°C, the equilibrium mixture contained 0.01 moles of $I_2(g)$.
The equation is:
$$H_2(g) + I_2(g) \rightleftharpoons 2HI(g)$$
(a) How many moles of $I_2(g)$ reacted?
(b) How many moles of $H_2(g)$ were present at equilibrium?
(c) How many moles of $HI(g)$ were present at equilibrium?
(d) Calculate the equilibrium constant, K_c for this reaction.

1 values: 47.6; 46.7; 46.9; 45.1. average = 46.6
2 (a) CH_3COOH, 0.5/V mol dm^{-3}; C_2H_5OH, 1.5/V mol dm^{-3}; $CH_3COOC_2H_5$, 1.5/V mol dm^{-3}; H_2O, 1.5/V mol dm^{-3}.
(b) $K_c = 3$.
3 (a) 0.17 mol (b) 0.33 mol (c) 0.34 mol (d) 35.

2.2 The equilibrium constant, K_p

After studying this section you should be able to:

- *understand and use the terms mole fraction and partial pressure*
- *calculate from data the partial pressures present at equilibrium*
- *deduce, for homogeneous reactions, K_p in terms of partial pressures*
- *calculate from data the value of K_p, including determination of units*

LEARNING SUMMARY

Partial pressure

AQA	M4	SALTERS	M5
EDEXCEL	M4	WJEC	CH5
OCR	M6	NICCEA	M4
NUFFIELD	M4		

Equilibria involving gases are usually expressed in terms of K_p, the equilibrium constant in terms of partial pressures.

> In a gas mixture, the *partial pressure* of a gas, p, is the contribution that a gas makes towards the total pressure, P.
>
> **KEY POINT**

The air is a gas mixture with approximate molar proportions of 80% $N_2(g)$ and 20% $O_2(g)$.

- The partial pressure of $N_2(g)$ is 80% of the total pressure.
- The partial pressure of $O_2(g)$ is 20% of the total pressure.

> The mole fraction of a gas is the same as its proportion by volume.

> For a gas **A** in a gas mixture:
>
> the mole fraction of **A**, $x_A = \dfrac{\text{number of moles of } \textbf{A}}{\text{total number of moles in gas mixture}}$
>
> partial pressure of **A**, p_A = mole fraction of **A** × total pressure = $x_A \times P$
>
> **KEY POINT**

> As with all fractions, the sum of the mole fractions in a mixture must equal ONE.

> The sum of the partial pressures in a mixture must equal the total pressure.

The mole fractions in air are: $\quad x_{N_2} = \dfrac{80}{100} = 0.8; \quad x_{O_2} = \dfrac{20}{100} = 0.2$

At normal atmospheric pressure, 100 kPa,

$p_{N_2} = 0.8 \times 100 = 80$ kPa; $\qquad p_{O_2} = 0.2 \times 100 = 20$ kPa

Progress check

1 A gas mixture at a total pressure of 300 kPa contains 3 moles of $N_2(g)$ and 1 mole of $O_2(g)$.
   (a) What is the mole fraction of each gas?
   (b) What is the partial pressure of each gas?

The equilibrium constant, K_p

AQA	M4	SALTERS	M5
EDEXCEL	M4	WJEC	CH5
OCR	M6	NICCEA	M4
NUFFIELD	M4		

K_p is written in a similar way to K_c but with partial pressures replacing concentration terms.

For the equilibrium: $\quad H_2(g) + I_2(g) \rightleftharpoons 2HI(g)$

the equilibrium constant in terms of partial pressures, K_p, is given by:

$$K_p = \frac{p_{HI}^2}{p_{H_2} \times p_{I_2}}$$

> Don't use [ ] in K_p.
> This is a common exam mistake.

- p means the equilibrium partial pressure

> K_p only includes gases. Ignore any other species.

- suitable units for partial pressures are kilopascals (kPa) or pascals (Pa) – but the same unit must used for all gases

- the power to which the partial pressures is raised is the *balancing number* in the chemical equation.

Working out K_p

An equilibrium mixture contains 13.5 mol $N_2(g)$, 3.6 mol $H_2(g)$, and 1.0 mol $NH_3(g)$. The total equilibrium pressure is 200 kPa. Calculate K_p.

Find the mole fractions and partial pressures

Total number of gas moles = 13.5 + 3.6 + 1.0 = 18.1 mol

> Check that the partial pressures add up to give the total pressure:
> 149 + 40 + 11 = 200 kPa

$$p_{N_2} = \frac{13.5}{18.1} \times 200 = 149 \text{ kPa} \qquad p_{H_2} = \frac{3.6}{18.1} \times 200 = 40 \text{ kPa}$$

$$p_{NH_3} = \frac{1.0}{18.1} \times 200 = 11 \text{ kPa}$$

Calculate K_p

For the reaction: $N_2(g) + 3H_2(g) \rightleftharpoons 2NH_3(g)$,

$$K_p = \frac{p_{NH_3}{}^2}{p_{N_2} \times p_{H_2}{}^3}$$

$$\therefore K_p = \frac{11^2}{149 \times 40^3} = 1.27 \times 10^{-5} \text{ kPa}^{-2}$$

The units for K_p are found by replacing each partial pressure value in the K_p expression by its units:

substituting units: $K_p = \dfrac{(kPa)^2}{(kPa)\,(kPa)^3}$ $\quad \therefore$ units of K_p are: kPa^{-2}

Progress check

1 For the following equilibria, write an expression for K_p and work out the value for K_p including units.
   (a) $2HI(g) \rightleftharpoons H_2(g) + I_2(g)$
      partial pressures: $HI(g)$, 56 kPa; $H_2(g)$, 22 kPa; $I_2(g)$, 22 kPa
   (b) $2NO_2(g) \rightleftharpoons 2NO(g) + O_2(g)$
      partial pressures: $NO_2(g)$, 45 kPa; $NO(g)$, 60 kPa; $O_2(g)$, 30kPa

2 In the equilibrium: $2SO_2(g) + O_2(g) \rightleftharpoons 2SO_3(g)$
   2.0 mol of $SO_2(g)$ were mixed with 1.0 mol $O_2(g)$. The mixture was allowed to reach equilibrium at constant temperature and a constant pressure of 900 kPa in the presence of a catalyst. At equilibrium, 0.5 mol of the $SO_2(g)$ had reacted.
   (a) How many moles of SO_2, O_2 and SO_3 were in the equilibrium mixture?
   (b) What were the mole fractions and partial pressures of SO_2, O_2 and SO_3 in the equilibrium mixture?
   (c) Calculate K_p

1 (a) 0.15 (b) 53 kPa.
2 (a) 1.5 mol SO_2; 0.75 mol O_2; 0.5 mol SO_3.
(b) $x(SO_2) = 0.55$; $x(O_2) = 0.27$; $x(SO_3) = 0.18$.
$p(SO_2) = 495$ kPa; $p(O_2) = 243$ kPa; $p(SO_3) = 162$ kPa.
(c) 4.4×10^{-4} kPa^{-1}

2.3 Acids and bases

After studying this section you should be able to:

- *describe what is meant by Brønsted–Lowry acids and bases*
- *understand conjugate acid–base pairs*
- *understand the difference between a strong and weak acid*
- *define the acid dissociation constant, K_a*

Key point from AS

- **Calculations in acid–base titrations**
  Revise AS page 40
- **Acids and bases**
  Revise AS pages 109–110

During the study of acids and bases at GCSE, you learnt that the pH scale can be used to measure the strength of acids and bases. You also learnt the reactions of acids with metals, carbonates and alkalis. During AS Chemistry, you revisited these reactions and found out how titrations can be used to measure the concentration of an unknown acid or base. For A2 Chemistry, you will study acids in terms of proton transfer, the strength of acids, pH and buffers.

Brønsted–Lowry acids and bases

AQA	M4	SALTERS	M5
EDEXCEL	M4	WJEC	CH5
OCR	M6	NICCEA	M4
NUFFIELD	M4		

An acid–base reaction involves proton, H^+, transfer. This idea was first developed by Brønsted and Lowry.

- A Brønsted–Lowry **acid** is a proton donor.
- A Brønsted–Lowry **base** is a proton acceptor.
- An **alkali** is a base that dissolves in water forming $OH^-(aq)$ ions.

Key points from AS

- **Acids, bases and alkalis**
  Revise AS pages 109–110

An acid is a proton donor.

A base is a proton acceptor.

This means that a molecule of an acid contains a hydrogen atom that can be released as a positive hydrogen ion or proton, H^+.

Acid–base pairs

Acids and bases are linked by H^+ as **conjugate pairs**:

- the **conjugate acid** donates H^+
- the **conjugate base** accepts H^+.

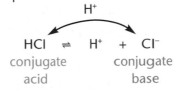

$$HCl \rightleftharpoons H^+ + Cl^-$$

conjugate   conjugate
acid   base

Examples of some conjugate acid–base pairs are shown below.

	acid			base
hydrochloric acid	HCl	$\rightleftharpoons$	H^+ +	Cl^-
sulphuric acid	H_2SO_4	$\rightleftharpoons$	H^+ +	HSO_4^-
ethanoic acid	CH_3COOH	$\rightleftharpoons$	H^+ +	CH_3COO^-

An acid needs a base

An acid can only donate a proton if there is a base to accept it. Most reactions of acids take place in aqueous conditions with water acting as the base. By mixing an acid with a base, an equilibrium is set up comprising **two** acid–base **conjugate pairs**.

The equilibrium system in aqueous hydrochloric acid is shown below.

proton donor ⟵ H^+ ⟶ proton acceptor

		ACID 1				BASE 1
$H_2O(l)$	+	$HCl(aq)$	$\rightleftharpoons$	$H_3O^+(aq)$	+	$Cl^-(aq)$
BASE 2				ACID 2		

proton acceptor ⟵ H^+ ⟶ proton donor

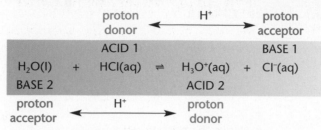

You should be able to identify acid–base pairs in equations such as this.

- In a reaction involving an aqueous acid such as hydrochloric acid, the 'active' species is the **oxonium** ion, H_3O^+ (ACID 2 below).
- Formation of the oxonium ion requires **both** an acid **and** water.

In equations, the oxonium ion H_3O^+ is usually shown as $H^+(aq)$.

Progress check

1 Identify the acid–base pairs in the acid–base equilibria below.
(a) $HNO_3 + H_2O \rightleftharpoons H_3O^+ + NO_3^-$
(b) $NH_3 + H_2O \rightleftharpoons NH_4^+ + OH^-$
(c) $H_2SO_4 + H_2O \rightleftharpoons HSO_4^- + H_3O^+$

2 Write equations for the following acid–base equilibria:
(a) hydrochloric acid and hydroxide ions
(b) ethanoic acid, CH_3COOH, and water.

1 (a) acid 1: HNO_3, base 1: NO_3^-; acid 2: H_3O^+, base 2: H_2O
(b) acid 1: H_2O, base 1: OH^-; acid 2: NH_4^+, base 2: NH_3
(c) acid 1: H_2SO_4, base 1: HSO_4^-; acid 2: H_3O^+, base 2: H_2O
2 (a) $HCl + OH^- \rightleftharpoons H_2O + Cl^-$
(b) $CH_3COOH + H_2O \rightleftharpoons H_3O^+ + CH_3COO^-$

Strength of acids and bases

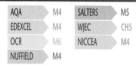

AQA	M4	SALTERS	M5
EDEXCEL	M4	WJEC	CH5
OCR	M6	NICCEA	M4
NUFFIELD	M4		

The acid–base equilibrium of an acid, HA, in water is shown below.

$$HA(aq) + H_2O(l) \rightleftharpoons H_3O^+(aq) + A^-(aq)$$

To emphasise the loss of H^+, this can be shown more simply as **dissociation** of the acid HA:

$$HA(aq) \rightleftharpoons H^+(aq) + A^-(aq).$$

The **strength** of an acid shows the extent of dissociation into H^+ and A^-.

A strong acid is completely dissociated.

Strong acids

A **strong** acid, such as nitric acid, HNO_3, is a **good** proton donor.

- The equilibrium position lies well over to the right.
- There is almost **complete** dissociation.

$$\xrightarrow{\text{equilibrium}}$$
$$HNO_3(aq) \rightleftharpoons H^+(aq) + NO_3^-(aq)$$

- Virtually all of the potential acidic power has been released as $H^+(aq)$.
- At equilibrium, $[H^+(aq)]$ is much greater than $[HNO_3(aq)]$.

A weak acid only partially dissociates.

Weak acids

A **weak** acid, such as ethanoic acid, CH_3COOH, is a **poor** proton donor.

- The equilibrium position lies well over to the left.
- There is only **partial** dissociation.

$$\xleftarrow{\text{equilibrium}}$$
$$CH_3COOH(aq) \rightleftharpoons H^+(aq) + CH_3COO^-(aq)$$

Only a small proportion of the potential acidic power has been released as $H^+(aq)$. At equilibrium, $[CH_3COOH(aq)]$ is much greater than $[H^+(aq)]$.

K_a is just a special equilibrium constant K_c for equilibria showing the dissociation of acids.

[HA(aq)], [H⁺(aq)] and [A⁻(aq)] are equilibrium concentrations.

The acid dissociation constant, K_a

The extent of acid dissociation is shown by an equilibrium constant called the **acid dissociation constant, K_a**.

For the reaction: HA(aq) $\rightleftharpoons$ H⁺(aq) + A⁻(aq),

$$K_a = \frac{[H^+(aq)]\,[A^-(aq)]}{[HA(aq)]}$$

units: $K_a = \dfrac{(mol\ dm^{-3})^2}{(mol\ dm^{-3})} = mol\ dm^{-3}$.

- A **large K_a** value shows that the extent of **dissociation is large** – the acid is strong.
- A **small K_a** value shows that the extent of **dissociation is small** – the acid is weak.

Acid strength and concentration

*Concentrated and dilute are terms used to describe the **amount** of dissolved acid in a solution.*

*Strong and weak are terms used to describe the degree of **dissociation** of an acid.*

The distinction between the strength and concentration of an acid is important.

> **Concentration** is the **amount** of an acid dissolved in 1 dm³ of solution.
> - Concentration is measured in mol dm⁻³.
>
> **Strength** is the extent of **dissociation** of an acid.
> - Strength is measured as K_a in units determined from the equilibrium equation.

KEY POINT

Progress check

1 For each of the following acid–base equilibria, write down the expression for K_a. State the units of K_a for each reaction.
(a) HCOOH(aq) $\rightleftharpoons$ H⁺(aq) + HCOO⁻(aq)
(b) C₆H₅COOH(aq) $\rightleftharpoons$ H⁺(aq) + C₆H₅COO⁻(aq)

2 Samples of two acids, hydrochloric acid and ethanoic acid, have the same concentration: 0.1 mol dm⁻³. Explain why one 'dilute acid' is *strong* whereas the other 'dilute acid' is *weak*!!

and it is a weak acid.
However, ethanoic acid only donates a small proportion of its potential protons, its dissociation is incomplete
1 dm³ of solution and are dilute. Hydrochloric acid is strong because its dissociation is near to complete.
2 Concentration applies to the amount, in mol, in 1 dm³ of solution. Both solutions have 0.1 mol dissolved in
1 (a) $K_a = \dfrac{[H^+(aq)]\,[HCOO^-(aq)]}{[HCOOH(aq)]}$ units: mol dm⁻³
(b) $K_a = \dfrac{[H^+(aq)]\,[C_6H_5COO^-(aq)]}{[C_6H_5COOH(aq)]}$ units: mol dm⁻³

2.4 The pH scale

After studying this section you should be able to:

- define the terms pH, pK_a and K_w
- calculate pH from $[H^+(aq)]$
- calculate $[H^+(aq)]$ from pH
- understand the meaning of the ionisation product of water, K_w
- calculate pH for strong bases

LEARNING SUMMARY

pH and $[H^+(aq)]$

AQA	M4	SALTERS	M5
EDEXCEL	M4	WJEC	CH5
OCR	M6	NICCEA	M4
NUFFIELD	M4		

The concentrations of $H^+(aq)$ ions in aqueous solutions vary widely between about 10 mol dm^{-3} and about 1×10^{-15} mol dm^{-3}.

The **pH scale** is a logarithmic scale used to overcome the problem of using this large range of numbers and to ease the use of negative powers.

The pH scale

pH	$[H^+]$ / mol dm^{-3}
0	1
1	1×10^{-1}
2	1×10^{-2}
3	1×10^{-3}
4	1×10^{-4}
5	1×10^{-5}
6	1×10^{-6}
7	1×10^{-7}
8	1×10^{-8}
9	1×10^{-9}
10	1×10^{-10}
11	1×10^{-11}
12	1×10^{-12}
13	1×10^{-13}
14	1×10^{-14}

KEY POINT

pH is defined as: $\quad pH = -\log_{10} [H^+(aq)]$
- where $[H^+(aq)]$ is the concentration of hydrogen ions in aqueous solution.
- $[H^+(aq)]$ can be calculated from pH using:
$$[H^+(aq)] = 10^{-pH}$$

Notice how the value of pH is linked to the power of 10.

pH	2	9	3.6	10.3
$[H^+(aq)]$/mol dm^{-3}	10^{-2}	10^{-9}	$10^{-3.6}$	$10^{-10.3}$

What does a pH value mean?

- A **low** value of $[H^+(aq)]$ matches a **high** value of pH.
- A **high** value of $[H^+(aq)]$ matches a **low** value of pH.
- A change of pH by 1 changes $[H^+(aq)]$ by 10 times.
- An acid of pH 4 contains 10 times the concentration of $H^+(aq)$ ions as an acid of pH 5.

Calculating the pH of strong acids

AQA	M4	SALTERS	M5
EDEXCEL	M4	WJEC	CH5
OCR	M6	NICCEA	M4
NUFFIELD	M4		

For a strong acid, HA:

- we can assume complete dissociation
- the concentration of $H^+(aq)$ can be found directly from the acid concentration: $[H^+] = [HA]$.

HA is **monoprotic**: each HA molecule can donate **one** H$^+$ ion.

10^x

log

This is the key to use on your calculator when doing pH and $[H^+]$ calculations.

For 10^x, press the **SHIFT** or **INV** key first.

You should be able to convert pH into $[H^+(aq)]$ and *vice versa*

Example 1

A strong acid, HA, has a concentration of 0.010 mol dm^{-3}. What is the pH?

Complete dissociation. $\quad \therefore [H^+(aq)] = 0.010$ mol dm^{-3}

$$pH = -\log_{10} [H^+(aq)] = -\log_{10} (0.010) = \textbf{2.0}$$

Example 2

A strong acid, HA, has a pH of 3.4. What is the concentration of $H^+(aq)$?

Complete dissociation. $\quad \therefore [H^+(aq)] = 10^{-pH} = 10^{-3.4}$ mol dm^{-3}

$$\therefore [H^+(aq)] = \textbf{3.98} \times \textbf{10}^{-4} \textbf{ mol dm}^{-3}$$

Hints for pH calculations

Calculations involving pH are easy once you have learnt how to use your calculator properly.

- Try the examples above until you can remember the order to press the keys.
- Try reversing each calculation to go back to the original value. Repeat several times until you have mastered how to use **your** calculator for pH calculations.
- **Don't** borrow a calculator or you will get confused. Different calculators may need the keys to be pressed in a different order!
- Look at your answer and decide whether it looks sensible.

> Learn: $pH = -\log_{10}[H^+(aq)]$
>
> $[H^+(aq)] = 10^{-pH}$.

KEY POINT

Progress check

1 Calculate the pH of solutions with the following $[H^+(aq)]$ values.
  (a) 0.01 mol dm^{-3} (d) 2.50×10^{-3} mol dm^{-3}
  (b) 0.0001 mol dm^{-3} (e) 8.10×10^{-6} mol dm^{-3}
  (c) 1.0×10^{-13} mol dm^{-3} (f) 4.42×10^{-11} mol dm^{-3}

2 Calculate $[H^+(aq)]$ of solutions with the following pH values.
  (a) pH 3 (c) pH 2.8 (e) pH 12.2
  (b) pH 10 (d) pH 7.9 (f) pH 9.6

3 How many times more hydrogen ions are in an acid of pH 1 than an acid of pH 5?

4 How can a solution have a pH with a negative value?

4 A solution with $[H^+(aq)] > 1$ mol dm^{-3} has a negative pH value.

3 pH 1 has 10 000 times more H$^+$ ions than pH 5.

2 (a) 1×10^{-3} mol dm^{-3} (d) 1.26×10^{-8} mol dm^{-3}
(b) 1×10^{-10} mol dm^{-3} (e) 6.31×10^{-13} mol dm^{-3}
(c) 1.58×10^{-3} mol dm^{-3} (f) 2.51×10^{-10} mol dm^{-3}

1 (a) 2 (b) 4 (c) 13 (d) 2.60 (e) 5.09 (f) 10.4

Calculating the pH of weak acids

AQA	M4	SALTERS	M5
EDEXCEL	M4	WJEC	CH5
OCR	M6	NICCEA	M4
NUFFIELD	M4		

The pH of a weak acid HA can be calculated from:

- the **concentration** of the acid and
- the value of the acid dissociation constant, K_a.

Assumptions and approximations

Consider the equilibrium of a weak aqueous acid HA(aq):

$$HA(aq) \rightleftharpoons H^+(aq) + A^-(aq)$$

- Assuming that only a very small proportion of HA dissociates, the equilibrium concentration of HA(aq) will be very nearly the same as the concentration of undissociated HA(aq).

$$\therefore [HA(aq)]_{equilibrium} \approx [HA(aq)]_{start}$$

- Assuming that there is a negligible proportion of H$^+$(aq) from ionisation of water:

$$[H^+(aq)] \approx [A^-(aq)]$$

- Using these approximations

For calculations, use

$$K_a \approx \frac{[H^+(aq)]^2}{[HA(aq)]}$$

$$K_a = \frac{[H^+(aq)][A^-(aq)]}{[HA(aq)]} \qquad \therefore K_a \approx \frac{[H^+(aq)]^2}{[HA(aq)]}$$

Take care to learn this method.

Many marks are dropped on exam papers by students who have not done so!

Example

For a weak acid $[HA(aq)] = 0.100$ mol dm^{-3}, $K_a = 1.70 \times 10^{-5}$ mol dm^{-3} at 25°C. Calculate the pH.

$$K_a = \frac{[H^+(aq)][A^-(aq)]}{[HA(aq)]} \approx \frac{[H^+(aq)]^2}{[HA(aq)]}$$

$$\therefore 1.70 \times 10^{-5} = \frac{[H^+(aq)]^2}{0.100}$$

$$\therefore [H^+(aq)] = \sqrt{0.100 \times 1.70 \times 10^{-5}} = 0.00130 \text{ mol dm}^{-3}$$

$$pH = -\log_{10}[H^+(aq)] = -\log_{10}(0.00130) = \textbf{2.89}$$

K_a and pK_a

AQA	M4	SALTERS	M5
EDEXCEL	M4	WJEC	CH5
OCR	M6	NICCEA	M4
NUFFIELD	M4		

K_a and pK_a conversions are just like those between pH and H^+.

Values of K_a can be made more manageable if expressed in a logarithmic form, pK_a (see also page 38: $[H^+(aq)]$ and pH).

$$pK_a = -\log_{10}K_a$$
$$K_a = 10^{-pK_a}$$

- A **low** value of K_a matches a **high** value of pK_a
- A **high** value of K_a matches a **low** value of pK_a

The smaller the pK_a value, the stronger the acid.

Comparison of K_a and pK_a

acid		K_a / mol dm^{-3}	pK_a
methanoic acid	HCOOH	1.6×10^{-4}	$-\log_{10}(1.6 \times 10^{-4}) = 3.8$
benzoic acid	C_6H_5COOH	6.3×10^{-5}	$-\log_{10}(6.3 \times 10^{-5}) = 4.2$

Progress check

1 Find the pH of solutions of a weak acid HA ($K_a = 1.70 \times 10^{-5}$ mol dm^{-3}), with the following concentrations:
   (a) 1.00 mol dm^{-3}; (b) 0.250 mol dm^{-3}; (c) 3.50×10^{-2} mol dm^{-3}

2 Find values of K_a and pK_a for the following weak acids.
   (a) 1.0 mol dm^{-3} solution with a pH of 4.5
   (b) 0.1 mol dm^{-3} solution with a pH of 2.2
   (c) 2.0 mol dm^{-3} solution with a pH of 3.8.

2 (a) $K_a = 1 \times 10^{-9}$ mol dm^{-3}, $pK_a = 9$
   (b) K_a 3.98 × 10^{-4} mol dm^{-3}, $pK_a = 3.4$
   (c) K_a 1.26 × 10^{-8} mol dm^{-3}, $pK_a = 7.9$

1 (a) 2.38 (b) 2.69 (c) 3.11

The ionisation of water and K_w

AQA	M4	SALTERS	M5
EDEXCEL	M4	WJEC	CH5
OCR	M6	NICCEA	M4
NUFFIELD	M4		

In water, a very small proportion of molecules dissociates into $H^+(aq)$ and $OH^-(aq)$ ions. The position of equilibrium lies well to the left of the equation below, representing this dissociation.

$$\overset{\text{equilibrium}}{\longleftarrow}$$
$$H_2O(l) \rightleftharpoons H^+(aq) + OH^-(aq)$$

Treating water as a weak acid: $K_a = \dfrac{[H^+(aq)][OH^-(aq)]}{[H_2O(l)]}$

Rearranging gives:

$[H_2O(l)]$ is constant and is included within K_w

$$\underbrace{K_a \times [H_2O(l)]}_{\text{constant, } K_w} = [H^+(aq)][OH^-(aq)]$$

At 25°C,
$[H^+(aq)] \times [OH^-(aq)]$
$= 1 \times 10^{-14}$ mol^2 dm^{-6}.

> **KEY POINT**
>
> The constant K_w is called the **ionic product of water**
> - $K_w = [H^+(aq)][OH^-(aq)]$
> - At 25°C, $K_w = 1.0 \times 10^{-14}$ mol^2 dm^{-6}.

In water, the concentrations of H^+(aq) and OH^-(aq) ions are the same.

- $[H^+(aq)] = [OH^-(aq)] = 10^{-7}$ mol dm^{-3} ($10^{-14} = 10^{-7} \times 10^{-7}$)

Hydrogen ion and hydroxide ion concentrations

All aqueous solutions contain H^+(aq) and OH^-(aq) ions. The proportions of these ions in a solution are determined by the pH.

In water	$[H^+(aq)] = [OH^-(aq)]$
In **acidic** solutions	$[H^+(aq)] > [OH^-(aq)]$
In **alkaline** solutions	$[H^+(aq)] < [OH^-(aq)]$

The concentrations of H^+(aq) and OH^-(aq) are linked by K_w.

- At 25°C $1.0 \times 10^{-14} = [H^+(aq)] [OH^-(aq)]$
- The indices of $[H^+(aq)]$ and $[OH^-(aq)]$ add up to -14.

Linking [H⁺] and [OH⁻]

$10^{-14} = [H^+] [OH^-]$

Water: pH = 7,
$[H^+] = 10^{-7}$ mol dm^{-3}
$10^{-14} = 10^{-7} \times 10^{-7}$

An acid: pH = 3
$[H^+] = 10^{-3}$ mol dm^{-3}
$10^{-14} = 10^{-3} \times 10^{-11}$

An alkali: pH = 10
$[H^+] = 10^{-10}$ mol dm^{-3}
$10^{-14} = 10^{-10} \times 10^{-4}$

Calculating the pH of strong alkalis

The pH of a strong alkali can be found using K_w.

For a strong alkali, BOH:

- we can assume complete dissociation
- the concentration of OH^-(aq) can be found directly from the alkali concentration: $[BOH] = [OH^-]$.

To find the pH of an alkali, first find $[H^+]$ using K_w and $[OH^-]$.

Example

A strong alkali, BOH, has a concentration of 0.50 mol dm^{-3}.

What is the pH?

Complete dissociation. $\therefore [OH^-(aq)] = 0.50$ mol dm^{-3}

$$K_w = [H^+(aq)] [OH^-(aq)] = 1 \times 10^{-14} \text{ mol}^2 \text{ dm}^{-6}$$

$$\therefore [H^+(aq)] = \frac{K_w}{[OH^-(aq)]} = \frac{1 \times 10^{-14}}{0.50} = 2 \times 10^{-13} \text{ mol dm}^{-3}$$

$$pH = -\log_{10} [H^+(aq)] = -\log_{10} (2 \times 10^{-13}) = \textbf{12.7}$$

Progress check

1 Find the $[H^+(aq)]$ and pH of the following alkalis at 25°C:
   (a) 1×10^{-3} mol dm^{-3} OH^-(aq)
   (b) 3.5×10^{-2} mol dm^{-3} OH^-(aq).

2 Find the pH of the following solutions of strong bases at 25°C:
   (a) 0.01 mol dm^{-3} KOH (aq)
   (b) 0.20 mol dm^{-3} NaOH (aq).

2 (a) pH = 12; (b) pH = 13.3.
(b) $[H^+(aq)] = 2.86 \times 10^{-13}$ mol dm^{-3}; pH = 12.5.
1 (a) $[H^+(aq)] = 1 \times 10^{-11}$ mol dm^{-3}; pH = 11

2.5 pH changes

After studying this section you should be able to:

- *understand how an indicator changes colour at different pHs*
- *recognise the shapes of titration curves for acids and bases with different strengths*
- *explain the choice of suitable indicators for acid–base titrations*
- *state what is meant by a buffer solution*
- *explain how pH is controlled by each component in a buffer solution*
- *calculate the pH of a buffer solution*

LEARNING SUMMARY

Indicators and end-points

AQA	M4	SALTERS	M5
EDEXCEL	M4	WJEC	CH5
OCR	M6	NICCEA	M4
NUFFIELD	M4		

Indicators as weak acids

An acid–base indicator is a weak acid, represented simply as HInd.

- The weak acid, HInd, and its conjugate base, Ind⁻, have different colours.

E.g. for methyl orange:

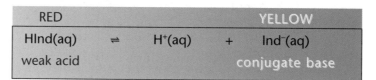

RED				YELLOW
HInd(aq)	$\rightleftharpoons$	H⁺(aq)	+	Ind⁻(aq)
weak acid				conjugate base

At the end-point of a titration:

- **HInd** and **Ind⁻** are present in **equal** concentrations.

When using methyl orange as indicator:

- the colour at the end-point is orange, from equal concentrations of the weak acid HInd (red) and its conjugate base Ind⁻ (yellow)
- the pH of the end-point is called the pK_{ind} of the indicator.

> Each indicator has its own pK_{ind} value at which its colour changes.

pH ranges for common indicators

An indicator changes colour over a range of about 2 pH units within which is the pK_{ind} value of the indicator. The pH range for the indicators methyl orange and phenolphthalein are shown below.

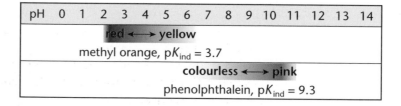

pH	0	1	2	3	4	5	6	7	8	9	10	11	12	13	14

red ⟷ yellow
methyl orange, pK_{ind} = 3.7

colourless ⟷ pink
phenolphthalein, pK_{ind} = 9.3

Titration curves

AQA	M4	SALTERS	M4
EDEXCEL	M4	WJEC	CH5
OCR	M6	NICCEA	M4
NUFFIELD	M4		

Choosing an indicator using titration curves

A titration curve shows the changes in pH during a titration.
In the titration curves on the next page:

- different combinations of strong and weak acids have been used
- the pK_{ind} values are shown for the indicators methyl orange (**MO**) and phenolphthalein (**P**).

strong acid/strong alkali

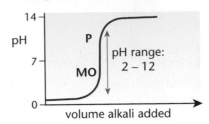

both indicators suitable

strong acid/weak alkali

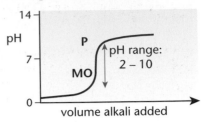

*methyl orange (**MO**) suitable*
*phenolphthalein (**P**) unsuitable*

weak acid/strong alkali

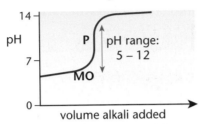

*phenolphthalein (**P**) suitable*
*methyl orange (**MO**) unsuitable*

weak acid/weak alkali

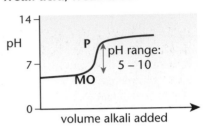

neither indicator suitable

Choose the correct indicator

Strong acid/strong alkali
phenolphthalein ✓
methyl orange ✓

Strong acid/weak alkali
phenolphthalein ✗
methyl orange ✓

Weak acid/strong alkali
phenolphthalein ✓
methyl orange ✗

Weak acid/weak alkali
phenolphthalein ✗
methyl orange ✗

Key features of titration curves

- The pH changes rapidly at the near vertical portion of the titration curve. This is the **end-point** or **equivalence point** of the titration.
- The sharp change in pH is brought about by a very small addition of alkali, typically the addition of one drop.
- The indicator is only suitable if its pK_{ind} value is within the pH range of the near vertical portion of the titration curve.

Titrations of weak acids/weak alkalis

- The pH changes slowly through the end point as alkali is added – there is **no** near vertical portion to the titration curve.
- The colour change of the indicator is gradual.
- It is very difficult to see a sharp colour change.
- The pH change is often followed using a pH meter instead of an indicator.

Progress check

1 The indicator, bromophenol blue, has an HInd form that is a yellow colour and an Ind⁻ form that is a blue colour.
   (a) Write an equation to link these two forms of the indicator.
   (b) Explain, with reasons, the colour of bromophenol blue
       (i)   in a strong acid
       (ii) in a strong alkali
       (iii) at the end-point of a titration.

1 (a) HInd ⇌ H⁺ + Ind⁻
(b) (i) Bromophenol blue is yellow. The high concentration of H⁺(aq) moves the equilibrium to the left, forming HInd(aq).
(ii) Bromophenol blue is blue. The high concentration of OH⁻(aq) removes H⁺(aq) and moves the equilibrium to the right, forming Ind⁻(aq).
(iii) Bromophenol blue is green. At the end-point, there are equal concentrations of HInd and Ind⁻, blue and yellow make green.

Buffer solutions

A buffer solution minimises changes in pH during the addition of an acid or an alkali. The buffer solution maintains a near constant pH by removing most of any acid or alkali that is added to the solution.

A **buffer solution** is a mixture of:

- a **weak acid**, HA, and
- its **conjugate base**, A^-:

$$HA(aq) \rightleftharpoons H^+(aq) + A^-(aq)$$
weak acid conjugate base

In a buffer solution, the concentration of hydrogen ions, $[H^+(aq)]$, is very small compared with the concentrations of the weak acid $[HA(aq)]$ or the conjugate base $[A^-(aq)]$.

$$[H^+(aq)] \ll [HA(aq)] \quad \text{and} \quad [H^+(aq)] \ll [A^-(aq)].$$

> We can explain how a buffer minimises pH changes by using le Chatelier's principle.

How does a buffer act?

Addition of an acid, $H^+(aq)$, to a buffer

On addition of an acid:

- $[H^+(aq)]$ is increased
- the pH change is opposed and the **equilibrium moves to the left, removing** $[H^+(aq)]$ and forming HA(aq).
- the **conjugate base $A^-(aq)$** removes most of any added $[H^+(aq)]$.

> A^- removes most of any added acid.

add acid, $H^+(aq)$

$$HA(aq) \rightleftharpoons H^+(aq) + A^-(aq)$$
weak acid conjugate base

$H^+(aq)$ is removed

> This idea is very similar to that explaining the action of indicators, see p. 42.

Addition of an alkali, $OH^-(aq)$, to a buffer

On addition of an alkali:

- the added $OH^-(aq)$ reacts with the small concentration of $H^+(aq)$:

$$H^+(aq) + OH^-(aq) \longrightarrow H_2O(l)$$

- the pH change is opposed – the **equilibrium moves to the right, restoring** $[H^+(aq)]$ as HA(aq) dissociates
- the **weak acid HA** restores most of any $[H^+(aq)]$ that has been removed.

add alkali, $OH^-(aq)$

$$HA(aq) \rightleftharpoons H^+(aq) + A^-(aq)$$
weak acid conjugate base

$H^+(aq)$ is restored

> HA removes added alkali.

KEY POINT Although the two components in a buffer solution react with added acid and alkali, they **cannot stop** the pH from changing. They do however **minimise** pH changes.

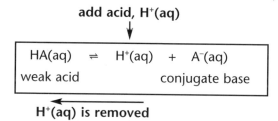

Remember these buffers:
CH_3COOH/CH_3COONa
and
NH_4Cl/NH_3

Common buffer solutions

An acidic buffer

A common acidic buffer is an aqueous solution containing a mixture of ethanoic acid, CH_3COOH, and the ethanoate ion, CH_3COO^-.

- Ethanoic acid acts as the **weak acid**, CH_3COOH.
- Sodium ethanoate, $CH_3COO^-Na^+$, acts as a source of the **conjugate base**, CH_3COO^-.

An alkaline buffer

A common alkaline buffer is an aqueous solution containing a mixture of the ammonium ion, NH_4^+, and ammonia, NH_3.

- Ammonium chloride, $NH_4^+Cl^-$, acts as source of the **weak acid**, NH_4^+.
- Ammonia acts as the **conjugate base**, NH_3.

Calculations involving buffer solutions

AQA	M4	SALTERS	M5
EDEXCEL	M4	WJEC	CH5
OCR	M6	NICCEA	M4
NUFFIELD	M4		

For buffer calculations, learn:

$$[H^+(aq)] = K_a \times \frac{[HA(aq)]}{[A^-(aq)]}$$

The pH of a buffer can be controlled by the acid/base ratio.

The pH of a buffer solution depends upon:

- the acid dissociation constant, K_a, of the buffer system
- the **ratio** of the weak **acid** and its conjugate **base**.

For a buffer comprising the weak acid, HA, and its conjugate base, A^-,

$$K_a = \frac{[H^+(aq)]\,[A^-(aq)]}{[HA(aq)]}$$

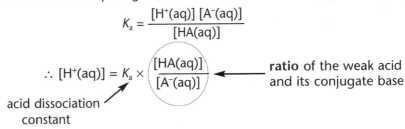

$\therefore [H^+(aq)] = K_a \times \dfrac{[HA(aq)]}{[A^-(aq)]}$

acid dissociation constant

ratio of the weak acid and its conjugate base

Example

Calculate the pH of a buffer comprising 0.30 mol dm^{-3} $CH_3COOH(aq)$ ($K_a = 1.7 \times 10^{-5}$ mol dm^{-3}) and 0.10 mol dm^{-3} $CH_3COO^-(aq)$.

$$[H^+(aq)] = K_a \times \frac{[HA(aq)]}{[A^-(aq)]}$$

$$\therefore [H^+(aq)] = 1.7 \times 10^{-5} \times \frac{0.30}{0.10} = 5.1 \times 10^{-5} \text{ mol dm}^{-3}$$

$$\therefore pH = -\log_{10}[H^+(aq)] = -\log_{10}(5.1 \times 10^{-5})$$

$\therefore$ pH of the buffer solution = **4.3**

Progress check

1 Three buffer solutions are made from benzoic acid, C_6H_5COOH and sodium benzoate, C_6H_5COONa with the following compositions:

> Buffer A: 0.10 mol dm^{-3} C_6H_5COOH and 0.10 mol dm^{-3} C_6H_5COONa
> Buffer B: 0.75 mol dm^{-3} C_6H_5COOH and 0.25 mol dm^{-3} C_6H_5COONa
> Buffer C: 0.20 mol dm^{-3} C_6H_5COOH and 0.80 mol dm^{-3} C_6H_5COONa

(a) Write the equation for the equilibrium in these buffers.
(b) Write an expression for K_a of benzoic acid.
(c) Calculate the pH of each of the buffer solutions
(K_a for $C_6H_5COOH = 6.3 \times 10^{-5}$ mol dm^{-3}.)

1 (a) $C_6H_5COOH \rightleftharpoons H^+(aq) + C_6H_5COO^-(aq)$
(b) $K_a = \dfrac{[H^+(aq)]\,[C_6H_5COO^-(aq)]}{[C_6H_5COOH(aq)]}$
(c) Buffer A: 4.2 Buffer B: 3.7 Buffer C: 4.8

Sample question and model answer

Blood is an example of a buffered solution. Human blood is slightly basic and has a pH of approximately 7.40. If the pH falls, a condition known as acidosis can occur. Death may arise if the pH drops below 6.80.

(a) (i) What do you understand by the term a *buffered solution*?

A solution that resists a change in pH. ✓

> Make sure that your calculator skills are good and show your working. A correct method will secure most available marks.

(ii) Calculate the approximate hydrogen ion concentration, $[H^+(aq)]$, of normal human blood.

$[H^+] = 10^{-pH}$ ✓ ∴ $[H^+] = 10^{-7.40} = 3.98 \times 10^{-8}$ ✓ mol dm^{-3}

(iii) How many times greater is the hydrogen ion concentration in blood at pH 6.8 compared to blood at pH 7.4?

At pH 6.8, $[H^+] = 10^{-6.80} = 1.58 \times 10^{-7}$ ✓ mol dm^{-3}

$[H^+]$ is $\dfrac{1.58 \times 10^{-7}}{3.98 \times 10^{-8}}$ times greater ✓ = 3.97 times greater.

[5]

(b) Acidosis may occur as a result of strenuous exercise when glucose is converted into lactic acid, $CH_3CH(OH)CO_2H$. Lactic acid is a weak acid with an acid dissociation constant, K_a, of 8.40×10^{-4} mol dm^{-3}.

> Notice the use throughout of 3 significant figures (in the question and in answers).

(i) Write an equation to show the dissociation of lactic acid into its ions.

$CH_3CH(OH)CO_2H \rightleftharpoons CH_3CH(OH)CO_2^- + H^+$ ✓

(ii) Write an expression for the acid dissociation constant, K_a, of lactic acid.

$K_a = \dfrac{[CH_3CH(OH)CO_2^-(aq)]\,[H^+(aq)]}{CH_3CH(OH)CO_2H(aq)]}$ ✓

(iii) Calculate the pH of a 0.100 mol dm^{-3} solution of lactic acid.

$CH_3CH(OH)CO_2^- \approx H^+$

equilibrium conc of weak acid ≈ undissociated concentration of weak acid.

> These approximations simplify the K_a expression to:
>
> $K_a \approx \dfrac{[H^+]^2}{[HA]}$

∴ $K_a \approx \dfrac{[H^+(aq)]^2}{[CH_3CH(OH)COOH(aq)]}$

$8.4 \times 10^{-4} \dfrac{[H^+(aq)]^2}{[CH_3CH(OH)COOH(aq)]} = \dfrac{[H^+(aq)]^2}{0.100}$ ✓

$[H^+(aq)] = \sqrt{0.100 \times 8.40 \times 10^{-4}} = 9.17 \times 10^{-3}$ ✓ mol dm^{-3}

pH $= -\log(9.17 \times 10^{-3}) = 2.04$. ✓

[5]

(c) Oxygen is carried in the blood attached to haemoglobin (*Hb*) in the form HbO_2 and is transported around the body. The haemoglobin is involved in a series of equilibria whose overall result can be represented by the following equation.

$HbH^+(aq) + O_2(aq) \rightleftharpoons HbO_2(aq) + H^+(aq)$

Show how an increase in the concentration of lactic acid influences this equilibrium and suggest a possible consequence of this within the body.

> Notice that 3 marks means 3 key points.

H^+ is added to right-hand side of equilibrium. ✓
The equilibrium moves to the left. ✓
This reduces the ability to transport oxygen. ✓

[3]

[Total: 13]

Cambridge How Far, How Fast? Q2 June 1996

Practice examination questions

1

Consider the esterification reaction:
$$CH_3COOH + CH_3CH_2OH \rightleftharpoons CH_3COOCH_2CH_3 + H_2O$$

(a) (i) Give the expression for K_c. [1]
   (ii) An equilibrium mixture from the above reaction contains 0.90 mol of ethyl ethanoate, 0.90 mol of water, 0.10 mol of ethanoic acid and 2.1 mol of ethanol. It has a volume of 235 cm³. Calculate K_c to 2 significant figures, having regard for the units. [3]
   (iii) State, giving a reason, what would happen to the equilibrium composition if more ethanoic acid were to be added to the equilibrium mixture. [2]

(b) The reaction employs an acid catalyst, usually a little sulphuric acid. State, with a reason, the effect on the equilibrium position if the concentration of the acid catalyst were to be increased. [2]

[Total: 8]

London Module Test 2 Q1 (a)-(b) January 1999

2

In the Contact Process for the production of sulphuric acid, highly purified sulphur dioxide and oxygen react together to form sulphur trioxide. The process is carried out in the presence of vanadium(V) oxide, at about 700 K and at a pressure of 120 kPa according to the equation:

$$2SO_2(g) + O_2(g) \rightleftharpoons 2SO_3(g) \qquad \Delta H^\ominus = -190 \text{ kJ mol}^{-1}$$

Under these conditions, the partial pressures of sulphur dioxide and sulphur trioxide at equilibrium are 33 kPa and 39 kPa, respectively.

(a) What would be the effect on the yield of SO_3 of increasing the temperature? Explain your answer. [3]
(b) Determine the partial pressure and hence the mole fraction of oxygen in the equilibrium mixture. [3]
(c) (i) Write an expression for the equilibrium constant, K_p, for the reaction shown.
   (ii) Calculate the value of the equilibrium constant, K_p, and state its units. [4]

[Total: 10]

NEAB Further Physical Chemistry Q3 June 1996

3

An equilibrium mixture of N_2O_4 and NO_2 appears as brown fumes.

$$N_2O_4(g) \rightleftharpoons 2NO_2(g)$$
pale yellow   dark brown

A transparent glass syringe was filled with the gaseous mixture of N_2O_4 and NO_2 and its tip sealed. When the piston of the syringe was rapidly pushed well into the body of the syringe, thereby compressing the gas mixture considerably, the colour of the gas became momentarily darker but then became lighter again.

(a) Suggest why compressing the gases causes the mixture to darken. [1]
(b) Explain why the mixture turns lighter on standing. [2]
(c) Write an expression for the equilibrium constant, K_p, for this equilibrium. [1]
(d) 1.0 mole of N_2O_4 was allowed to reach equilibrium at 400 K. At equilibrium the partial pressure of N_2O_4 was found to be 0.15 atm. Given that the equilibrium constant K_p for this reaction is 48 atm, calculate the partial pressure of NO_2 in the equilibrium mixture. [3]

[Total: 7]

Edexcel Module Test 3 Q3(b) modified June 2000

Practice examination questions (continued)

4

(a) Define the term Brønsted–Lowry acid. [1]

(b) Write an equation for the reaction between gaseous hydrogen chloride and water. State the role of water in this reaction, using the Brønsted–Lowry definition. [2]

(c) Write an equation for the reaction between gaseous ammonia and water. State the role of water in this reaction, using the Brønsted–Lowry definition. [2]

(d) The ion $H_2NO_3^+$ is formed in the first stage of a reaction between concentrated nitric acid and an excess of concentrated sulphuric acid. In this first stage the two acids react in a 1 : 1 molar ratio. In the second stage, the $H_2NO_3^+$ ion decomposes to form the nitronium ion, NO_2^+. Write equations for these two reactions and state the role of nitric acid in the first reaction. [3]

(e) (i)   Explain the term weak acid.

(ii) Write an expression for the acid dissociation constant, K_a of HA, a weak monoprotic acid.

(iii) The value of the acid dissociation constant for the monoprotic acid HX is 144 mol dm^{-3}. What does this suggest about the concentration of undissociated HX in dilute aqueous solution?

(iv) State whether HX should be classified as a strong acid or a weak acid. Justify your answer. [5]

[Total:13]

NEAB Equilibria and Inorganic Chemistry Q1 June 2000

5

(a) (i)   Define pH.

(ii) Define the term 'weak acid' as applied to methanoic acid, HCOOH. [3]

(b) Calculate the pH of the following solutions:
(The ionic product of water, $K_w = 1.00 \times 10^{-14}$ mol^2 dm^{-6} at 25°C.
The acid dissociation constant for methanoic acid is 1.78×10^{-4} mol dm^{-3}.)
(i)   a solution of hydrochloric acid of concentration 0.152 mol dm^{-3}
(ii) a solution of sodium hydroxide of concentration 0.747 mol dm^{-3}
(iii) a solution of methanoic acid of concentration 0.152 mol dm^{-3}. [9]

(c) (i)   What is the principle property of a buffer solution?

(ii) The acid dissociation constant, K_a for ethanoic acid is 1.80×10^{-5} mol dm^{-3}. Calculate the pH of a buffer solution which has a concentration of 0.105 mol dm^{-3} with respect to ethanoic acid and 0.342 mol dm^{-3} with respect to sodium ethanoate. [5]

[Total: 17]

Edexcel Specimen Unit Test 4 Q1 2000

Chapter 3

Energy changes in chemistry

The following topics are covered in this chapter:

- Enthalpy changes
- Electrochemical cells
- Predicting redox reactions

3.1 Enthalpy changes

After studying this section you should be able to:

- *explain and use the term 'lattice enthalpy'*
- *construct Born–Haber cycles to calculate the lattice enthalpy of simple ionic compounds*
- *explain the effect of ionic charge and ionic radius on the numerical magnitude of a lattice enthalpy*
- *calculate enthalpy changes of solution for ionic compounds from enthalpy changes of hydration and lattice enthalpies*

LEARNING SUMMARY

Key points from AS

- **Enthalpy changes**
  Revise AS pages 87–95

During the study of enthalpy changes in AS Chemistry, you learnt how to:

- calculate enthalpy changes directly from experiments using the relationship
  $Q = mc\Delta T$
- calculate enthalpy changes indirectly using Hess's law.

These key principles are built upon by considering the enthalpy changes that bond together an ionic lattice.

Lattice enthalpy

AQA	M5	SALTERS	M5
EDEXCEL	M4	WJEC	CH5
OCR	M5.1	NICCEA	M4
NUFFIELD	M6		

Key points from AS

- **Ionic bonding**
  Revise AS pages 45–47

> Each ion is surrounded by oppositely-charged ions, forming a giant ionic lattice.

Ionic bonding is the electrostatic attraction between oppositely charged ions. This attraction acts in all directions, resulting in a giant ionic lattice with hundreds of thousands of ions (depending upon the size of the crystal). Lattice enthalpy indicates the strength of the ionic bonds in an ionic lattice.

Part of the sodium chloride lattice

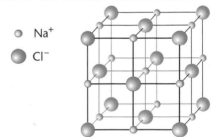

- Na⁺
- Cl⁻

- Each Na⁺ ion surrounds 6 Cl⁻ ions
- Each Cl⁻ ion surrounds 6 Na⁺ ions

> **The lattice enthalpy ($\Delta H^{\ominus}_{L.E.}$)** of an ionic compound is the enthalpy change that accompanies the formation of 1 mole of an ionic compound from its constituent gaseous ions. ($\Delta H^{\ominus}_{L.E.}$ is **exothermic**.)
>
> **KEY POINT**

$$Na^+(g) + Cl^-(g) \longrightarrow Na^+Cl^-(s)$$

The term *lattice enthalpy* is also sometimes called the **enthalpy change of lattice formation**.

The opposite change, to break up the lattice, is called the **enthalpy change of lattice dissociation**.

$$Na^+Cl^-(s) \longrightarrow Na^+(g) + Cl^-(g)$$

Key points from AS

• **Indirect determination of enthalpy changes**
  Revise AS pages 91–92

Determination of lattice enthalpies

Lattice enthalpies cannot be determined directly and must be calculated indirectly using Hess's Law from other enthalpy changes that can be found experimentally. The energy cycle used to calculate a lattice enthalpy is the **Born–Haber cycle**.

The basis of the Born–Haber cycle is the formation of an ionic lattice from its elements. For sodium chloride, one route is the single stage process corresponding to its enthalpy change of formation.

ROUTE 1 $\qquad Na(s) + \frac{1}{2}Cl_2(g) \longrightarrow Na^+Cl^-(s)$

A second route can be broken down into separate steps resulting in:

ROUTE 2
- *production of gaseous atoms*
  $Na(s) \longrightarrow Na(g)$
  $\frac{1}{2}Cl_2(g) \longrightarrow Cl(g)$
- *production of gaseous ions*
  $Na(g) \longrightarrow Na^+(g) + e^-$
  $Cl(g) + e^- \longrightarrow Cl^-(g)$
- *production of the solid ionic lattice*
  $Na^+(g) + Cl^-(g) \longrightarrow Na^+Cl^-(s).$

The Born–Haber cycle

Definitions for these enthalpy changes are shown on p. 51.

The complete Born–Haber cycle for sodium chloride is shown below.

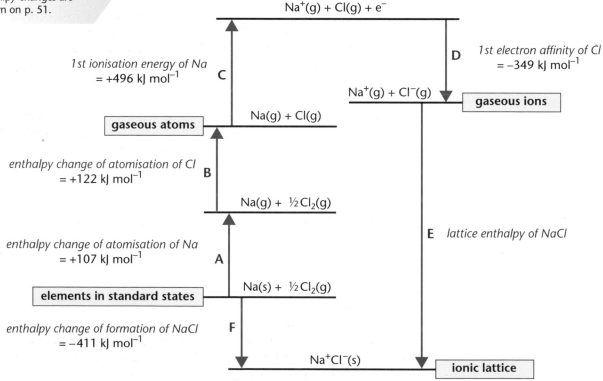

• the 1st ionisation energy of Na is **endothermic**
• the 1st electron affinity of Cl is **exothermic**.

Route 1: $\qquad$ **A + B + C + D + E**

Route 2: $\qquad$ **F**

Using Hess's Law: $\qquad$ **A + B + C + D + E = F**

$\Delta H^{\ominus}_{at} Na(g) + \Delta H^{\ominus}_{at} Cl(g) + \Delta H^{\ominus}_{I.E.} Na(g) + \Delta H^{\ominus}_{E.A.} Cl(g) + E = \Delta H^{\ominus}_f Na^+Cl^-(s)$

$\qquad \therefore +107 + 122 + 496 + (-349) + E = -411$

Hence, the lattice energy of $Na^+Cl^-(s)$, **E = −787** kJ mol^{-1}

Definitions for enthalpy changes

The enthalpy changes involved in these two routes are shown below.

> **The standard enthalpy change of formation** ($\Delta H^{\ominus}_f$) is the enthalpy change that takes place when one mole of a compound in its standard state is formed from its constituent elements in their standard states under standard conditions.
>
> $$Na(s) + \tfrac{1}{2}Cl_2(g) \longrightarrow Na^+Cl^-(s) \qquad \Delta H^{\ominus}_f = -411 \text{ kJ mol}^{-1}$$
>
> **KEY POINT**

Be careful when using $\Delta H^{\ominus}_{at}$ involving diatomic elements.

$\Delta H^{\ominus}_{at}$ relates to the formation of 1 mole of atoms. For chlorine, this involves $\tfrac{1}{2}Cl_2$ only.

> **The standard enthalpy change of atomisation** ($\Delta H^{\ominus}_{at}$) of an element is the enthalpy change that accompanies the formation of 1 mole of gaseous atoms from the element in its standard state.
>
> $$Na(s) \longrightarrow Na(g) \qquad \Delta H^{\ominus}_{at} = +107 \text{ kJ mol}^{-1}$$
> $$\tfrac{1}{2}Cl_2(g) \longrightarrow Cl(g) \qquad \Delta H^{\ominus}_{at} = +122 \text{ kJ mol}^{-1}$$
>
> **KEY POINT**

For gaseous molecules this enthalpy change can be determined from the **bond dissociation enthalpy** (B.D.E.) – the enthalpy change required to break and separate 1 mole of bonds so that the resulting gaseous atoms exert no forces upon each other.

$$Cl–Cl(g) \longrightarrow 2Cl(g) \qquad \Delta H^{\ominus}_{B.D.E.} = +244 \text{ kJ mol}^{-1}$$
$$\tfrac{1}{2}Cl–Cl(g) \longrightarrow Cl(g) \qquad \Delta H^{\ominus}_{at} = +122 \text{ kJ mol}^{-1}$$

> **The first ionisation energy** ($\Delta H^{\ominus}_{I.E.}$) of an element is the enthalpy change that accompanies the removal of 1 electron from each atom in 1 mole of gaseous atoms to form 1 mole of gaseous 1+ ions.
>
> $$Na(g) \longrightarrow Na^+(g) + e^- \qquad \Delta H^{\ominus}_{I.E} = +496 \text{ kJ mol}^{-1}$$
>
> **KEY POINT**

Be careful using electron affinities. $\Delta H^{\ominus}_{E.A.}$ relates to the formation of 1 mole of 1– ions. For chlorine, this involves $Cl(g)$ only.

> **The first electron affinity** ($\Delta H^{\ominus}_{E.A.}$) of an element is the enthalpy change that accompanies the addition of 1 electron to each atom in 1 mole of gaseous atoms to form 1 mole of gaseous 1–ions.
>
> $$Cl(g) + e^- \longrightarrow Cl^-(g) \qquad \Delta H^{\ominus}_{E.A.} = -349 \text{ kJ mol}^{-1}$$
>
> **KEY POINT**

Key points from AS

- **Charge density**
  Revise AS page 54

Factors affecting the size of lattice enthalpies

The strength of an ionic lattice and the value of its lattice enthalpy depend upon:

- ionic size
- ionic charge.

Effect of ionic size

The effect of increasing ionic size can be seen by comparing the lattice enthalpies of sodium halides. From Cl^- to I^-, the increasing size of the negative halide ion produces a smaller charge density. This results in weaker attraction between ions and a less negative lattice enthalpy.

Lattice energy has a negative value. You should use the term '*becomes less/more negative*' instead of '*becomes bigger/smaller*' to describe any trend in lattice energy.

compound	lattice enthalpy / kJ mol^{-1}	ions	effect of size of halide ion
NaCl	−787	$\oplus\ominus$	ionic size increases:
NaBr	−751	$\oplus\ominus$	• charge density decreases
			• attraction between ions decreases
NaI	−705	$\oplus\ominus$	• lattice energy becomes less negative.

Cations are smaller than their atoms.

Anions are larger than their atoms.

Effect of ionic charge

The strongest ionic lattices with the most negative lattice enthalpies contain **small, highly charged ions**.

The diagram below shows the change in ionic size and ionic charge across Period 3 in the Periodic Table. Notice how the effects of ionic size and charge affect the size of the attraction between ions.

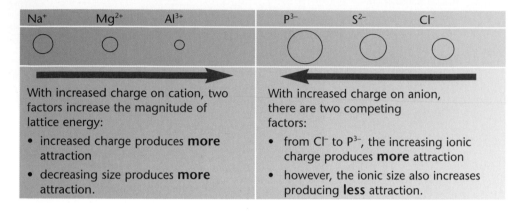

Na^+	Mg^{2+}	Al^{3+}	P^{3-}	S^{2-}	Cl^-

With increased charge on cation, two factors increase the magnitude of lattice energy:
- increased charge produces **more** attraction
- decreasing size produces **more** attraction.

With increased charge on anion, there are two competing factors:
- from Cl^- to P^{3-}, the increasing ionic charge produces **more** attraction
- however, the ionic size also increases producing **less** attraction.

Key points from AS

- **Polarisation**
  Revise AS page 54
- **The thermal stability of s-block compounds**
  Revise AS pages 74–75

Limitations of lattice enthalpies

Theoretical lattice enthalpies can be calculated by considering each ion as a perfect sphere. Any difference between this theoretical lattice enthalpy and lattice enthalpy calculated using a Born–Haber cycle indicates a degree of covalent bonding caused by polarisation. Polarisation is greatest between:

- a small densely charged cation and
- a large anion.

From the ions shown in the diagram above, we would expect the largest degree of covalency between Al^{3+} and P^{3-}.

Progress check

1 Using the information below:
 (a) name each enthalpy change **A** to **E**
 (b) construct a Born–Haber cycle for sodium bromide and calculate the lattice
     enthalpy (enthalpy change of lattice formation) of sodium bromide.

enthalpy change	equation	$\Delta H^{\ominus}/kJ\ mol^{-1}$
A	$Na(s) + \frac{1}{2}Br_2(l) \longrightarrow NaBr(s)$	−361
B	$Br(g) + e^- \longrightarrow Br^-(g)$	−325
C	$Na(s) \longrightarrow Na(g)$	+107
D	$Na(g) \longrightarrow Na^+(g) + e^-$	+496
E	$\frac{1}{2}Br_2(l) \longrightarrow Br(g)$	+112

1 (a) A enthalpy change of formation of sodium bromide
       B 1st electron affinity of bromine
       C enthalpy change of atomisation of sodium
       D 1st ionisation energy of sodium
       E enthalpy change of atomisation of bromine
   (b) −751 kJ mol⁻¹

Enthalpy change of solution

AQA   M5   SALTERS   M5
EDEXCEL   M4   WJEC   CH5
NUFFIELD   M6

An ionic lattice dissolves in **polar** solvents (e.g. water). In this process, the giant ionic lattice is broken up by polar water molecules which surround each ion in solution. The diagrams below compare the ionic lattice of sodium chloride with aqueous sodium and chloride ions.

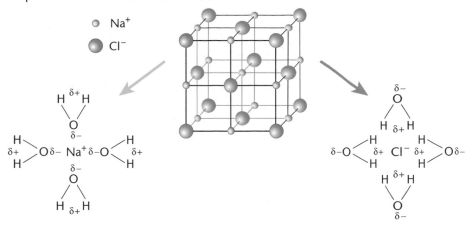

Enthalpy changes

When sodium chloride is dissolved in water, the enthalpy change of this process can be measured directly as the enthalpy change of solution.

See also complex ions p. 73.

breaking of the lattice $\longrightarrow$ hydrated ions

$Na^+Cl^-(s) + aq \longrightarrow Na^+(aq) + Cl^-(aq)$ **ROUTE 1**

This change can be broken down into separate steps:

- breaking of the lattice $\longrightarrow$ gaseous ions

  $Na^+Cl^-(s) \longrightarrow Na^+(g) + Cl^-(g)$

- hydration of gaseous ions $\longrightarrow$ hydrated ions   **ROUTE 2**

  $Na^+(g) + aq \longrightarrow Na^+(aq)$

  $Cl^-(g) + aq \longrightarrow Cl^-(aq)$

Key points from AS

- **Indirect determination of enthalpy changes**
  Revise AS pages 91–92

The energy cycle

The complete energy cycle for dissolving of sodium chloride in water is shown below. Definitions of these enthalpy changes are shown on page 54.

Lattice enthalpy **forms** the ionic lattice. Notice that the energy change to break 1 mole of the ionic lattice = – (lattice enthalpy).

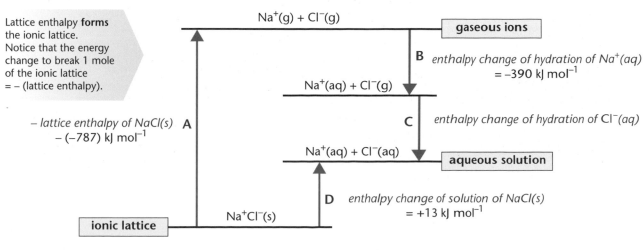

Route 1:   **A + B + C**

Route 2:   **D**

The production of hydrated ions comprises two changes:

- hydration of $Na^+(g)$
- hydration of $Cl^-(g)$.

Using Hess's Law: $A + B + C = D$

To determine the enthalpy change of hydration of $Cl^-(aq)$, **C**,

$$-\Delta H^{\ominus}_{L.E.} NaCl(s) + \Delta H^{\ominus}_{hyd} Na^+(g) + C = \Delta H^{\ominus}_{solution} NaCl(s)$$

$$\therefore\ -(-787) + (-390) + C = +13$$

Hence, the enthalpy change of hydration of $Cl^-(aq)$, **C = −384 kJ mol^{-1}**

Definitions for enthalpy changes

The enthalpy changes involved in these two routes are shown below.

> **KEY POINT**
>
> The **standard enthalpy change of solution** ($\Delta H^{\ominus}_{solution}$) is the enthalpy change that accompanies the dissolving of 1 mole of a solute in a solvent to form an infinitely dilute solution under standard conditions.
>
> $Na^+Cl^-(s) + aq \longrightarrow Na^+(aq) + Cl^-(aq)$ $\Delta H^{\ominus}_{solution} = +13$ kJmol^{-1}

> **KEY POINT**
>
> The **standard enthalpy change of hydration** ($\Delta H^{\ominus}_{hyd}$) of an ion is the enthalpy change that accompanies the hydration of 1 mole of gaseous ions to form 1 mole of hydrated ions in an infinitely dilute solution under standard conditions.
>
> $Na^+(g) + aq \longrightarrow Na^+(aq)$ $\Delta H^{\ominus}_{hyd} = -390$ kJmol^{-1}
> $Cl^-(g)\ + aq \longrightarrow Cl^-(aq)$ $\Delta H^{\ominus}_{hyd} = -384$ kJmol^{-1}

Factors affecting the magnitude of hydration enthalpy

ion	Na^+	Mg^{2+}	Al^{3+}	Cl^-	Br^-	I^-
$\Delta H^{\ominus}_{hyd}$/kJ mol^{-1}	−390	−1891	−4613	−384	−351	−307
effect of charge	• increasing ionic charge					
	• greater attraction for water					
ionic radius/nm	0.102	0.072	0.053	0.180	0.195	0.215
effect of size	• decreasing ionic size			• increasing ionic size		
	• greater attraction for water			• less attraction for water		

The values obtained for each ion depend upon the size of the charge and the size of the ion (see also pages 51–52).

Progress check

1 You are provided with the following enthalpy changes.

$Na^+(g) + F^-(g) \longrightarrow NaF(s)$ $\Delta H = -918$ kJ mol^{-1}
$Na^+(g) + aq \longrightarrow Na^+(aq)$ $\Delta H = -390$ kJ mol^{-1}
$F^-(g) + aq \longrightarrow F^-(aq)$ $\Delta H = -457$ kJ mol^{-1}

(a) Name each of these three enthalpy changes.
(b) Write an equation, including state symbols, for which the enthalpy change is the enthalpy change of solution of NaF.
(c) Calculate the enthalpy change of solution of NaF.

1 (a) lattice enthalpy of NaF
enthalpy change of hydration of Na^+
enthalpy change of hydration of F^-
(b) NaF(s) + aq $\longrightarrow Na^+(aq) + F^-(aq)$
(c) +71 kJ mol^{-1}

3.2 Electrochemical cells

After studying this section you should be able to:

- understand the principles of an electrochemical cell
- define the term 'standard electrode (redox) potential', $E^{\ominus}$
- describe how to measure standard electrode potentials
- calculate a standard cell potential from standard electrode potentials

Key points from AS

- **Redox reactions**
  Revise AS pages 68–70

During the study of redox in AS Chemistry you learnt:

- that oxidation and reduction involve transfer of electrons in a redox reaction
- the rules for assigning oxidation states
- how to construct an overall redox equation from two half-equations.

These key principles are built upon by considering electrochemical cells in which redox reactions provide electrical energy.

Electricity from chemical reactions

AQA	M5	SALTERS	M4
EDEXCEL	M5	WJEC	CH5
OCR	M5.6	NICCEA	M4
NUFFIELD	M6		

In a redox reaction, electrons are transferred between the reacting chemicals.

For example, zinc react with aqueous copper(II) ions in a redox reaction.

$$Zn(s) + Cu^{2+}(aq) \longrightarrow Zn^{2+}(aq) + Cu(s)$$

oxidation: $Zn(s) \longrightarrow Zn^{2+}(aq) + 2e^-$
reduction: $Cu^{2+}(aq) + 2e^- \longrightarrow Cu(s)$

If this reaction is carried out simply by mixing zinc with aqueous copper(II) ions:

- heat energy is produced and, unless used as useful heat, this is often wasted.

An electrochemical cell controls the transfer of electrons:

- electrical energy is produced and can be used for electricity.

Electrochemical cells are the basis of all batteries. Their true value is as a portable form of electricity.

The zinc-copper electrochemical cell

An electrochemical cell comprises two **half-cells**. In each half-cell, the **oxidation** and **reduction** processes take place **separately**.

A zinc-copper cell has two half-cells:

A half-cell contains the oxidised and reduced species from a half-equation.

The species can be converted into one another by addition or removal of electrons.

- an oxidation half-cell containing reactants and products from the oxidation half-reaction: $Zn^{2+}(aq)$ and $Zn(s)$
- a reduction half-cell containing reactants and products from the reduction half-reaction: $Cu^{2+}(aq)$ and $Cu(s)$.

The circuit is completed using:

- a connecting **wire** which transfers **electrons** between the half-cells
- a **salt bridge** which transfers **ions** between the half-cells.

The salt bridge consists of filter paper soaked in aqueous KNO_3 or NH_4NO_3.

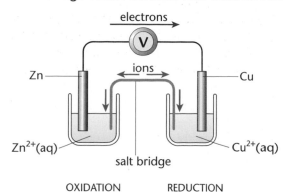

Remember these charge carriers:

ELECTRONS through the wire;

IONS through the salt bridge.

oxidation half-cell: Zn^{2+}(aq) and Zn(s)
- **Zn is oxidised to Zn^{2+}:** $Zn(s) \longrightarrow Zn^{2+}(aq) + 2e^-$
- electrons are **supplied** to the external circuit
- the polarity is **negative**.

reduction half-cell: Cu^{2+}(aq) and Cu(s)
- **Cu^{2+} is reduced to Cu:** $Cu^{2+}(aq) + 2e^- \longrightarrow Cu(s)$
- electrons are **taken** from the external circuit
- the polarity is **positive**.

The measured cell e.m.f. of 1.10 V indicates that there is a potential difference of 1.10 V between the copper and zinc half-cells.

Standard electrode potentials

AQA	M5	SALTERS	M4
EDEXCEL	M5	WJEC	CH5
OCR	M5.6	NICCEA	M4
NUFFIELD	M6		

A **hydrogen half-cell** is used as the standard for the measurement of standard electrode potentials.

The standard hydrogen half-cell is based upon the half-reaction below.

$$H^+(aq) + e^- \rightleftharpoons \tfrac{1}{2}H_2(g)$$

A hydrogen half-cell typically comprises

- 1 mol dm^{-3} hydrochloric acid as the source of H^+(aq)

- a supply of hydrogen gas, H_2(g) at 100 kPa

- an inert platinum electrode on which the half-reaction above takes place.

> **KEY POINT**
>
> The **standard electrode potential** of a half-cell, $E^\ominus$, is the e.m.f. of a half-cell compared with a standard hydrogen half-cell.
> All measurements are at **298 K** with solution **concentrations of 1 mol dm^{-3}** and gas **pressures of 100 kPa.**

Measuring standard electrode potentials

Half-cells comprising a metal and a metal ion

The diagram below shows how a hydrogen half-cell can be used to measure the standard electrode potential of a copper half-cell.

A high-resistance voltmeter is used to minimise the current that flows.

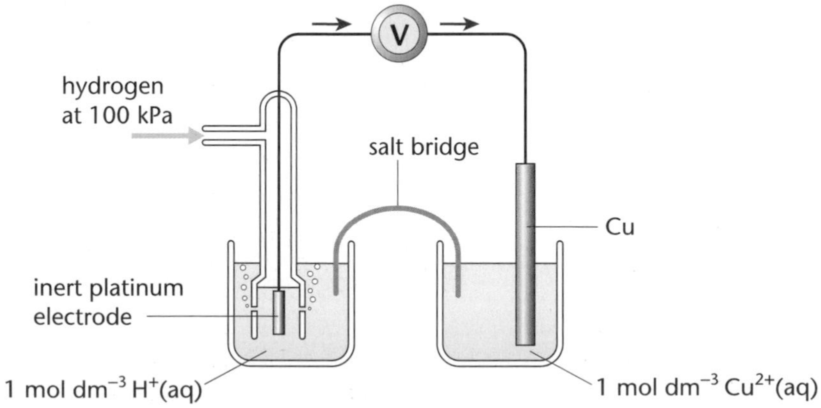

Hydrogen half-cell

> **KEY POINT**
>
> - The standard electrode potential of a half-cell represents its contribution to the cell e.m.f.
> - The contribution made by the hydrogen half-cell to the cell e.m.f. is defined as 0 V.
> - The **sign** of the standard electrode potential of a half-cell indicates its **polarity** compared with the hydrogen half-cell.

As with the hydrogen half-cell, the platinum electrode provides a surface on which the half-reaction can take place.

Half-cells comprising ions of different oxidation states

The half-reaction taking place in a half-cell can be between aqueous ions with different oxidation states. To allow electrons to pass into the half-cell, an inert electrode of platinum is used.

For example, a half-cell containing $Fe^{3+}(aq)$ and $Fe^{2+}(aq)$ ions is based upon the half-reaction below.

$$Fe^{3+}(aq) \rightleftharpoons Fe^{2+}(aq) + e^-$$

A **standard** $Fe^{3+}(aq)/Fe^{2+}(aq)$ half-cell requires:

- a solution containing **both**
  - 1 mol dm^{-3} $Fe^{3+}(aq)$ **and**
  - 1 mol dm^{-3} $Fe^{2+}(aq)$
- an inert platinum electrode with a connecting wire.

Pt (inert electrode)

$Fe^{2+}(aq), 1 \text{ mol dm}^{-3}$
$Fe^{3+}(aq), 1 \text{ mol dm}^{-3}$

The electrochemical series

A standard electrode potential indicates the availability of electrons from a half-reaction.

Metals are reducing agents.

- Metals typically react by donating electrons in a redox reaction.
- Reactive metals have the greatest tendency to donate electrons and their half-cells have the most negative electrode potentials.

Non-metals are oxidising agents.

- Non-metals typically react by accepting electrons in a redox reaction.
- Reactive non-metals have the greatest tendency to accept electrons and their half-cells have the most positive electrode potentials.

The **electrochemical series** shows electrode reactions listed in order of their standard electrode potentials. The electrochemical series below shows standard electrode potentials listed with the **most positive** $E^\ominus$ value at the top. This enables oxidising and reducing agents to be easily compared.

Key points from AS

- **Reactivity of s-block elements**
  Revise AS page 72
- **The relative reactivity of the halogens as oxidising agents**
  Revise AS page 77

This version of the electrochemical series show standard electrode potentials listed with the **most positive** $E^\ominus$ value at the top.

You will find versions of the electrochemical series listed with the **most negative** $E^\ominus$ value at the top.

This does not matter. The important point is that the potentials are listed **in order** and **can be compared**.

		electrode reaction			$E^\ominus$/ V	
strongest oxidising agent	$F_2(g)$	$+$	$2e^-$	$\rightleftharpoons$	$2F^-(aq)$	$+2.87$
	$Cl_2(g)$	$+$	$2e^-$	$\rightleftharpoons$	$2Cl^-(aq)$	$+1.36$
	$Br_2(l)$	$+$	$2e^-$	$\rightleftharpoons$	$2Br^-(aq)$	$+1.07$
	$Ag^+(aq)$	$+$	e^-	$\rightleftharpoons$	$Ag(s)$	$+0.80$
	$Cu^{2+}(aq)$	$+$	$2e^-$	$\rightleftharpoons$	$Cu(s)$	$+0.34$
	$H^+(aq)$	$+$	e^-	$\rightleftharpoons$	$\frac{1}{2}H_2(g)$	0
	$Fe^{2+}(aq)$	$+$	$2e^-$	$\rightleftharpoons$	$Fe(s)$	-0.44
	$Cr^{3+}(aq)$	$+$	$3e^-$	$\rightleftharpoons$	$Cr(s)$	-0.77
	$K^+(aq)$	$+$	e^-	$\rightleftharpoons$	$K(s)$	-2.92

strongest reducing agent

> **KEY POINT**
> In the electrochemical series:
> - the **oxidised** form is on the **left-hand side** of the half-equation
> - the **reduced** form is on the **right-hand side** of the half-equation.

Standard cell potentials and cell reactions

AQA	M5	SALTERS	M4
EDEXCEL	M5	WJEC	CH5
OCR	M5.6	NICCEA	M4
NUFFIELD	M6		

The **standard cell potential** of a cell is the e.m.f. acting between the two half-cells making up the cell under standard conditions. A standard cell potential can be determined using standard electrode potentials.

The **cell reaction** is:

- the overall process taking place in the cell
- the sum of the reduction and oxidation half-reactions taking place in each half-cell.

Key points from AS

- **Combining half-equations**
  Revise AS page 70

Calculating the standard cell potential of a silver-iron(II) cell

Identify the two relevant half-reactions and the polarity of each electrode.

The more positive of the two systems is the positive terminal of the cell.

$Ag^+(aq) + e^- \rightleftharpoons Ag(s)$ $E^\ominus = +0.80$ V *positive terminal*
$Fe^{2+}(aq) + 2e^- \rightleftharpoons Fe(s)$ $E^\ominus = -0.44$ V *negative terminal*

The standard electrode potential is the difference between the $E^\ominus$ values:

> $E^\ominus_{cell}$ is the difference between the standard electrode potentials.

Simply subtract the $E^\ominus$ of the negative terminal from the $E^\ominus$ of the positive terminal:

$$E^\ominus_{cell} = E^\ominus \text{ (positive terminal)} - E^\ominus \text{ (negative terminal)}$$
$$\therefore E^\ominus_{cell} = 0.080 - (-0.44) = \mathbf{1.24 \text{ V}}$$

Determination of the cell reaction in a silver-iron(II) cell

Work out the actual direction of the half-equations.

> The more negative half-equation is reversed. The negative terminal is then on the left-hand side.

The **half-equation** with the **more negative** $E^\ominus$ value provides the electrons and proceeds to the left. This half-equation is reversed so that the two half-reactions taking place can be clearly seen and compared.

$Ag^+(aq) + e^- \longrightarrow Ag(s)$ *reaction at positive terminal*
$Fe(s) \longrightarrow Fe(aq) + 2e^-$ *reaction at negative terminal*

The overall cell reaction can be found by adding the half-equations.

> This is discussed in detail in the AS Chemistry Study Guide, p. 70.

The silver half-equation must first be multiplied by '2' to balance the electrons.
The two half-equations are then added.

$2Ag^+(aq) + Fe(s) \longrightarrow 2Ag(s) + Fe^{2+}(aq)$

Progress check

1 Use the standard electrode potentials on page 57, to answer the questions which follow.
   (a) Calculate the standard cell potential for the following cells.
      (i) F_2/F^- and Ag^+/Ag
      (ii) Ag^+/Ag and Cr^{3+}/Cr
      (iii) Cu^{2+}/Cu and Cr^{3+}/Cr.
   (b) For each cell in (a) determine which half-reactions proceed and hence construct the cell reaction.

1 (a) (i) 2.07 V (ii) 1.57 V (iii) 1.11 V.
   (b) $F_2 + 2Ag \longrightarrow 2F^- + 2Ag^+$
   $3Ag^+ + Cr \longrightarrow 3Ag + Cr^{3+}$
   $3Cu^{2+} + 2Cr \longrightarrow 3Cu + 2Cr^{3+}$

3.3 Predicting redox reactions

After studying this section you should be able to:

- *use standard cell potentials to predict the feasibility of a reaction*
- *recognise the limitations of such predictions*

LEARNING SUMMARY

During AS Chemistry, you learnt how to combine two half-equations to construct a full redox equation.

Standard electrode potentials can be used to predict the direction of any aqueous redox reaction from the relevant half-reactions.

Using standard electrode potentials to make predictions

AQA	M5	SALTERS	M4
EDEXCEL	M5	WJEC	CH5
OCR	M5.6	NICCEA	M4
NUFFIELD	M6		

By studying standard electrode potentials, predictions can be made about the likely feasibility of a reaction.

The table below shows three redox systems.

Key points from AS

- **Redox reactions**
  Revise AS pages 68–70

	redox system	$E^{\ominus}$ /V
A	$H_2O_2(aq) + 2H^+ + 2e^- \rightleftharpoons 2H_2O(l)$	+1.77
B	$I_2(aq) + 2e^- \rightleftharpoons 2I^-(aq)$	+0.54
C	$Fe^{3+}(aq) + 3e^- \rightleftharpoons Fe(s)$	–0.04

We can predict the redox reactions that may take place between these species by treating the redox systems as if they are half-cells.

Comparing the redox systems A and B

- The I_2/I^- system has the less positive $E^{\ominus}$ value and its half-reaction will proceed to the left, donating electrons:

$$H_2O_2(aq) + 2H^+(aq) + 2e^- \longrightarrow 2H_2O(l) \qquad E^{\ominus} = +1.77 \text{ V}$$
$$I_2(aq) + 2e^- \longleftarrow 2I^-(aq) \qquad E^{\ominus} = +0.54 \text{ V}$$

We would therefore predict that **H_2O_2 and H⁺** would react with **I⁻**.

Notice that a reaction takes place between reactants from different sides of each half equation.

Comparing the redox systems B and C

- The Fe^{3+}/Fe system has the more negative $E^{\ominus}$ value and its half-reaction will proceed to the left, donating electrons:

$$I_2(aq) + 2e^- \longrightarrow 2I^-(aq) \qquad E^{\ominus} = +0.54 \text{ V}$$
$$Fe^{3+}(aq) + 3e^- \longleftarrow Fe(s) \qquad E^{\ominus} = -0.04 \text{ V}$$

We would therefore predict that **I_2** would react with **Fe**.

Comparing the redox systems A and C

- The Fe^{3+}/Fe system has the more negative $E^{\ominus}$ value and its half-reaction will proceed to the left, donating electrons:

$$H_2O_2(aq) + 2H^+(aq) + 2e^- \longrightarrow 2H_2O(l) \qquad E^{\ominus} = +1.77 \text{ V}$$
$$Fe^{3+}(aq) + 3e^- \longleftarrow Fe(s) \qquad E^{\ominus} = -0.04 \text{ V}$$

We would therefore predict that **H_2O_2 and H⁺** would react with **Fe**.

> When comparing two redox systems, we can predict that a reaction may take place between:
> - the stronger reducing agent with the more negative $E^{\ominus}$, on the right-hand side of the redox system
> - the stronger oxidising agent with the more positive $E^{\ominus}$, on the left-hand side of the redox system.

KEY POINT

Limitations of standard electrode potentials

AQA	M5	SALTERS	M4
EDEXCEL	M5	WJEC	CH5
OCR	M5.6	NICCEA	M4
NUFFIELD	M6		

Electrode potentials and ionic concentration

Non-standard conditions alter the value of an electrode potential.

The half-equation for the copper half-cell is shown below.

$$Cu^{2+}(aq) + 2e^- \rightleftharpoons Cu(s)$$

Using le Chatelier's principle, by increasing the concentration of $Cu^{2+}(aq)$:

- the equilibrium opposes the change by moving to the right
- electrons are removed from the equilibrium system
- the electrode potential becomes more positive.

> A change in electrode potential resulting from concentration changes means that predictions made on the basis of the **standard** value may not be valid.

Will a reaction actually take place?

Remember that these are equilibrium processes.

- Predictions can be made but these give no indication of the reaction rate which may be extremely slow, caused by a large activation energy.

- The actual conditions used may be different from the standard conditions used to record $E^\ominus$ values. This will affect the value of the electrode potential (see above).

- Standard electrode potentials apply to aqueous equilibria. Many reactions take place that are not aqueous.

As a general working rule:

- the larger the difference between $E^\ominus$ values, the more likely that a reaction will take place

- if the difference between $E^\ominus$ values is less than 0.4 V, then a reaction is unlikely to take place.

Progress check

1 For each of the three redox systems on page 59, construct full redox equations.

2 Use the standard electrode potentials below to answer the questions which follow.

$Fe^{3+}(aq) + e^-$	$\rightleftharpoons$	$Fe^{2+}(aq)$	$E^\ominus = +0.77$ V
$Br_2(l) + 2e^-$	$\rightleftharpoons$	$2Br^-(aq)$	$E^\ominus = +1.07$ V
$Ni^{2+}(aq) + 2e^-$	$\rightleftharpoons$	$Ni(s)$	$E^\ominus = -0.25$ V
$O_2(g) + 4H^+(aq) + 4e^-$	$\rightleftharpoons$	$2H_2O(l)$	$E^\ominus = +1.23$ V

(a) Arrange the reducing agents in order with the most powerful first.
(b) Predict the reactions that could take place.

O_2 and H^+ with Ni.
Br_2 with Fe^{2+}; Fe^{3+} with Ni; O_2 and H^+ with Fe^{2+}; Br_2 with Ni;
2 Ni(s), $Fe^{2+}(aq)$, $Br^-(l)$, $H_2O(l)$.
A and C: $3H_2O_2(aq) + 6H^+(aq) + 2Fe^{3+}(s) \longrightarrow 2Fe^{3+}(aq) + 6H_2O(l)$
B and C: $2Fe(s) + 3I_2(aq) \longrightarrow 2Fe^{3+}(aq) + 6I^-(aq)$
1 A and B: $H_2O_2(aq) + 2H^+(aq) + 2I^-(aq) \longrightarrow I_2(aq) + 2H_2O(l)$

Sample question and model answer

Take care to include correct state symbols – essential when showing the equations representing these enthalpy changes.

(a) The lattice energy of copper(II) oxide can be found using a Born–Haber cycle. Write equations, including state symbols, which correspond to:

(i) the enthalpy change of formation of copper(II) oxide

$Cu(s) + \frac{1}{2}O_2(g) \longrightarrow CuO(s)$ formulae ✓ state symbols ✓

(ii) the lattice energy of copper(II) oxide

$Cu^{2+}(g) + O^{2-}(g) \longrightarrow CuO(s)$ formulae ✓ state symbols ✓

(iii) the first ionisation energy of copper.

$Cu(g) \longrightarrow Cu^+(g) + e^-$ formulae ✓ state symbols ✓ [6]

(b) Use the data below to construct a Born–Haber cycle and hence calculate a value for the lattice energy of copper(II) oxide.

enthalpy change	/ kJ mol^{-1}
atomisation of copper	+339
1st ionisation energy of copper	+745
2nd ionisation energy of copper	+1960
atomisation of oxygen, i.e. $\frac{1}{2}O_2(g) \rightarrow O(g)$	+248
1st electron affinity of oxygen	−141
2nd electron affinity of oxygen	+791
formation of copper(II) oxide	−155

This Born–Haber cycle includes all the enthalpy changes that you may encounter in a Born–Haber cycle.

Most cycles in exams are simpler.

Note the different signs for the 1st and 2nd electron affinities.

copper part of cycle ✓
oxygen part of cycle ✓
CuO part of cycle ✓

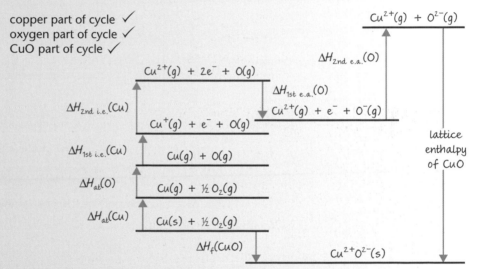

By Hess's Law.

$\Delta H_{at}(Cu) + \Delta H_{at}(O) + \Delta H_{1st\ i.e.}(Cu) + \Delta H_{2nd\ i.e.}(Cu) + \Delta H_{1st\ e.a.}(O) + \Delta H_{2nd\ e.a.}(O) + L.E.(CuO) = \Delta H_f(CuO)$

∴ $(+339) + (+248) + (+745) + (+1960) + (−141) + (+791) + L.E.(CuO)$
   $= (−155)$

∴ Lattice energy CuO $= −4097$ kJ mol^{-1} correct value correct sign ✓✓

[Total: 11]

Cambridge How Far, How Fast? Q7 Modified June 1998

Practice examination questions

1

(a) Construct a Born–Haber cycle for the formation of sodium oxide, Na_2O, and use the data given below to calculate the second electron affinity of oxygen.

$Na(s)$	$\longrightarrow$	$Na(g)$	$\Delta H^{\ominus} = +107$ kJ mol^{-1}
$Na(g)$	$\longrightarrow$	$Na^+(g) + e^-$	$\Delta H^{\ominus} = +496$ kJ mol^{-1}
$O_2(g)$	$\longrightarrow$	$2O(g)$	$\Delta H^{\ominus} = +249$ kJ mol^{-1}
$O(g) + e^-$	$\longrightarrow$	$O^-(g)$	$\Delta H^{\ominus} = -141$ kJ mol^{-1}
$O^-(g) + e^-$	$\longrightarrow$	$O^{2-}(g)$	$\Delta H^{\ominus}$ to be calculated
$2Na^+(g) + O^{2-}(g)$	$\longrightarrow$	$Na_2O(s)$	$\Delta H^{\ominus} = -2478$ kJ mol^{-1}
$2Na(s) + \frac{1}{2}O_2(g)$	$\longrightarrow$	$Na_2O(s)$	$\Delta H^{\ominus} = -414$ kJ mol^{-1} [9]

(b) Explain why the second electron affinity of oxygen has a large positive value. [2]

(c) Explain, by reference to steps from relevant Born–Haber cycles, why sodium forms a stable oxide consisting of Na^+ and O^{2-} ions but not oxides consisting of Na^+ and O^- or Na^{2+} and O^{2-} ions. [4]

[Total: 15]

Assessment and Qualifications Alliance Further Physical Chemistry Q2 June 2000

2

A cell was set up as shown below.

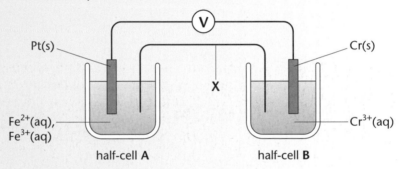

(a) What species are involved in the equilibria in the separate half-cells? [2]

(b) What name is given to the part of the cell labelled X? [1]

(c) The standard electrode potentials for the half-cells above are:

$$Fe^{3+} + e^- \rightleftharpoons Fe^{2+} \qquad E^{\ominus} = +0.77 \text{ V}$$
$$Cr^{3+} + 3e^- \rightleftharpoons Cr \qquad E^{\ominus} = -0.74 \text{ V}$$

(i) Calculate the standard cell potential of the complete cell.

(ii) Draw an arrow on the diagram above to show the direction of electron flow in the external circuit. [2]

(d) (i) Write equations for the reaction at each electrode. [2]

(ii) Hence write an equation to show the overall reaction. [1]

[Total: 8]

Cambridge How Far, How Fast? Q2 March 1998

Practice examination questions (continued)

3

(a) Use the standard electrode potentials given below to answer the questions which follow.

			$E^\ominus$/V
$Fe^{3+}(aq) + e^-$	$\rightleftharpoons$	$Fe^{2+}(aq)$	+0.77
$I_2(aq) + 2e^-$	$\rightleftharpoons$	$2I^-(aq)$	+0.54
$[Fe(CN)_6]^{3-}(aq) + e^-$	$\rightleftharpoons$	$[Fe(CN)_6]^{4-}(aq)$	+0.34

(i) Which of the ions $Fe^{3+}(aq)$ and $[Fe(CN)_6]^{3-}(aq)$ is the stronger oxidising agent?

(ii) State, with an equation and a reason, which of the iron-containing species shown above would be capable of liberating iodine when added to an aqueous solution containing iodide ions. [4]

(b) Consider the following standard electrode potentials.

					$E^\ominus$/V
$Sn^{2+}(aq)$	+	$2e^-$	$\rightleftharpoons$	$Sn(s)$	–0.14
$Fe^{2+}(aq)$	+	$2e^-$	$\rightleftharpoons$	$Fe(s)$	–0.44
$Zn^{2+}(aq)$	+	$2e^-$	$\rightleftharpoons$	$Zn(s)$	–0.76

Some cars are made from coated steel. Use the above data to explain why zinc, rather than tin, is used for this coating. [2]

[Total: 6]

NEAB Paper 1B Q5 June 1993 modified

4

(a) Name the standard reference electrode against which electrode potentials are measured and, for this electrode, state the conditions to which the term standard refers. [4]

(b) The standard electrode potentials for two electrode reactions are given below.

$S_2O_8^{2-}(aq) + 2e^- \longrightarrow 2SO_4^{2-}(aq)$ $E^\ominus = +2.01$ V

$Ag^+(aq) + e^- \longrightarrow Ag(s)$ $E^\ominus = +0.80$ V

(i) A cell is produced when these two half-cells are connected. Deduce the cell potential, $E^\ominus$, for this cell and write an equation for the spontaneous reaction.

(ii) State how, if at all, the electrode potential of the $S_2O_8^{2-}/SO_4^{2-}$ equilibrium would change if the concentration of SO_4^{2-} ions was increased. Explain your answer. [6]

[Total: 10]

Assessment and Qualifications Alliance Further Physical Chemistry Q1 March 2000

Chapter 4

The Periodic Table

The following topics are covered in this chapter:

- Oxides of the Period 3 elements
- Chlorides of the Period 3 elements
- The transition elements
- Transition element complexes
- Ligand substitution of complex ions
- Redox reactions of transition metal ions
- Catalysis
- Reactions of metal aqua-ions

Key points from AS

- **Bonding and structure**
  Revise AS pages 44–61

During AS Chemistry, you learnt about physical trends in properties of **elements** across Period 3. Across Period 3, **compounds** of the elements also show characteristic trends. For A2 Chemistry, physical and chemical trends in the properties of Period 3 oxides and chlorides are explained in terms of bonding and structure.

4.1 Oxides of the Period 3 elements

After studying this section you should be able to:

- describe and explain the periodic variation of the formulae and boiling points of the main oxides of Period 3
- describe the reactions of Period 3 elements with oxygen
- describe the action of water on Period 3 oxides
- describe these reactions in terms of structure and bonding

LEARNING SUMMARY

The formulae of the Period 3 oxides

AQA	M5	SALTERS	M5
EDEXCEL	M4	WJEC	CH5
OCR	M5.1	NICCEA	M4
NUFFIELD	M6		

Key points from AS

- **The modern Periodic Table**
  Revise AS pages 64–67
- **Redox reactions**
  Revise AS pages 68–70

The highest oxidation state of an element is usually the group number.

Across Period 3, the formulae of compounds show a regular pattern which depends upon the **number of electrons** in the outer shell.

The **oxidation state** of an element indicates the **number of electrons** involved in **bonding**. The relationship of the formulae of the Period 3 oxides with electronic structure and oxidation state is shown below.

element	Na	Mg	Al	Si	P	S
electrons in outer shell	1	2	3	4	5	6
formula of oxide	Na_2O	MgO	Al_2O_3	SiO_2	P_4O_6 P_4O_{10}	SO_2 SO_3
oxidation state of element in oxide	+1	+2	+3	+4	+3 +5	+4 +6

increase in highest oxidation state →

The amount of oxygen atoms per mole of the element increases steadily across Period 3.

For each element in the Period, there is an increase of 0.5 mole of O per mole of element.

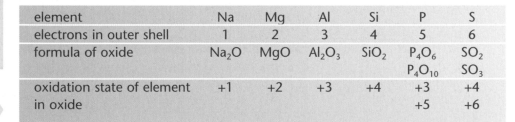

$$Na_1O_{0.5} \quad Mg_1O_1 \quad Al_1O_{1.5} \quad Si_1O_2 \quad P_1O_{2.5} \quad S_1O_3$$

Periodicity means a repeating trend. This is one of the important ideas of chemistry – a trend across one period is repeated across other periods.

This periodicity is also seen across Period 2 and 4:

Period 2	Li_2O	BeO	B_2O_3	CO_2
Period 3	Na_2O	MgO	Al_2O_3	SiO_2
Period 4	K_2O	CaO	Ga_2O_3	GeO_2

Preparation of oxides from the elements

AQA	M5	SALTERS	M5
EDEXCEL	M4	WJEC	CH5
OCR	M5.1	NICCEA	M4
NUFFIELD	M6		

Oxides of the Period 3 elements Na to Cl can be prepared by heating each element in oxygen. These are redox reactions:

- the Period 3 element is oxidised
- oxygen is reduced.

Metal oxides

Equations for the reactions of oxygen with the metals sodium, magnesium and aluminium are shown below.

$$4Na(s) + O_2(g) \xrightarrow{\text{yellow flame}} 2Na_2O(s)$$

$$2Mg(s) + O_2(g) \xrightarrow{\text{white flame}} 2MgO(s)$$

$$4Al(s) + 3O_2(g) \xrightarrow{\text{white flame}} 2Al_2O_3(s)$$

At room temperature and pressure:
Na_2O, MgO and Al_2O_3 are white solids.

- Na_2O and MgO are **ionic compounds**.
- Al_2O_3 has bonding **intermediate** between ionic and covalent.

Non-metal oxides

Phosphorus and sulphur exist as P_4 and S_8 molecules.
Sulphur, S_8, is usually simplified in equations as S.

Equations for the reactions of phosphorus and sulphur with oxygen are shown below.

$$P_4(s) + 5O_2(g) \xrightarrow{\text{white flame}} P_4O_{10}(s)$$

$$S(s) + O_2(g) \xrightarrow{\text{blue flame}} SO_2(g)$$

At room temperature and pressure:
- P_4O_{10} is a white solid
- SO_2 is a colourless gas
- SO_3 is a colourless liquid.

SO_2 reacts further with oxygen in the presence of a catalyst (V_2O_5) see page 83:

$$2SO_2(g) + O_2(g) \longrightarrow 2SO_3(l)$$

- P_4O_{10}, SO_2 and SO_3 are **covalent compounds**.

Trend in the boiling points of the Period 3 oxides

AQA	M5	SALTERS	M5
EDEXCEL	M4	WJEC	CH5
OCR	M5.1	NICCEA	M4
NUFFIELD	M6		

The table below compares the boiling points of some Period 3 oxides with their structure and bonding.

Key point from AS

- **Bonding, structure and properties**
  Revise AS pages 58–61
- **Trends in boiling points of Period 3 elements**
  Revise AS pages 66–67

oxide	Na_2O	MgO	Al_2O_3	SiO_2	P_4O_{10}	SO_3
boiling point /°C	1275	2827	2017	1607	580	33
structure	giant lattice				simple molecules	
bonding	ionic		mixed		covalent	
	strong forces between ions		strong forces between atoms		weak forces between molecules	

Giant structures and boiling point

During your study of AS Chemistry, you learnt that the boiling points of the elements and compounds are dependent upon bonding and structure.

The same principles can be applied many times in chemistry covalent bonds.

The oxides with **giant** structures ($Na_2O \rightarrow SiO_2$):

- have **high boiling points**
- require a large amount of energy to break the **strong forces** between their particles.

The particles making up a giant lattice can be different.

- Na_2O and MgO have ionic bonding and the **strong** forces are electrostatic attractions acting between positive and negative **ions**.
- Al_2O_3 has bonding intermediate between ionic and covalent. **Strong** forces act between particles **intermediate** between ions and atoms.
- SiO_2 has covalent bonding and the **strong** forces are 'shared pairs of electrons' acting between the **atoms**.

Simple molecular structures and boiling point

The oxides with **simple molecular** structures ($P_4O_{10} \rightarrow SO_3$):

- have low boiling points
- require a small amount of energy to break the **weak van der Waals' forces** acting between the **molecules**.

The action of water on the Period 3 oxides

AQA	M5
EDEXCEL	M4
OCR	M5.1
NUFFIELD	M6

SALTERS	M5
WJEC	CH5
NICCEA	M4

Reactions of sodium and magnesium oxides

Metal oxides form alkalis in water.

$$Na_2O(s) + 2H_2O(l) \longrightarrow 2NaOH(aq) \qquad pH = 14$$

$$MgO(s) + 2H_2O(l) \longrightarrow Mg(OH)_2(aq) \qquad pH = 11$$

Each hydroxide dissociates in water, releasing OH^- ions into solution:

$$NaOH(aq) \longrightarrow Na^+(aq) + OH^-(aq)$$

> Although MgO reacts with water, the magnesium hydroxide formed is only slightly soluble in water forming a weak alkaline solution only.

Key points from AS

- **The Group 1 elements**
  Revise AS pages 72–73
- **The Group 2 elements**
  Revise AS pages 73–74

Oxides of aluminium and silicon

Al_2O_3 and SiO_2 have very strong lattices which cannot be broken down by water. Consequently these compounds are **insoluble** in water.

Reactions of non-metal oxides

Non-metal oxides form acids in water. Equations for the reactions of $P_4O_{10}(s)$, $SO_2(g)$ and $SO_3(l)$ with water are shown below.

$$P_4O_{10}(s) + 6H_2O(l) \longrightarrow 4H_3PO_4(aq) \qquad \textit{phosphoric acid}$$

$$SO_2(g) + H_2O(l) \longrightarrow H_2SO_3(aq) \qquad \textit{sulphurous acid}$$

$$SO_3(l) + H_2O(l) \longrightarrow H_2SO_4(aq) \qquad \textit{sulphuric acid}$$

Each acid dissociates in the water, releasing H^+ ions into solution.

E.g. $\quad H_2SO_4(aq) \longrightarrow H^+(aq) + HSO_4^-(aq)$

> Metal oxides are basic oxides.
>
> Non-metal oxides are acidic oxides.

KEY POINT

An important general rule is:

- **metal oxides** are **basic**
  When soluble, metal oxides form alkaline solutions in water
- **non-metal oxides** are **acidic**
  When soluble, non-metal oxides form acidic solutions in water.

The action of water on Period 3 oxides is summarised below.

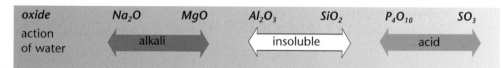

oxide	Na_2O	MgO	Al_2O_3	SiO_2	P_4O_{10}	SO_3
action of water	alkali		insoluble		acid	

Progress check

1 (a) Across the Periodic Table, what pattern is shown in the formulae of the Period 3 oxides?

(b) Why do elements in the **same** group form compounds with **similar** formulae?

2 The reactions of elements with oxygen are redox reactions. Use oxidation numbers to show that the reaction of aluminium with oxygen is a redox reaction.

O changes oxidation state from 0 to −2 (reduction)
Al changes oxidation number from 0 to +3 (oxidation)
2 $4Al(s) + 3O_2(g) \longrightarrow 2Al_2O_3(s)$
(b) Although the outer shell is different, there is the same number of electrons in the outer shell.
1 (a) For each element in the period, there is an increase of 0.5 mole of O per mole of the element.

4.2 Chlorides of the Period 3 elements

After studying this section you should be able to:

- *describe and explain the periodic variation of the formulae and boiling points of the main chlorides of Period 3*
- *describe the reactions of the Period 3 elements with chlorine*
- *describe the action of water on Period 3 chlorides*
- *describe these reactions in terms of structure and bonding*

LEARNING SUMMARY

The formulae of the Period 3 chlorides

AQA	M5	SALTERS	M5
EDEXCEL	M4	WJEC	CH5
OCR	M5.1	NICCEA	M4
NUFFIELD	M6		

Key points from AS

- **The modern Periodic Table**
  Revise AS pages 64–67
- **Redox reactions**
  Revise AS pages 68–70

The oxidation state of the element in its chloride is equal to the number of chlorine atoms bonded to the element.

The relationship of the formulae of some Period 3 chlorides with electron structure and oxidation state is shown below.

element	Na	Mg	Al	Si	P
electrons in outer shell	1	2	3	4	5
formula of chloride	NaCl	$MgCl_2$	Al_2Cl_6	$SiCl_4$	PCl_3 PCl_5
oxidation state of element in chloride	+1	+2	+3	+4	+3 +5

increase in highest oxidation state →

The amount of chlorine atoms per mole of the element increases steadily across Period 3.

For each element in the Period, there is an increase of 1 mole of Cl per mole of element.

NaCl $MgCl_2$ Al_2Cl_6 $SiCl_6$ $SiCl_4$ PCl_5

The same trend is seen across each Period of the Periodic Table.

The reason is the same – the number of electrons in the outer shell increases steadily across each period.

As with the Period 3 oxides (see page 64), this periodicity in formulae is also seen as a repeating trend across Periods 2 and 4. The table below shows the formulae of the chlorides from Period 2 to Period 4 of the Periodic Table.

Period 2	LiCl	$BeCl_2$	BCl_3	CCl_4
Period 3	NaCl	$MgCl_2$	$AlCl_3$	$SiCl_4$
Period 4	KCl	$CaCl_2$	$GaCl_3$	$GeCl_4$

Preparation of chlorides from the elements

AQA	M5	SALTERS	M5
EDEXCEL	M4	WJEC	CH5
OCR	M5.1	NICCEA	M4
NUFFIELD	M6		

Chlorides of Period 3 can be prepared by heating the elements in chlorine. As with the Period 3 oxides (see page 65), these are redox reactions.

Metal chlorides

Equations for the reactions of chlorine with the metals sodium, magnesium and aluminium are shown below.

$$2Na(s) + Cl_2(g) \longrightarrow 2NaCl(s)$$
$$Mg(s) + Cl_2(g) \longrightarrow MgCl_2(s)$$
$$2Al(s) + 3Cl_2(g) \longrightarrow Al_2Cl_6(s)$$

NaCl and $MgCl_2$ are white **ionic** compounds.

Al_2Cl_6 is a white covalent compound.

At room temperature and pressure:

- NaCl and $MgCl_2$ are white **ionic compounds** with **giant** structures
- Al_2Cl_6 is a white compound but, surprisingly, it has **covalent** bonding and a **simple molecular** structure.

Although aluminium has the characteristic properties of a metal, its compounds often show covalent or intermediate bonding (as in Al_2O_3 on page 65). This property results from the large polarising effect of the very small Al^{3+} ion.

Non-metal chlorides

Equations for the reactions of silicon and phosphorus with chlorine are shown below.

$$Si(s) + 2Cl_2(g) \longrightarrow SiCl_4(l)$$

$$P_4(s) + 6Cl_2(g) \longrightarrow 4PCl_3(l)$$

- $SiCl_4$ and PCl_3 are **covalent compounds.**

Phosphorus pentachloride: ionic or covalent?

Phosphorus pentachloride, PCl_5, exists as a covalent molecule in the gaseous state. However phosphorus pentachloride exists in the solid state in the **ionic** form, $[PCl_6]^- [PCl_4]^+$.

Thus solid PCl_3 and PCl_5 have different structures with PCl_5 having a much higher melting point and boiling points than PCl_3.

compound	PCl_3	PCl_5
structure	simple molecular: PCl_3 molecules	ionic lattice: $[PCl_6]^- [PCl_4]^+$
state at room temperature	liquid	solid
melting point /°C	−112	162
forces holding solid lattice together	weak van der Waals' forces	strong ionic bonds

Trend in the boiling points of the Period 3 chlorides

The table below compares the boiling points of some Period 3 chlorides with structure and bonding.

chloride	NaCl	MgCl₂	Al₂Cl₆	SiCl₄	PCl₅
boiling point /°C	1413	1412	178	58	56
structure	giant lattice →		← simple molecules		
bonding	← ionic →		← covalent →		
forces	strong forces between ions		weak forces between molecules		

Giant structures and boiling point

The chlorides with **giant** structures (NaCl $\longrightarrow$ MgCl₂)

- have **high boiling points**
- require a large amount of energy to break the **strong forces** between their particles
- NaCl and MgCl₂ have ionic bonding and the **strong** forces are electrostatic attractions acting between positive and negative **ions.**

The chlorides with **simple molecular** structures (Al₂Cl₆ $\longrightarrow$ PCl₃)

- have low boiling points
- require a small amount of energy to break the weak **van der Waals' forces** acting between the **molecules.**

Comparison between the oxides and chlorides of Period 3

The structure and bonding of the Period 3 oxides and chlorides are compared below.

For **oxides**, bonding and structure change at **different** points.

- Bonding changes from ionic to covalent between MgO and SiO_2 with Al_2O_3 having intermediate bonding.
- Structure changes from giant lattice to simple molecular between SiO_2 and P_4O_{10}.

For **chlorides**, both bonding and structure change at the **same** point.

Between $MgCl_2$ and Al_2Cl_6:

- bonding changes from ionic to covalent **and**
- structure changes from giant to simple molecular.

The action of water on the chlorides of Period 3

AQA	M5	SALTERS	M5
EDEXCEL	M4	WJEC	CH5
OCR	M5.1	NICCEA	M4
NUFFIELD	M6		

Dissolving ionic chlorides

The ionic chlorides **dissolve** in water forming a **neutral** or very weakly acidic solution.

$$NaCl(s) \quad + aq \longrightarrow Na^+(aq) \quad + Cl^-(aq) \qquad pH = 7$$
$$MgCl_2(s) \quad + aq \longrightarrow Mg^{2+}(aq) + 2Cl^-(aq) \qquad pH = 6$$

> This periodicity of properties is repeated across each period.

Reactions of covalent chlorides

The covalent chlorides are **hydrolysed** by water in **a vigorous reaction**. Strong acid solutions are formed containing hydrochloric acid.

Equations for the reactions of $Al_2Cl_6(s)$, $SiCl_4(l)$ and $PCl_5(s)$ with water are shown below.

$$Al_2Cl_6(s) \quad + 6H_2O(l) \longrightarrow 2Al(OH)_3(s) \quad + 6HCl(aq)$$
$$SiCl_4(l) \quad + 2H_2O(l) \longrightarrow SiO_2(s) \quad + 4HCl(aq)$$
$$PCl_5(s) \quad + 4H_2O(l) \longrightarrow H_3PO_4(aq) \quad + 5HCl(aq)$$

> **KEY POINT**
>
> An important general rule is:
> - **ionic chlorides** form **neutral** solutions in water
> - **covalent chlorides** form **acidic** solutions in water.

The action of water on the Period 3 oxides is summarised below.

chloride	NaCl	MgCl₂	AlCl₃	SiCl₄	PCl₅
action of water	neutral	very weak acid		strong acid	

Progress check

1 (a) What is the trend in pH shown by the chlorides of Period 3?
(b) Water is added to sodium chloride and phosphorus trichloride to form aqueous solutions with different pH values. Explain, with equations, why these solutions have different pH values.

1 (a) neutral to acidic
(b) $NaCl(s) + aq \longrightarrow Na^+(aq) + Cl^-(aq)$
$PCl_3(s) + 3H_2O(l) \longrightarrow H_3PO_3(aq) + 3HCl(aq)$
NaCl(aq) simply dissolves in water and does not generate excess H^+ or OH^- ions in solution. ∴ NaCl(aq) is neutral.
PCl_3 reacts with water forming H_3PO_3 and HCl both generating H^+ ions in solution. ∴ the solution is acidic.

4.3 The transition elements

After studying this section you should be able to:

- *explain what is meant by the terms d-block element and transition element*
- *deduce the electronic configurations of atoms and ions of the d-block elements Sc → Zn*
- *know the typical properties of a transition element*

LEARNING SUMMARY

Key points from AS

- **The Periodic Table**
  Revise AS page 64–86

During AS Chemistry, you studied the metals in the s-block of the Periodic Table. For A2 Chemistry, this understanding is extended with a detailed study of the elements scandium to zinc in the d-block of the Periodic Table. Many of the principles introduced during AS Chemistry will be revisited, especially redox chemistry, and structure and bonding.

The d-block elements

AQA	M5	SALTERS	M4
EDEXCEL	M5	WJEC	CH5
OCR	M5.1	NICCEA	M5
NUFFIELD	M6		

The d-block is found in the centre of the Periodic Table. For A2 Chemistry, you will be studying mainly the d-block elements in Period 4 of the Periodic Table: Sc, Ti, V, Cr, Mn, Fe, Co, Ni, Cu and Zn.

Electronic configurations of d-block elements

The diagram below shows the sub-shell being filled across Period 4 of the Periodic Table.

4s		3d									4p					
K	Ca	Sc	Ti	V	Cr	Mn	Fe	Co	Ni	Cu	Zn	Ga	Ge	As	Se	Br	Kr

Key points from AS

- **Sub-shells and orbitals**
  Revise AS pages 21–22
- **Filling the sub-shells**
  Revise AS pages 23–24
- **Sub-shells and the Periodic Table**
  Revise AS page 24

> The 3d sub-shell is at a higher energy and fills **after** the 4s sub-shell.

> See also the stability from a half-full p sub-shell.

- The elements Sc ⟶ Zn have their highest energy electrons in the 4s and 3d sub-shells.

- Across the d-block, electrons are filling the 3d sub-shell. The diagram below shows the outermost electron shell for Sc ⟶ Zn.

Notice how the orbitals are filled:

- the orbitals in the 3d sub-shell are first occupied singly to prevent any repulsion caused by pairing.

Notice chromium and copper:

- chromium has one electron in each orbital of the 4s and 3d sub-shells ⟶ extra stability

- copper has a full 3d sub-shell ⟶ extra stability.

		4s	3d				
Sc	[Ar] $3d^1 4s^2$	↑↓	↑				
Ti	[Ar] $3d^2 4s^2$	↑↓	↑	↑			
V	[Ar] $3d^3 4s^2$	↑↓	↑	↑	↑		
Cr	[Ar] $3d^5 4s^1$	↑	↑	↑	↑	↑	↑
Mn	[Ar] $3d^5 4s^2$	↑↓	↑	↑	↑	↑	↑
Fe	[Ar] $3d^6 4s^2$	↑↓	↑↓	↑	↑	↑	↑
Co	[Ar] $3d^7 4s^2$	↑↓	↑↓	↑↓	↑	↑	↑
Ni	[Ar] $3d^8 4s^2$	↑↓	↑↓	↑↓	↑↓	↑	↑
Cu	[Ar] $3d^{10} 4s^1$	↑	↑↓	↑↓	↑↓	↑↓	↑↓
Zn	[Ar] $3d^{10} 4s^2$	↑↓	↑↓	↑↓	↑↓	↑↓	↑↓

Transition elements

The elements in the d-block, Ti→Cu, are called transition elements.

> - A transition element has **at least one ion** with a **partially filled d sub-shell**.

KEY POINT

Although scandium and zinc are d-block elements, they are **not** *transition elements*. Neither element forms an ion with a partially filled d sub-shell.

Scandium forms one ion only, Sc^{3+} with the 3d sub-shell **empty**.

$$Sc \ [Ar]3d^14s^2 \xrightarrow{\textit{loss of 3 electrons}} Sc^{3+} \text{ only:} \quad [Ar]$$

Zinc forms one ion only, Zn^{2+} with the 3d sub-shell **full**.

$$Zn \ [Ar]3d^{10}4s^2 \xrightarrow{\textit{loss of 2 electrons}} Zn^{2+} \text{ only:} \quad [Ar]3d^{10}$$

d–block elements

Sc	Ti	V	Cr	Mn	Fe	Co	Ni	Cu	Zn

transition elements

Key points from AS

- **The s-block elements**
  Revise AS pages 71–75

Properties of transition elements

1 high density

2 high m. and b. pt.

3 moderate to low reactivity

4 two or more oxidation states

5 coloured ions

6 complex ions with ligands

7 catalysts

Typical properties of transition elements

The transition elements are all metals – they are good conductors of heat and electricity. In addition, some general properties distinguish the transition elements from the s-block metals.

Density

- The transition elements are more **dense** than other metals.
  - They have smaller atoms than the metals in Groups 1 and 2. The small atoms are able to pack closely together $\longrightarrow$ high density.

Melting point and boiling point

- The transition elements have **higher melting and boiling points** than other metals.
  - Within the metallic lattice, ions are **smaller** than those of the s-block metals – the metallic bonding between the metal ions and the delocalised electrons is strong.

Reactivity

- The transition elements have **moderate to low reactivity**.
  - Unlike the s-block metals, they do **not** react with cold water. Many transition metals react with dilute acids although some, such as gold, silver and platinum, are extremely unreactive.

Oxidation states

- Most transition elements have compounds with **two or more oxidation states**.

Colour of compounds

- The transition elements have **coloured compounds and ions**.
  - Most transition metals have at least one oxidation state that is coloured.

Complex ions

- The ions of transition elements form **complex ions** with ligands.

Catalysis

- Many transition elements can act as **catalysts**.

Formation of ions

AQA	M5	SALTERS	M4
EDEXCEL	M5	WJEC	CH5
OCR	M5,1	NICCEA	M5
NUFFIELD	M6		

The energy levels of the 4s and 3d sub-shells are very close together.

> **KEY POINT**
>
> - Transition elements are able to form **more than one ion**, each with a different oxidation state, by losing the **4s** electrons and different numbers of **3d** electrons.
> - When forming ions, the 4s electrons are **lost first**, before the 3d electrons.

The Fe^{3+} ion, [Ar]3d^5 is more stable than the Fe^{2+} ion, [Ar]3d^{5}4s^1.

Fe^{3+} has a half-full 3d sub-shell $\longrightarrow$ stability.

The 4s and 3d energy levels are very close together and both are involved in bonding.

When forming ions, the 4s electrons are **lost before** the 3d electrons.

Note that this is different from the order of **filling** – 4s before 3d.

Once the 3d sub-shell starts to fill, the 4s electrons are repelled to a higher energy level and are lost first.

Forming Fe^{2+} and Fe^{3+} ions from iron

Iron has the electronic configuration [Ar]3d^{6}4s^2

Iron forms two common ions, Fe^{2+} and Fe^{3+}.

An Fe atom loses **two 4s electrons** $\longrightarrow$ Fe^{2+} ion.

An Fe atom loses **two 4s electrons** and **one 3d electron** $\longrightarrow$ Fe^{3+} ion.

The variety of oxidation states

The table below summarises the stable oxidation states of the elements scandium to zinc. Those in bold type represent the commonest oxidation numbers of the elements in their compounds.

Sc	Ti	V	Cr	Mn	Fe	Co	Ni	Cu	Zn
	+1	+1	+1	+1	+1	+1	+1	+1	
	+2	+2	+2	**+2**	**+2**	**+2**	**+2**	**+2**	**+2**
+3	+3	+3	**+3**	+3	**+3**	+3	+3	+3	
	+4	+4	+4	+4	+4	+4	+4		
		+5	+5	+5	+5	+5			
			+6	+6	+6				
				+7					

Refer to the electronic configurations of the d-block elements, shown on p. 70.

- The variety of oxidation states results from the close similarity in energy of the 4s and the 3d electrons.
- The number of available oxidation states for the element **increases** from Sc $\rightarrow$ Mn.
  - All of the available 3d and 4s electrons may be used for bond formation.
- The number of available oxidation states for the element **decreases** from Mn $\rightarrow$ Zn.
  - There is a decreasing number of unpaired d electrons available for bond formation.
- Higher oxidation states involve covalency because of the high charge densities involved.
  - In practice, for oxidation states of +4 and above, the bonding electrons are involved in covalent bond formation.

Progress check

1 Write down the electronic configuration of the following ions:
(a) Ni^{2+} (b) Cr^{3+} (c) Cu$^+$ (d) V^{3+}.

2 What is the oxidation state of the metal in each of the following:
(a) MnO$_2$ (b) CrO$_3$ (c) MnO$_4^-$ (d) VO$_2^+$?

2 (a) +4 (b) +6 (c) +7 (d) +5.

1 (a) [Ar]3d^8 (b) [Ar]3d^3 (c) [Ar]3d^{10} (d) [Ar]3d^2.

4.4 Transition element complexes

After studying this section you should be able to:

- *explain what is meant by the terms 'complex ion' and 'ligand'*
- *predict the formula and possible shape of a complex ion*
- *know that ligands can be unidentate, bidentate and multidentate*
- *know that transition metal ions can be identified by their colour*
- *know that electronic transitions are responsible for colour*
- *explain how colorimetry can be used to determine the concentration and formula of a complex ion*

LEARNING SUMMARY

Ligands and complex ions

AQA	M5	SALTERS	M4
EDEXCEL	M5	WJEC	CH5
OCR	M5.1	NICCEA	M5
NUFFIELD	M6		

Transition metal ions are small and densely charged. They strongly attract electron-rich species called **ligands** forming **complex ions**.

A ligand has a lone pair of electrons.

> **KEY POINT**
>
> A ligand is a molecule or ion that bonds to a metal ion by:
> - forming a coordinate (dative covalent) bond
> - donating a lone pair of electrons into a vacant d-orbital.
>
> Common ligands include: $H_2O\colon$, $\colon Cl^-$, $\colon NH_3$, $\colon CN^-$

> **KEY POINT**
>
> A **complex ion** is a central metal ion surrounded by ligands.
>
> The **coordination number** is the total **number** of **coordinate bonds** from ligands to the central transition metal ion of a complex ion.

Key points from AS

- **Electron pair repulsion theory**
  Revise AS pages 50–51

For AS, you learnt that electron pair repulsion is responsible for the shape of a molecule or ion.

Shapes of complex ions

An important factor in deciding the geometry of the complex ion is the size of the ligand.

Six coordinate complexes

Six water molecules are able to fit around a Cu^{2+} ion to form:
- the complex ion $[Cu(H_2O)_6]^{2+}$
- with a coordination number of **6**.

The **six** electron pairs surrounding the central Cu^{2+} ion in $[Cu(H_2O)_6]^{2+}$ repel one another as far apart as possible forming a complex ion with an **octahedral** shape.

The hexaaquacopper(II) complex ion, $[Cu(H_2O)_6]^{2+}$

The size of a ligand helps to decide the geometry of the complex ion.

$[Cu(H_2O)_6]^{2+}$
octahedral
6 coordinate

$$\left[\begin{array}{c} H_2O \\ H_2O\cdots \overset{\displaystyle 90°}{\underset{H_2O}{\overset{\displaystyle |}{Cu}}}\overset{OH_2}{\underset{OH_2}{}} \\ H_2O \end{array}\right]^{2+}$$

Four coordinate complexes

Chloride ions are larger than water molecules and it is only possible for **four** chloride ions to fit around the central copper(II) ion to form:
- the complex ion $[CuCl_4]^{2-}$
- with a coordination number of **4**.

The **four** electron pairs surrounding the central Cu^{2+} ion in $[CuCl_4]^{2-}$ repel one another as far apart as possible to form a complex ion with a **tetrahedral** shape.

The tetrachlorocuprate(II) complex ion [CuCl₄]²⁻

Notice the overall charge on the complex ion [CuCl₄]²⁻ is 2−.
Cu²⁺ and 4 Cl⁻ → 2− ions.

$[CuCl_4]^{2-}$
tetrahedral
4 coordinate

$$\begin{bmatrix} & Cl & \\ & | & 109.5° \\ & Cu & \\ Cl & & Cl \\ & Cl & \end{bmatrix}^{2-}$$

General rules for deciding the shape of a complex ion

Although there are exceptions, the following general rules are useful.

> **KEY POINT**
>
> Complex ions with small ligands such as H_2O and NH_3 are usually **6-coordinate** and **octahedral**.
>
> Complex ions with large Cl⁻ ligands are usually **4-coordinate** and **tetrahedral**.

The table below compares complex ions formed from cobalt(II) and iron(III) ions.

ligand	complex ions from Co²⁺	complex ions from Fe³⁺	shape
H_2O	$[Co(H_2O)_6]^{2+}$	$[Fe(H_2O)_6]^{3+}$	octahedral
NH_3	$[Co(NH_3)_6]^{2+}$	$[Fe(NH_3)_6]^{3+}$	octahedral
Cl⁻	$[CoCl_4]^{2-}$	$[FeCl_4]^{-}$	tetrahedral

- Some complex ions contain more than one type of ligand. For example, copper(II) forms a complex ion with a mixture of water and ammonia ligands, $[Cu(NH_3)_4(H_2O)_2]^{2+}$.

- Ag⁺ forms linear complexes that are 2-coordinate:
  e.g. $[Ag(H_2O)_2]^+$, $[Ag(NH_3)_2]^+$ and $[AgCl_2]^-$.

Ligands have teeth

AQA	M5	SALTERS	M4
OCR	M5.6	NICCEA	M5
NUFFIELD	M6		

Ligands such as H_2O, NH_3 and Cl⁻ are called **unidentate** ligands. They form only **one** coordinate bond to the central metal ion.

A molecule or ion with more than one oxygen or nitrogen atom may form more than one coordinate bond to the central metal ion.

Monodentate means **one** tooth.

Bidentate ligands have **two** teeth.

Multidentate ligands, **many** teeth.

Each 'tooth' is a coordinate bond.

Ligands that can form **two** coordinate bonds to the central metal ion are called **bidentate** ligands. Examples of bidentate ligands are ethane-1,2-diamine, $NH_2CH_2CH_2NH_2$ and the ethanedioate ion, $(COO^-)_2$.

Multidentate ligands can form **many** coordinate bonds to the central metal ion. For example, the **hexa**dentate ligand, edta⁴⁻, is able to form **six** coordinate bonds to the central metal ion. The diagram of edta⁴⁻ shows four oxygen atoms and two nitrogen atoms able to form coordinate bonds.

edta⁴⁻

Ligands as Lewis bases

| AQA | M5 |

A Lewis acid is an electron pair acceptor

A Lewis base is an electron pair donor

A ligand behaves as:
- a **Lewis base** by **donating** an electron pair to the central metal ion of a complex ion.

The transition metal ion acts as:
- a **Lewis acid** by **accepting** the electron pair from the ligand.

These definitions of an acid and a base extend acid–base chemistry to reactions which do not involve transfer of protons.

Colour and complex ions

AQA	M5	SALTERS	M4
EDEXCEL	M5	WJEC	CH5
OCR	M5.1	NICCEA	M5
NUFFIELD	M6		

Complex ions have different colours

Many transition metal ions have different colours. These colours indicate the possible identity of a transition metal ion. The table below shows the colours of some **aqua** complex ions of d-block elements. Notice that different oxidation states of a d-block element can have different colours.

Sc	Ti	V	Cr	Mn	Fe	Co	Ni	Cu	Zn
								+1 colourless	
				+2 pale pink	+2 pale green	+2 pink	+2 green	+2 blue	+2 colourless
+3 colourless	+3 violet	+3 blue	+3 ruby		+3 red-brown				
		+5 yellow							
			+6 yellow or orange						
				+7 purple					

Why are transition element ions coloured?

| AQA | M5 | OCR | M5.6 |
| EDEXCEL | M5 | WJEC | CH5 |

Splitting the 3d energy level

All five d orbitals in the outer d sub-shell of an **isolated** transition element ion have the same energy. However, when ligands bond to the central ion, the d orbitals move to two different energy levels. The gap between the energy levels depends upon:

• the ligand
• the coordination number
• the transition metal ion.

This diagram shows how water ligands split the energy levels of the five 3d orbitals in a copper(II) ion.

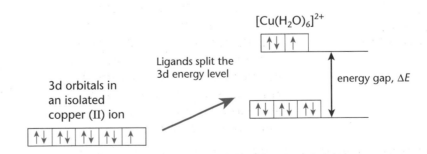

Absorption of light energy

An electron can be promoted from the lower 3d energy level by absorbing energy **exactly** equal to the energy gap ΔE. This energy is provided by radiation from the visible and ultraviolet regions of the spectrum.

Energy and frequency

The difference in energy between the d-orbitals, ΔE, and the frequency of the absorbed radiation, f, are linked by the following relationship:

$\Delta E = hf$

h = Planck's constant, 6.63×10^{-34} J s.

Radiation with a frequency f has an energy hf, where h = Planck's constant, 6.63×10^{-34} J s.

So to promote an electron through as energy gap ΔE, a quantity of light energy is needed given by:

$$\Delta E = hf$$

The diagram below shows what happens when a complex ion absorbs light energy.

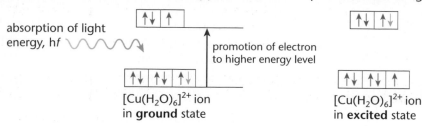

absorption of light energy, hf

promotion of electron to higher energy level

$[Cu(H_2O)_6]^{2+}$ ion in **ground** state

$[Cu(H_2O)_6]^{2+}$ ion in **excited** state

Only those ions that have a partially filled d subshell are coloured.
Cu^{2+} is $[Ar]3d^9 4s^2$

The blue colour of $[Cu(H_2O)_6]^{2+}$ results from:

- **absorption** of light energy in the red, yellow and green regions of the spectrum. This absorbed radiation provides the energy for an electron to be excited to a higher energy level
- **reflection** of blue light only, giving the copper(II) hexaaqua ion its characteristic blue colour. The blue region of the spectrum is **not** absorbed.

Copper has another ion, Cu^+, with the electronic configuration: $[Ar]3d^{10}$.

The Cu^+ ion is colourless because:

- the 3d sub-shell is full, $[Ar]3d^{10}$, preventing electron transfer.

Ag^+ complexes are also colourless because the 3d sub-shell is full.

> **KEY POINT**
> The colour of a transition metal complex ion results from the **transfer** of an **electron** between the orbitals of an **unfilled** d sub-shell.

Using colorimetry

AQA	M5	SALTERS	M4
OCR	M5.1	WJEC	CH5
NUFFIELD	M4, M6		

A colorimeter measures the amount of light absorbed by a coloured solution. The concentration and formula of a complex ion can be determined from the intensity of light absorbed by the colorimeter.

Finding an unknown concentration

A calibration curve for the colorimeter is first plotted using solutions of a complex ion of known concentration.

The solution of the complex ion with an unknown concentration is then placed in the colorimeter and the absorbance measured. The concentration can be simply determined from the calibration curve.

Progress check

1. Predict the formula of the following complex ions:
   (a) iron(III) with water ligands
   (b) vanadium(III) with chloride ligands
   (c) nickel(II) with ammonia ligands.

2. Explain the origin of colour in a complex ion that is yellow.

1 (a) $[Fe(H_2O)_6]^{3+}$ (b) $[VCl_4]^-$ (c) $[Ni(NH_3)_6]^{2+}$.

2 The 3d orbitals are split between two energy levels. A 3d electron from the lower 3d energy level absorbs visible light in the blue and red regions of the spectrum. The absorbed energy allows the d electron to be promoted to a vacancy in the orbitals of the higher 3d energy level. With the red and blue regions of the spectrum absorbed, only yellow light is reflected, giving the complex ion its colour.

4.5 Ligand substitution of complex ions

After studying this section you should be able to:

- *describe what is meant by ligand substitution*
- *understand that ligand exchange may produce changes in colour and coordination number*
- *understand stability of complex ions in terms of stability constants*

LEARNING SUMMARY

Exchange between ligands

AQA	M5	SALTERS	M4
EDEXCEL	M5	WJEC	CH5
OCR	M5.1	NICCEA	M5
NUFFIELD	M6		

Ligand substitution usually produces a change in colour.

A **ligand substitution** reaction takes place when a ligand in a complex ion exchanges for another ligand. A change in ligand usually changes the energy gap between the 3d energy levels. With a different ΔE value, light with a different frequency is absorbed, producing a colour change.

Exchange between H_2O and NH_3 ligands

The **similar sizes** of water and ammonia molecules ensure that ligand exchange takes place with **no change in coordination number**.

For example, addition of an excess of concentrated aqueous ammonia to aqueous nickel(II) ions results in ligand substitution of ammonia ligands for water ligands. Both complex ions have an octahedral shape with a coordination number of six.

Remember that the **size** of the ligand is a major factor in deciding the coordination number (see p. 73):

- H_2O and NH_3 have similar sizes – same coordination number
- H_2O and Cl^- have different sizes – different coordination numbers.

octahedral 6 coordinate — **same** coordinate number → octahedral 6 coordinate

green solution — blue solution

$$[Ni(H_2O)_6]^{2+} + 6NH_3 \longrightarrow [Ni(NH_3)_6]^{2+} + 6H_2O$$

Exchange between H_2O and Cl^- ligands

The **different sizes** of water molecules and chloride ions ensure that ligand exchange takes place **with a change in coordination number**.

For example, addition of an excess of concentrated hydrochloric acid (as a source of Cl^- ions) to aqueous cobalt(II) ions results in ligand substitution of chloride ligands for water ligands. The complex ions have different shapes with different coordination numbers.

You should also be able to construct a balanced equation for any ligand exchange reaction.

octahedral 6 coordinate — **change** in coordinate number → tetrahedral 4 coordinate

pink solution — blue solution

$$[Co(H_2O)_6]^{2+} + 4Cl^- \longrightarrow [CoCl_4]^{2-} + 6H_2O$$

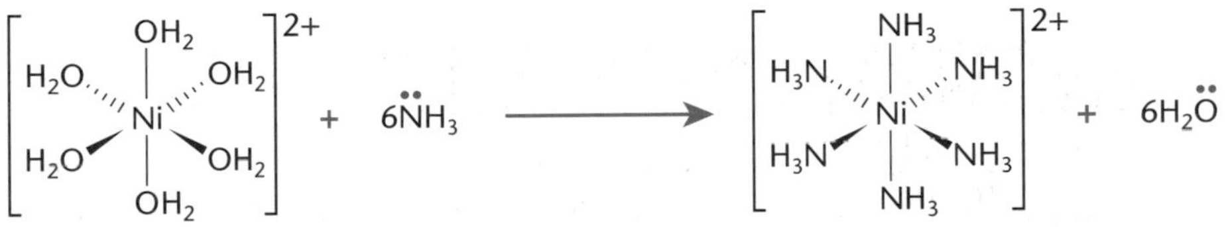

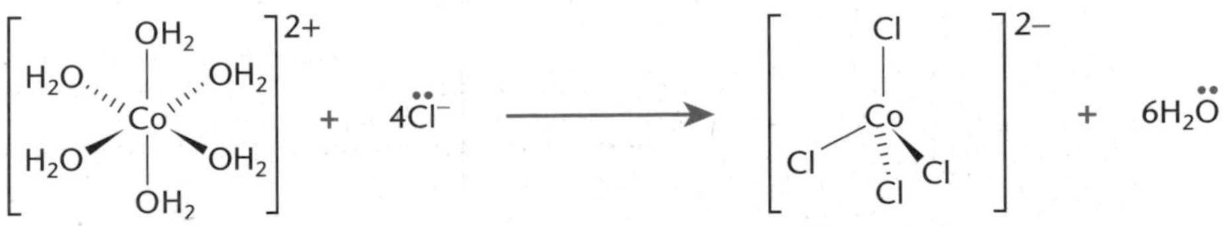

Ligand substitutions of copper(II) and cobalt(II)

Ligand substitutions of copper(II) and cobalt(II) complexes are shown below.

Concentrated hydrochloric acid is used as a source of Cl^- ligands.

Concentrated aqueous ammonia is used as a source of NH_3 ligands.

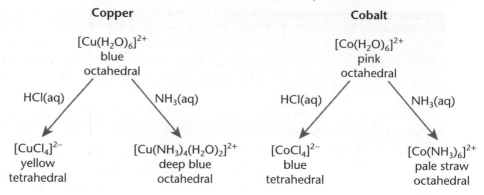

Copper

$[Cu(H_2O)_6]^{2+}$
blue
octahedral

HCl(aq) / NH₃(aq)

$[CuCl_4]^{2-}$
yellow
tetrahedral

$[Cu(NH_3)_4(H_2O)_2]^{2+}$
deep blue
octahedral

Cobalt

$[Co(H_2O)_6]^{2+}$
pink
octahedral

HCl(aq) / NH₃(aq)

$[CoCl_4]^{2-}$
blue
tetrahedral

$[Co(NH_3)_6]^{2+}$
pale straw
octahedral

Incomplete substitution

Substitution may be incomplete.

This reaction is used as a test for the Fe^{3+} ion – the deep red-brown colour of $[Fe(H_2O)_5(SCN)]^{2+}$ is intense and extremely small traces of Fe^{3+} ions can be detected using this ligand substitution reaction.

- Aqueous ammonia only exchanges **four** from the six water ligands in $[Cu(H_2O)_6]^{2+}$.

$$[Cu(H_2O)_6]^{2+} + 4NH_3 \longrightarrow [Cu(NH_3)_4(H_2O)_2]^{2+} + 4H_2O$$
blue solution   deep blue solution

- Aqueous thiocyanate ions, SCN^-, only exchanges **one** from the six water ligands in $[Fe(H_2O)_6]^{3+}$.

$$[Fe(H_2O)_6]^{3+} + SCN^-(aq) \longrightarrow [Fe(H_2O)_5(SCN)]^{2+} + H_2O$$
pale brown solution   deep red solution

Stability of complex ions

NUFFIELD M6   NICCEA M5
SALTERS M4

The stability of a complex ion depends upon the ligands. For example, a complex ion of copper(II) is more stable with ligands of ammonia than with water ligands.

The concentration of water is virtually constant. Constant terms are not included in expressions for equilibrium constants.

The stability of a complex ion is measured as a **stability constant**, K_{stab}:

$$[Cu(H_2O)_6]^{2+} + 4NH_3(aq) \rightleftharpoons [Cu(NH_3)_4(H_2O)_2]^{2+} + 4H_2O(l)$$

$$K_{stab} = \frac{[Cu(NH_3)_4(H_2O)_2]^{2+}}{[Cu(H_2O)_6]^{2+}[NH_3(aq)]^4} = 1 \times 10^{12} \text{ dm}^{12} \text{ mol}^{-4} \text{ at } 25°C$$

Stability constants usually measure the stability of a complex compared with the metal aqua-ion. The greater the value of K_{stab}, the more stable the complex ion.

The data is expressed on a logarithmic scale because of the very large range of values.

complex ion	log K_{stab}	
$[Fe(H_2O)_5(SCN)]^{2+}$	3.9	
$[CuCl_4]^{2-}$	5.6	increased
$[Ni(NH_3)_6]^{2+}$	8.6	stability
$[Ni(edta)]^{2-}$	19.3	

Progress check

1 Write equations for the following ligand substitutions:
   (a) $[Ni(H_2O)_6]^{2+}$ with six ammonia ligands
   (b) $[Fe(H_2O)_6]^{3+}$ with four chloride ligands
   (c) $[Cr(H_2O)_6]^{3+}$ with six cyanide ligands, CN^-.

1 (a) $[Ni(H_2O)_6]^{2+} + 6NH_3 \longrightarrow [Ni(NH_3)_6]^{2+} + 6H_2O$
   (b) $[Fe(H_2O)_6]^{3+} + 4Cl^- \longrightarrow [FeCl_4]^- + 6H_2O$
   (c) $[Cr(H_2O)_6]^{3+} + 6CN^- \longrightarrow [Cr(CN)_6]^{3-} + 6H_2O$.

4.6 Redox reactions of transition metal ions

After studying this section you should be able to:

- describe redox behaviour in transition elements
- construct redox equations, using relevant half-equations
- perform calculations involving simple redox titrations

LEARNING SUMMARY

Redox reactions

AQA	M5	SALTERS	M4
EDEXCEL	M5	WJEC	CH5
OCR	M5.1	NICCEA	M5
NUFFIELD	M6		

Transition elements have variable oxidation states with characteristic colours (see page 75). Many **redox** reactions take place in which transition metals ions change their oxidation state by **gaining** or **losing** electrons.

The iron(II) – manganate(VII) reaction

Iron(II) ions are oxidised by acidified manganate(VII) ions.

- Acidified manganate(VII) ions are reduced to manganate(II) ions.

Key points from AS

- **Redox reactions**
  Revise AS pages 68–70

reduction: $MnO_4^-(aq) + 8H^+(aq) + 5e^- \longrightarrow Mn^{2+}(aq) + 4H_2O(l)$
 +7 $\phantom{\longrightarrow\ Mn^{2+}}$ +2
 purple $\longrightarrow$ very pale pink

The reaction is easy to see.

The deep purple MnO_4^- ions are reduced to the very pale pink Mn^{2+} ions, which are virtually colourless in solution.

- Iron(II) ions are oxidised to iron(III) ions.

oxidation: $Fe^{2+}(aq) \longrightarrow Fe^{3+}(aq) + e^-$
$\phantom{oxidation:\ Fe^{2+}}$ +2 +3

To give the overall equation:

- the electrons are balanced

$$MnO_4^-(aq) + 8H^+(aq) + 5e^- \longrightarrow Mn^{2+}(aq) + 4H_2O(l)$$
$$5Fe^{2+}(aq) \longrightarrow 5Fe^{3+}(aq) + 5e^-$$

- the half-equations are added

$$5Fe^{2+}(aq) + MnO_4^-(aq) + 8H^+(aq) \longrightarrow 5Fe^{3+}(aq) + Mn^{2+}(aq) + 4H_2O(l)$$

Redox titrations

AQA	M5	SALTERS	M4
EDEXCEL	M5	WJEC	CH5
OCR	M5.1	NICCEA	M5
NUFFIELD	M6		

Redox titrations can be used in analysis.

Essential requirements are:

- an oxidising agent
- a reducing agent
- an easily-seen colour change (with or without an indicator).

It is important to use an acid that does **not** react with either the oxidising or reducing agent – dilute sulphuric acid is usually used.

For example, hydrochloric acid cannot be used because HC is oxidised to Cl_2 by MnO_4^-.

The commonest redox titration encountered at A Level is that between manganate(VII) and acidified iron(II) ions.

Redox titrations using the acidified manganate(VII)

- $KMnO_4$ is reacted with a reducing agent such as Fe^{2+} under acidic conditions (see above).
- Purple $KMnO_4$ is added from the burette to a measured amount of the acidified reducing agent.
- At the end-point, the colour changes from colourless to pale pink, showing that all the reducing agent has exactly reacted. The pale-pink colour indicates the first trace of an excess of purple manganate(VII) ions.
- This titration is **self-indicating** – no indicator is required, the colour change from the reduction of MnO_4^- ions being sufficient to indicate that reaction is complete.

Key points from AS

• Calculations in acid–base titrations
  Revise AS page 40

Calculations for redox titrations follow the same principles as those used for acid–base titrations, studied during AS Chemistry.

The concentration and volume of $KMnO_4$ are known so the amount of MnO_4^- ions can be found.

Using the equation determine the number of moles of the second reagent.

If there is sampling of the original solution, you will need to scale.

Example

Five iron tablets with a combined mass of 0.900 g were dissolved in acid and made up to 100 cm³ of solution. In a titration, 10.0 cm³ of this solution reacted exactly with 10.4 cm³ of 0.0100 mol dm⁻³ potassium manganate(VII). What is the percentage by mass of iron in the tablets?

As with all titrations, we must consider **five** pieces of information:

• the balanced equation
  $$5Fe^{2+}(aq) + MnO_4^-(aq) + 8H^+(aq) \rightarrow 5Fe^{3+}(aq) + Mn^{2+}(aq) + 4H_2O(l)$$

• the concentration c_1 and reacting volume V_1 of $KMnO_4(aq)$

• the concentration c_2 and reacting volume V_2 of $Fe^{2+}(aq)$.

From the titration results, the amount of $KMnO_4$ can be calculated:

$$\text{amount of } KMnO_4 = c \times \frac{V}{1000} = 0.0100 \times \frac{10.4}{1000} = 1.04 \times 10^{-4} \text{ mol}$$

From the equation, the amount of Fe^{2+} can be determined:

$$5Fe^{2+}(aq) + MnO_4^-(aq) + 8H^+(aq) \rightarrow 5Fe^{3+}(aq) + Mn^{2+}(aq) + 4H_2O(l)$$

5 mol   1 mol *(balancing numbers)*

∴ 5 x 1.04×10^{-4} mol Fe^{2+} reacts with 1.04×10^{-4} mol MnO_4^-

∴ amount of Fe^{2+} that reacted = 5.20×10^{-4} mol

Find the amount of Fe^{2+} in the solution prepared from the tablets:

10.0 cm³ of $Fe^{2+}(aq)$ contains 5.20×10^{-4} mol Fe^{2+} (aq)
The 100 cm³ solution of iron tablets contains $10 \times (5.20 \times 10^{-4})$
$$= 5.20 \times 10^{-3} \text{ mol } Fe^{2+}$$

Find the percentage of Fe^{2+} in the tablets (A$_r$: Fe, 55.8)

5.20×10^{-3} mol Fe^{2+} has a mass of $5.20 \times 10^{-3} \times 55.8 = 0.290$ g

$$\therefore \% \text{ of } Fe^{2+} \text{ in tablets} = \frac{\text{mass of } Fe^{2+}}{\text{mass of tablets}} \times 100 = \frac{0.290}{0.900} \times 100 = 32.2\%$$

Further redox titrations

Providing there is a visible colour change, many other redox reactions can be used for redox titrations.

For example, acidified dichromate(VI) ions is an oxidising agent which can also oxidise Fe^{2+} ions to Fe^{3+}.

reduction: $Cr_2O_7^{2-}(aq) + 14H^+(aq) + 6e^- \longrightarrow 2Cr^{3+}(aq) + 7H_2O(l)$
oxidation: $Fe^{2+}(aq) \longrightarrow Fe^{3+}(aq) + e^-$
overall: $6Fe^{2+}(aq) + Cr_2O_7^{2-}(aq) + 14H^+(aq) \rightarrow 6Fe^{3+}(aq) + 2Cr^{3+}(aq) + 7H_2O(l)$

Further examples of redox reactions

AQA	M5	OCR	M5.6	
EDEXCEL	M5	WJEC	M5	

Vanadium

Vanadium(V) can be reduced to vanadium(II) using zinc in acid solution as the reducing agent. The formation of different oxidation states of vanadium can be followed by colour changes.

oxidation state	+2	+3	+4	+5
species	V^{2+}	V^{3+}	VO^{2+}	VO_2^+
	(purple)	(green)	(blue)	(yellow)

Zn/HCl ←

Chromium

The oxidation states of chromium can be interchanged using:

- hydrogen peroxide in alkaline solution as the oxidising agent
- zinc in acid solution as the reducing agent.

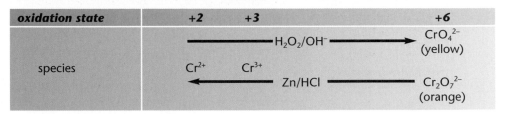

oxidation state	+2	+3		+6
species	Cr^{2+}	Cr^{3+}	⟵———— H_2O_2/OH^- ————⟶	CrO_4^{2-} (yellow)
			⟵———— Zn/HCl —————	$Cr_2O_7^{2-}$ (orange)

The dichromate(VI) – chromate(VI) conversion

The table above shows that the dichromate(VI) ion $Cr_2O_7^{2-}$ and the chromate(VI) ion CrO_4^{2-} have the **same** oxidation state. These ions can be interconverted at different pH values.

$$2H^+(aq) + 2CrO_4^{2-}(aq) \underset{alkali}{\overset{acid}{\rightleftharpoons}} Cr_2O_7^{2-}(aq) + H_2O(l)$$
$$\text{yellow} \qquad\qquad \text{orange}$$

This conversion can be explained by applying le Chatelier's principle.

Addition of acid increases $[H^+(aq)]$ – the equilibrium moves to the right, counteracting this change.

Added alkali reacts with $H^+(aq)$ ions present in the equilibrium system. $[H^+(aq)]$ decreases – the equilibrium moves to the left, counteracting this change.

Cobalt

Cobalt(II) can be oxidised to Co(III) using:

- hydrogen peroxide in alkaline solution or
- air in the presence of ammonia as an oxidising agent.

oxidation state	+2		+3
species	$[Co(NH_3)_6]^{2+}$ (light straw)	H_2O_2/OH^- *or* NH_3/air ⟶	$[Co(NH_3)_6]^{3+}$ (brown)

Useful oxidising and reducing agents

> **KEY POINT**
>
> For moving **up** oxidation states:
> - H_2O_2/OH^- is a good general oxidising agent.
>
> For moving **down** oxidation states:
> - Zn/HCl is a good general reducing agent.

Progress check

1 Write a full equation from the following pairs of half-equations. For each pair, identify the changes in oxidation number, what has been oxidised and what has been reduced.
  (a) $Zn \longrightarrow Zn^{2+} + 2e^-$
    $VO_2^+ + 4H^+ + 3e^- \longrightarrow V^{2+} + 2H_2O$
  (b) $Cr^{3+} + 8OH^- \longrightarrow CrO_4^{2-} + 4H_2O + 3e^-$
    $H_2O_2 + 2e^- \longrightarrow 2OH^-$

2 In a redox titration, 25.0 cm³ of an acidified solution containing $Fe^{2+}(aq)$ ions reacted exactly with 21.8 cm³ of 0.0200 mol dm⁻³ potassium manganate(VII). Calculate the concentration of $Fe^{2+}(aq)$ ions.

2 0.0872 mol dm⁻³

Cr: +3 → +6; O: –1 → –2; Cr^{3+} oxidised, H_2O_2 reduced.
(b) $2Cr^{3+} + 3H_2O_2 + 10OH^- \longrightarrow 2CrO_4^{2-} + 8H_2O$
V: +5 → +2; Zn: 0 → +2; VO_2^+ reduced, Zn oxidised.
1 (a) $2VO_2^+ + 3Zn + 8H^+ \longrightarrow 2V^{2+} + 3Zn^{2+} + 4H_2O$

4.7 Catalysis

After studying this section you should be able to:

- *understand how a transition element acts as a catalyst*
- *explain how a catalyst acts in heterogeneous catalysis*
- *explain how a catalyst acts in homogeneous catalysis*

LEARNING SUMMARY

Key points from AS

- **How do catalysts work?**
  Revise AS page 102
- **Heterogeneous and homogeneous catalysis**
  Revise AS pages 102–103

The two different classes of catalyst, homogeneous and heterogeneous, were discussed in detail during AS Chemistry. Both types are used industrially but heterogeneous catalysis is much more common. The AS coverage is summarised below and expanded to include extra examples of transition element catalysis.

Transition elements as catalysts

AQA	M5	SALTERS	M4
EDEXCEL	M5	WJEC	CH5
OCR	M5.1	NICCEA	M5
NUFFIELD	M6		

An important use of transition metals and their compounds is as catalysts for many industrial processes. Nickel and platinum are extensively used in the petroleum and polymer industries.

> Transition metal ions are able to act as catalysts by changing their oxidation states. This is made possible by using the partially full d orbitals for gaining or losing electrons.

KEY POINT

Heterogeneous catalysis

AQA	M5	SALTERS	M4
EDEXCEL	M5	WJEC	CH5
OCR	M5.6	NICCEA	M5

A **hetero**geneous catalyst has a **different** phase from the reactants.

- Many examples of heterogeneous catalysts involve reactions between **gases**, catalysed by a **solid** catalyst which is often a transition metal or one of its compounds.

Key points from AS

- **Heterogeneous and homogeneous catalysis**
  Revise AS pages 102–103

The process involves

- **diffusion** of gas molecules onto the surface of the iron catalyst
- **adsorption** of the gases to the surface of the catalyst
- **weakening of bonds**, allowing a chemical reaction to take place
- **diffusion** of the product molecules from the surface of the catalyst, allowing more gas molecules to diffuse onto its surface.

Choice of a heterogeneous catalyst (AQA only)

A key requirement of a heterogeneous catalyst is to bind reactant molecules to the surface of the catalyst by adsorption.

The strength of this adsorption is critical.

Catalysts can become coated by impurities and can become 'poisoned' – efficiency is reduced as active sites are blocked.

- Tungsten, W, bonds very strongly to reactant molecules. The adsorption is so strong that molecules are unable to leave the surface and the catalyst is poisoned and rendered useless.
- Silver, Ag, bonds too weakly to reactant molecules to allow a reaction to take place.

Nickel and platinum are commonly used as heterogeneous catalysts:
- the bonding of reactant molecules is sufficient to allow a reaction to take place but
- weak enough to allow molecules to escape following reaction.

Transition metal ions as heterogeneous catalysts

The catalytic converter

- Rh/Pt/Pd catalysts are used in catalytic converters for the removal of polluting gases, such as CO and NO, produced in a car engine.
- The **solid** Rh/Pt/Pd catalyst is supported on a ceramic honeycomb, giving a large surface area for the catalyst which can be spread extremely thinly on the support. The high surface area of catalyst increases the chances of a reaction and less of the expensive catalysts are needed, keeping down costs.

Iron

Iron-containing catalysts are used in many industrial processes, the most important being to form ammonia from nitrogen and hydrogen in the Haber process.

Vanadium

A vanadium(V) oxide, V_2O_5, catalyst is used in the Contact process for sulphur trioxide production from which sulphuric acid is manufactured.

This catalysis proceeds via an **intermediate state**.

- The vanadium(V) oxide catalyst first oxidises sulphur dioxide to sulphur trioxide. The vanadium is **reduced** from the +5 to +4 oxidation state, forming vanadium(IV) oxide $V_2O_4(s)$ as the **intermediate state**.

$$SO_2(g) + V_2O_5(s) \longrightarrow SO_3(g) + V_2O_4(s) \qquad \textit{vanadium reduced}$$

$$+4 \xrightarrow{\hspace{3cm}} +6$$
$$+5 \xrightarrow{\hspace{3cm}} +4$$
$$+5 \xrightarrow{\hspace{3cm}} +4$$

> Notice that the oxidation number of each atom has been shown. Two V atoms have been reduced from +5 to +4.

- The intermediate V_2O_4 is then oxidised back to the +5 oxidation state by oxygen, forming vanadium(V) oxide.

$$\tfrac{1}{2}O_2(g) + V_2O_4(s) \longrightarrow V_2O_5(s) \qquad \textit{vanadium oxidised}$$

$$0 \xrightarrow{\hspace{3cm}} -2$$
$$+4 \xrightarrow{\hspace{2cm}} +5$$
$$+4 \xrightarrow{\hspace{2cm}} +5$$

This is a chain reaction because the vanadium(V) oxide is then able to oxidise further sulphur dioxide.

Note

- Although V_2O_5 has been involved in the mechanism, the overall reaction does not include V_2O_5.
  - V_2O_5 is unchanged at the end of the reaction.
- The equation for the overall reaction taking place is:
  $$SO_2(g) + \tfrac{1}{2}O_2(g) \rightleftharpoons SO_3(g) \qquad \textit{The Contact process}$$
- The vanadium is able to catalyse the reaction by interchanging its +5 and +4 oxidation states.

Homogeneous catalysis

AQA	M5	WJEC	CH5
EDEXCEL	M5	NICCEA	M5
NUFFIELD	M6		

Key points from AS

- **How do catalysts work**
  Revise AS page 102

A **homogeneous** catalyst has the **same** phase as the reactants.

- During homogeneous catalysis, the catalyst initially reacts with the reactants forming an **intermediate state** with **lower activation energy**.
- The catalyst is then regenerated as the reaction completes, allowing the catalyst to react with further reactants.
- Overall the catalyst is **not** used up, but it is actively involved in the process by a chain reaction.

Transition metal ions as homogeneous catalysts

The reaction between I^- and $S_2O_8^{2-}$, catalysed by $Fe^{2+}(aq)$ ions

The reaction between I^- and $S_2O_8^{2-}$ is very slow:

- it can be catalysed by Fe^{2+} ions
- the Fe^{2+} ions are recycled in a chain reaction.

chain reaction: $\quad S_2O_8^{2-}(aq) + 2Fe^{2+}(aq) \longrightarrow 2SO_4^{2-}(aq) + 2Fe^{3+}(aq)$

$\qquad\qquad\qquad 2Fe^{3+}(aq) + 2I^-(aq) \longrightarrow I_2(aq) + 2Fe^{2+}(aq)$

overall: $\qquad\qquad S_2O_8^{2-}(aq) + 2I^-(aq) \longrightarrow I_2(aq) + 2SO_4^{2-}(aq)$

The overall reaction between I^- and $S_2O_8^{2-}$ involves two negative ions. These negative ions will repel one another, making it difficult for a reaction to take place. Notice that, in each stage of the mechanism, the catalyst provides positive ions which attract, and react with, each negative ion in turn.

Autocatalysis by $Mn^{2+}(aq)$ ions

Notice that, as in the first example, this reaction is between two negative ions.

The mechanism of this reaction proceeds in a similar manner – the catalyst provides positive ions which react with each negative ion.

In warm acidic conditions, manganate(VII) ions, MnO_4^-, oxidise ethanedioate ions, $C_2O_4^{2-}$ to carbon dioxide.

$$2MnO_4^-(aq) + 5C_2O_4^{2-}(aq) + 8H^+(aq) \rightarrow 2Mn^{2+}(aq) + 10CO_2(g) + H_2O(l)$$

The purple manganate(VII) colour disappears very slowly at first, indicating a slow reaction. As soon as $Mn^{2+}(aq)$ ions start to form, the colour disappears immediately because the $Mn^{2+}(aq)$ ions catalyse the reaction. This is called **autocatalysis**.

Progress check

1 (a) State what is meant by homogeneous and heterogeneous catalysis.
(b) State an example of a homogeneous and heterogeneous catalyst.

2 How does a transition metal act as a catalyst?

2 Transition metal changes oxidation states by gaining or losing electrons from partially filled d orbitals.
Heterogeneous: iron catalysing the reaction between N_2 and H_2 for ammonia production (Haber process).
(b) Homogeneous: Fe^{2+} catalysing the reaction between I^- and $S_2O_8^{2-}$.
Heterogeneous: catalyst and reactants have different phases.
1 (a) Homogeneous: catalyst and reactants have the same phase.

4.8 Reactions of metal aqua-ions

After studying this section you should be able to:

- describe the precipitation reactions of metal aqua-ions with bases
- describe the acidity of transition metal aqua-ions
- describe the acid–base behaviour of metal hydroxides

Simple precipitation reactions of metal aqua-ions

AQA	M5	SALTERS	M4
EDEXCEL	M5	WJEC	CH5
OCR	M5.1	NICCEA	M5
NUFFIELD	M6		

Notice that NaOH(aq) and NH$_3$(aq) **both** act as bases ⟶ precipitate of the transition metal hydroxide.

A precipitation reaction takes place between **aqueous alkali** and an aqueous solution of a **metal(II)** or **metal(III) cation**.

This results in the formation of a **precipitate** of the **metal hydroxide**, often with a characteristic colour.

Suitable aqueous alkalis include aqueous sodium hydroxide, NaOH(aq), and aqueous ammonia, NH$_3$(aq).

> Aqueous ammonia provides OH$^-$(aq) ions.
> NH$_3$(aq) + H$_2$O(l) $\rightleftharpoons$ NH$_4^+$(aq) + OH$^-$(aq)

These precipitation reactions can be represented simply as follows.

$$Cu^{2+}(aq) + 2OH^-(aq) \longrightarrow Cu(OH)_2(s)$$
$$Cr^{3+}(aq) + 3OH^-(aq) \longrightarrow Cr(OH)_3(s)$$

> The hydroxides of all transition metals are insoluble in water.

The characteristic colour of the precipitate can help to identify the metal ion. The colours of some hydroxide precipitates are shown below.

hydroxide	Fe(OH)$_2$(s)	Fe(OH)$_3$(s)	Co(OH)$_2$(s)	Cu(OH)$_2$(s)	Cr(OH)$_3$(s)
colour	green	brown	turquoise	blue	green

Reaction of complex metal aqua-ions with aqueous alkali

Equations can be written using complex aqua-ions for the reactions above, each producing a **precipitate** of the **hydrated hydroxide**.

$$[Cu(H_2O)_6]^{2+}(aq) + 2OH^-(aq) \longrightarrow Cu(OH)_2(H_2O)_4(s) + 2H_2O(l)$$
blue precipitate

The precipitate has no charge.

$$[Cr(H_2O)_6]^{3+}(aq) + 3OH^-(aq) \longrightarrow Cr(OH)_3(H_2O)_3(s) + 3H_2O(l)$$
green precipitate

Excess aqueous sodium hydroxide

Metal(III) hydroxides **dissolve** in an **excess** of dilute aqueous sodium hydroxide forming charged **anions**.

$$Al(OH)_3(H_2O)_3(s) + OH^-(aq) \rightleftharpoons [Al(OH)_4]^-(aq) + 3H_2O(l)$$
aluminate ions

Species in solution are charged.

$$Cr(OH)_3(H_2O)_3(s) + 3OH^-(aq) \rightleftharpoons [Cr(OH)_6]^{3-}(aq) + 3H_2O(l)$$
chromate(III) ions

Excess aqueous ammonia

Excess ammonia usually results in **ligand exchange** forming a soluble ammine complex:

$$Cr(H_2O)_3(OH)_3(s) + 6NH_3(aq) \longrightarrow [Cr(NH_3)_6]^{3+}(aq) + 3H_2O(l) + 3OH^-(aq)$$
green solution

If aqueous ammonia is slowly added to an aqueous solution of a **metal(II)** or **metal(III) cation**:

- a precipitate of the metal hydroxide first forms: **acid–base reaction**
- the precipitate then dissolves in excess ammonia: **ligand substitution**.

Progress check

In exams, a great emphasis is placed on these reactions. You will need to learn the colours of all solutions and precipitates in your course.

1 Write equations for the following reactions of metal aqua-ions with aqueous alkali. What is the colour of each precipitate?
(a) $Co^{2+}(aq)$ (b) $Fe^{2+}(aq)$ (c) $[Ni(H_2O)_6]^{2+}$ (d) $[Fe(H_2O)_6]^{3+}$.

2 Write equations to show what happens when excess $NH_3(aq)$ is added to the following hydrated metal hydroxides.
(a) $Co(OH)_2(H_2O)_4$ (b) $Cu(OH)_2(H_2O)_4$ (c) $Cr(OH)_3(H_2O)_3$.

2 (a) $Co(OH)_2(H_2O)_4(s) + 6NH_3(aq) \longrightarrow [Co(NH_3)_6]^{2+}(aq) + 4H_2O(l) + 2OH^-(aq)$ (pale straw)
(b) $Cu(OH)_2(H_2O)_4(s) + 4NH_3(aq) \longrightarrow [Cu(NH_3)_4(H_2O)_2]^{2+}(aq) + 2H_2O(l) + 2OH^-(aq)$ (deep-blue)
(c) $Cr(OH)_3(H_2O)_3(s) + 6NH_3(aq) \longrightarrow [Cr(NH_3)_6]^{3+}(aq) + 3H_2O(l) + 3OH^-(aq)$ (green)

1 (a) $Co^{2+}(aq) + 2OH^-(aq) \longrightarrow Co(OH)_2(s)$ (blue-green)
(b) $Fe^{2+}(aq) + 2OH^-(aq) \longrightarrow Fe(OH)_2(s)$ (green)
(c) $[Ni(H_2O)_6]^{2+}(aq) + 2OH^-(aq) \longrightarrow Ni(OH)_2(H_2O)_4(s) + 2H_2O(l)$ (blue-green)
(d) $[Fe(H_2O)_6]^{3+}(aq) + 3OH^-(aq) \longrightarrow Fe(OH)_3(H_2O)_3(s) + 3H_2O(l)$ (brown)

Precipitation as acid–base equilibria

| AQA | M5 | WJEC | CH5 |
| EDEXCEL | M5 | NICCEA | M5 |

Fission of O–H bond

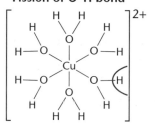

Metal(II) aqua-ions are weaker acids than metal(III) aqua-ions.

Examples of metal aqua-ions with 3+ cations include $[Al(H_2O)_6]^{3+}$, $[V(H_2O)_6]^{3+}$, $[Cr(H_2O)_6]^{3+}$ and $[Fe(H_2O)_6]^{3+}$.

Acidity of metal aqua-ions

Metal(II) and metal(III) aqua-ions behave as weak acids. This is achieved by:
- fission of the O–H bond in one of the water ligands of the complex ion
- donation of a proton to a water molecule.

Metal(II) cations

- With a metal(II) aqua-ion, the equilibrium lies well to the left-hand side: only a very weak acid is formed.

$$[Cu(H_2O)_6]^{2+}(aq) + H_2O(l) \rightleftharpoons [Cu(H_2O)_5(OH)]^+(aq) + H_3O^+(l)$$
very weakly acidic

Metal(III) cations

- Metal(III) aqua-ions are slightly more acidic than metal(II) aqua-ions: the equilibrium position has moved slightly to the right.

$$[Fe(H_2O)_6]^{3+}(aq) + H_2O(l) \rightleftharpoons [Fe(H_2O)_5(OH)]^{2+}(aq) + H_3O^+(aq)$$
weakly acidic

> The aqua-ions of metal(III) cations have a greater charge/size ratio and greater acidity than aqua-ions of metal(II) cations.   **KEY POINT**

Precipitation as acid–base equilibria

The precipitation reaction of a transition metal with hydroxide ions can be expressed as a series of acid–base equilibria.

The equilibrium set up in water by the hexaaqua ion of iron(III) is:

$$[Fe(H_2O)_6]^{3+}(aq) + H_2O(l) \rightleftharpoons [Fe(H_2O)_5(OH)]^{2+}(aq) + H_3O^+(aq)$$

OH^- reacts with H_3O^+:
$OH^- + H_3O^+ \longrightarrow 2H_2O$

On addition of aqueous hydroxide ions:
- H_3O^+ ions are removed from the equilibrium, which shifts to the right.
The $[Fe(H_2O)_5(OH)]^{2+}$ ion now sets up a second equilibrium:

$$[Fe(H_2O)_5(OH)]^{2+}(aq) + H_2O(l) \rightleftharpoons [Fe(H_2O)_4(OH)_2]^+(aq) + H_3O^+(aq)$$

With the weak alkali $Na_2CO_3(aq)$, metal(II) aqua-ions precipitate metal **carbonates**.

The more acidic metal(III) aqua-ions precipitate metal **hydroxides**.

Metal(III) carbonates do not exist.

- Again, the hydroxide ion removes H_3O^+ ions from the aqueous equilibrium, which shifts to the right.
- This process continues. Eventually **all charge** will be **removed** from the iron(III) complex. Then the uncharged **hydrated metal hydroxide** precipitates.

$$[Fe(H_2O)_4(OH)_2]^+(aq) + H_2O(l) \rightleftharpoons [Fe(H_2O)_3(OH)_3](s) + H_3O^+(aq)$$

Acid–base properties of metal hydroxides

| AQA | MS | WJEC | CH5 |
| EDEXCEL | M5 | NICCEA | M5 |

Many insoluble metal hydroxides are **amphoteric** – they can act as both an acid and a base.

Metal hydroxides as acids

See also:
'Reaction of complex metal aqua-ions with aqueous alkali' p. 85.

Some insoluble metal hydroxides can act as acids, donating protons to strong alkalis and forming soluble anions.

- Metal(III) hydroxides react with dilute aqueous sodium hydroxide.

$$Al(OH)_3(H_2O)_3(s) + OH^-(aq) \rightleftharpoons [Al(OH)_4]^-(aq) + 3H_2O(l)$$
aluminate ions

$$Cr(OH)_3(H_2O)_3(s) + 3OH^-(aq) \rightleftharpoons [Cr(OH)_6]^{3-}(aq) + 3H_2O(l)$$
chromate(III) ions

In general, metal(II) hydroxides form anions with the general formula $[M(OH)_4]^{-2}$.
Metal(III) hydroxides form anions with the general formula $[M(OH)_6]^{3-}$.

- Metal(II) hydroxides are weaker acids and react only with concentrated aqueous sodium hydroxide. E.g.

$$Cu(OH)_2(H_2O)_4(s) + 2OH^-(aq) \rightleftharpoons [Cu(OH)_4]^{2-}(aq) + 4H_2O(l)$$
deep blue cuprate(II) ions

Metal hydroxides as bases (AQA only)

All metal hydroxides can act as bases, removing protons from strong acids.

$$Cu(OH)_2(H_2O)_4(s) + 2H^+(aq) \rightleftharpoons [Cu(H_2O)_6]^{2+}(aq) + 2H_2O(l)$$
blue precipitate

This reaction reverses the direction of the acid-base equilibria that produce hydrated hydroxide precipitates (see p. 86)

$$Cr(OH)_3(H_2O)_3(s) + 3H^+(aq) \rightleftharpoons [Cr(H_2O)_6]^{3+}(aq) + 3H_2O(l)$$
green precipitate

Summary of reactions of metal aqua-ions with bases

AQA	M5	SALTERS	M4
EDEXCEL	M5	WJEC	CH5
OCR	M5.1	NICCEA	M5
NUFFIELD	M6		

	base				
	$OH^-(aq)$		$NH_3(aq)$		$CO_3^{2-}(aq)$
ion		**excess**		**excess**	
$[Fe(H_2O)_6]^{2+}$ green	$Fe(OH)_2(s)$ green	no change	$Fe(OH)_2(s)$ green	no change	$FeCO_3(s)$ white
$[Fe(H_2O)_6]^{3+}$ pale violet	$Fe(OH)_3(s)$ brown	no change	$Fe(OH)_3(s)$ brown	no change	$Fe(OH)_3(s)$ brown
$[Co(H_2O)_6]^{2+}$ pink	$Co(OH)_2(s)$ blue-green	$[Co(OH)_4]^{2-}(aq)$ deep blue	$Co(OH)_2(s)$ blue-green	$[Co(NH_3)_6]^{2+}(aq)$ pale straw	$CoCO_3(s)$ pink
$[Cu(H_2O)_6]^{2+}$ blue	$Cu(OH)_2(s)$ pale blue	$[Cu(OH)_4]^{2-}(aq)$ deep blue	$Cu(OH)_2(s)$ pale blue	$[Cu(NH_3)_4(H_2O)_2]^{2+}(aq)$ deep blue	$CuCO_3(s)$ green
$[Al(H_2O)_6]^{3+}$ colourless	$Al(OH)_3(s)$ white	$[Al(OH)_4]^-(aq)$ colourless	$Al(OH)_3(s)$ white	no change	$Al(OH)_3(s)$ white
$[Cr(H_2O)_6]^{3+}$ ruby	$Cr(OH)_3(s)$ green	$[Cr(OH)_6]^{3-}(aq)$ green	$Cr(OH)_3(s)$ green	$[Cr(NH_3)_6]^{3+}(aq)$ purple	$Cr(OH)_3(s)$ green

- The precipitate of $Fe(OH)_2(s)$ is oxidised in air forming a brown precipitate of $Fe(OH)_3(s)$.
- The complexion $[Co(NH_3)_6]^{2+}(aq)$ darkens in air as it is oxidised to $[Co(NH_3)_6]^{3+}(aq)$ (see page 81).

Sample question and model answer

(a) Scandium and zinc both appear in the d-block of the Periodic Table but their common compounds do not show the characteristic properties associated with transition element compounds.

Give the electronic configurations of each of these elements and deduce the oxidation state each has in its common compounds. [4]

Scandium [Ar] $3d^1 4s^2$ ✓   Zinc [Ar] $3d^{10} 4s^2$ ✓
Scandium has +3 oxidation state. ✓   Zinc has +2 oxidation state. ✓

(b) When concentrated hydrochloric acid is added to an aqueous solution of a cobalt(II) salt there is a change in colour. Account for this observation by giving the two cobalt species involved, their colours, their shapes and the co-ordination numbers of the cobalt they contain. Write an equation for the reaction and explain why the addition of dilute hydrochloric acid does not cause the same colour change.

This is standard bookwork but it also reinforces important principles:

H_2O forms octahedral complexes with a coordination number of 6

Cl^- forms tetrahedral complexes with a coordination number of 4.

Aqueous cobalt(II) contains $[Co(H_2O)_6]^{2+}$ ✓ which is pink. ✓ It has an octahedral shape ✓ and a coordination number of 6. ✓
After addition of concentrated hydrochloric acid, the complex ion $[CoCl_4]^{2-}$ ✓ is formed which is blue. ✓ It has an tetrahedral shape ✓ and a coordination number of 4. ✓

Equation: $[Co(H_2O)_6]^{2+} + 4Cl^- \longrightarrow CoCl_4^{2-} + 6H_2O$ ✓✓

The concentration of chloride ligands in dilute hydrochloric acid is much smaller than the concentration of water ligands. ✓ [11]

(c) A 0.570 g sample of hydrated iron(II) sulphate was dissolved in water. After acidification with dilute sulphuric acid, the solution was found to react exactly with 22.8 cm³ of 0.0180 M potassium manganate(VII) solution.

(i) Write half-equations for the oxidation of Fe^{2+} and the reduction of MnO_4^- and deduce the equation for the overall reaction.

Learn these important equations.

If your equation is wrong, you can still score in the next part provided that your method is built upon sound chemical principles.

Show your working!!

$MnO_4^-(aq) + 8H^+(aq) + 5e^- \longrightarrow Mn^{2+}(aq) + 4H_2O(l)$ ✓
$Fe^{2+}(aq) \longrightarrow Fe^{3+}(aq) + e^-$ ✓
$5Fe^{2+}(aq) + MnO_4^-(aq) + 8H^+(aq) \longrightarrow Mn^{2+}(aq) + 4H_2O(l)$ ✓

(ii) Use the above data to calculate the relative molecular mass of hydrated iron(II) sulphate. Hence calculate the number of moles of water of crystallisation present in one mole of hydrated iron(II) sulphate.

amount of $KMnO_4 = c \times \dfrac{V}{1000} = 0.0180 \times \dfrac{22.8}{1000} = 4.10 \times 10^{-4}$ mol ✓

From the equation, the amount of Fe^{2+} = 5 × amount of $KMnO_4$.
∴ amount of Fe^{2+} that reacted = $5 \times 4.10 \times 10^{-4}$ mol = 2.05×10^{-3} mol ✓

∴ $\dfrac{0.570}{2.05 \times 10^{-3}}$ g of hydrated iron(II) sulphate contains 1 mol Fe^{2+} ✓

M_r of hydrated iron(II) sulphate = 278 ✓
M_r of $FeSO_4 = 55.8 + 32.1 + 16 \times 4 = 151.9$ ✓

Number of waters of crystallisation = $\dfrac{278 - 151.9}{18} = \dfrac{126.1}{18} = 7$ ✓

(iii) State how the volume of potassium manganate(VII) solution used would have differed if the solution of iron(II) sulphate had been left for some time before being titrated. Explain your answer.

Less $KMnO_4$ solution would have been needed ✓ because some of the Fe^{2+} will have been oxidised by air to Fe^{3+}. ✓ [11]

[Total: 26]

NEAB Equilibria and Inorganic Chemistry Q6(a–c) June 1998

Practice examination questions

1

(a) Complete the following table.

	Na	Mg	Al	Si
Formula of anhydrous chloride				

[2]

(b) (i) Write equations, including state symbols, for the changes that take place when these chlorides are added to water:

sodium chloride   magnesium chloride

aluminium chloride   silicon chloride. [4]

    (ii) Interpret these changes on the basis of the bonding in the chlorides. [2]

    (iii) Suggest why the chloride of carbon, CCl_4, does not react with water. [4]

(c) Aluminium hydroxide is amphoteric. Write two ionic equations to show the meaning of this statement. [2]

[Total: 14]

Edexcel Specimen Unit Test 4 Q4 2000

2

(a) Write equations to show what happens when the following oxides are added to water and predict approximate values for the pH of the resulting solutions.

    (i) sodium oxide

    (ii) sulphur dioxide. [4]

(b) What is the relationship between bond type in the oxides of the Period 3 elements and the pH of the solutions which result from addition of the oxides to water? [2]

(c) Write equations to show what happens when the following chlorides are added to water and predict approximate values for the pH of the resulting solutions.

    (i) magnesium chloride

    (ii) silicon tetrachloride. [4]

[Total: 10]

Assessment and Qualifications Alliance Specimen Unit Test 5 Q4 2000

3

(a) Explain how a bond is formed between a metal ion and a ligand in a complex ion. [2]

(b) Consider the complex compound $[Co(NH_2CH_2CH_2NH_2)_3]Cl_3$

    (i) Name the ligand bonded to cobalt.

    (ii) What name is given to this type of ligand?

    (iii) What is the oxidation state of cobalt in this compound?

    (iv) What is the co-ordination number of cobalt in this compound?

    (v) Deduce the likely shape around cobalt in this complex. [5]

(c) From your knowledge of the reaction of cobalt(II) ions with ammonia, outline how you would prepare a solution of the complex $[Co(NH_2CH_2CH_2NH_2)_3]Cl_3$ starting from solid cobalt(II) chloride. [3]

[Total: 10]

Assessment and Qualifications Alliance Further Inorganic Chemistry Q1 June 1998

Practice examination questions (continued)

4

(a) Give the electronic configuration of a cobalt atom and a Co^{2+} ion. [2]

(b) Cobalt can be described both as a d-block element and as a transition element. State what is meant by each of the terms. [2]

(c) In aqueous solution water combines with the Co^{2+} ion to form the complex ion $[Co(H_2O)_6]^{2+}$ which gives a pink colour to the solution.

   (i) What feature of the water molecule allows it to form a complex ion with Co^{2+}? [1]

   (ii) What types of bond are present in the complex ion $[Co(H_2O)_6]^{2+}$? [2]

   (iii) Suggest the shape of the ion $[Co(H_2O)_6]^{2+}$. [1]

(d) Consider the following reactions:

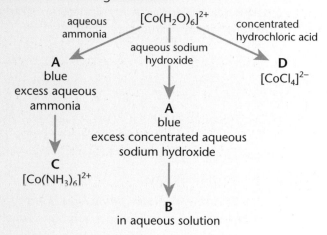

   (i) Give the name of the blue precipitate, **A**, and write an ionic equation for its formation from $[Co(H_2O)_6]^{2+}$. [2]

   (ii) What name is given to the type of reaction occurring in (i)? [1]

   (iii) Suggest a formula for the cobalt complex ion **B** present in the solution. [1]

   (iv) Give the name of the ion **C**. [1]

   (v) Write an equation for the formation of ion **D** from $[Co(H_2O)_6]^{2+}$ and suggest the type of reaction taking place. [2]

[Total: 15]

Edexcel Module Test 1 Q1 modified Jan 1999

5

The concentration of hydrogen peroxide in a solution can be determined by titrating an acidified solution against aqueous potassium manganate(VII) added from a burette. The potassium manganate(VII) reacts with the colourless aqueous solution of hydrogen peroxide as shown in the equation given below.

$$5H_2O_2 + 2MnO_4^- + 6H^+ \longrightarrow 5O_2 + 8H_2O + 2Mn^{2+}$$

(a) State the role of hydrogen peroxide in this reaction. [1]

(b) Identify a suitable acid for use in this titration. [1]

(c) State the colour change at the end-point of this reaction. [1]

(d) After acidification with a suitable acid, 25.0 cm³ of a dilute aqueous solution of hydrogen peroxide were found to react with 18.1 cm³ of 0.0200 M KMnO₄. Calculate the molar concentration of hydrogen peroxide in the solution. [4]

[Total: 7]

Assessment and Qualification Alliance Equilibria and Inorganic Chemistry Q6 March 1998

Chapter 5

Isomerism, aldehydes, ketones and carboxylic acids

The following topics are covered in this chapter:

- Isomerism and functional groups
- Aldehydes and ketones
- Carboxylic acids
- Esters
- Acylation

5.1 Isomerism and functional groups

After studying this section you should be able to:

- understand what is meant by structural isomerism and stereoisomerism
- explain the term chiral centre and identify any chiral centres in a molecule of given structural formula
- understand that many natural compounds are present as one optical isomer only
- understand that many pharmaceuticals are chiral drugs
- recognise common functional groups

LEARNING SUMMARY

Key points from AS

- **Basic concepts**
  Revise AS pages 113–117

During AS Chemistry, you learnt about the basic concepts used in organic chemistry. You should revise these thoroughly before you start the A2 part of this course.

Isomerism

AQA	M4	SALTERS	M4
EDEXCEL	M4	WJEC	CH4
OCR	M4	NICCEA	M4
NUFFIELD	M6		

Structural and *cis-trans* isomerism were introduced during AS Chemistry. These are reviewed below and *cis-trans* isomerism is discussed in the wider context of **stereoisomerism**.

Key points from AS

- **Structural isomerism**
  Revise AS pages 114–115
- **cis-trans isomerism**
  Revise AS page 122

Structural isomerism

Structural isomers are molecules with the same molecular formula but with different structural arrangements of atoms (structural formulae).

Two structural isomers of $C_2H_4O_2$ are shown below:

$$H_3C-C\diagup^{O}_{\diagdown OH} \qquad H-C\diagup^{O}_{\diagdown O-CH_3}$$

Stereoisomerism

Stereoisomers are molecules with the same structural formula but their atoms have different positions in space.

There are two types of stereoisomerism, each arising from a different structural feature:

- *cis-trans* isomerism about a **C=C** double bond
- **optical** isomerism about a **chiral** carbon centre.

cis-trans (or geometric) isomerism

cis-trans isomerism occurs in molecules with:

- a C=C double bond and
- two **different** groups attached to **each** carbon in the C=C bond.

The double bond prevents rotation.

E.g. *cis-trans* isomers of 1,2-dichloroethene, ClCH=CHCl.

> Each *cis-trans* isomer has the same structural formula.

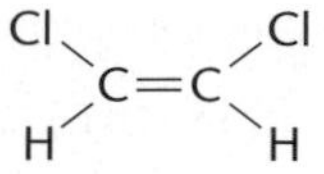

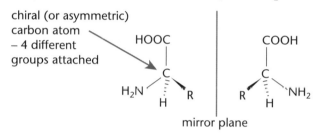

cis-isomer	*trans*-isomer
(groups on one side)	(groups on opposite sides)

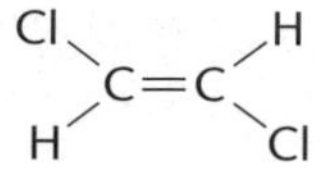

Optical isomers

> Optical isomerism occurs in the molecules of a compound with:
> - a **chiral** (or *asymmetric*) carbon atom.
>
> A chiral carbon atom has **four** different groups attached to it.
> - Optical isomers are non-superimposable mirror images of one another.

> Optical isomers are usually drawn as 3D diagrams. It is then easy to picture the mirror images.

E.g. the optical isomers of an amino acid, RCHNH$_2$COOH (see pages 123–125).

chiral (or asymmetric) carbon atom – 4 different groups attached

mirror plane

> Optical isomers are also called **enantiomers**.

Optical isomers

- are non-superimposable mirror images of one another
- rotate plane-polarised light in opposite directions
- are chemically identical.

> Society discovered the consequences of harmful side-effects from the 'wrong' optical isomer with the use of thalidomide. One optical isomer combated the effects of morning sickness in pregnant women. The other optical isomer was the cause of deformed limbs of unborn babies.
>
> Partly through this lesson, drugs are now often used as the optically pure form, comprising just the required optical isomer.

Chirality and drug synthesis

A synthetic amino acid, made in the laboratory, is optically inactive:
- it contains equal amounts of each optical isomer – a **racemic** mixture.

A natural amino acid, made by living systems, is optically active:
- it contains only one of the optical isomers.

The difference between the optical isomers present in natural and synthetic organic molecules has important consequences for drug design. The synthesis of pharmaceuticals often requires the production of chiral drugs containing a single optical isomer. Although one of the optical isomers may have beneficial effects, the other may be harmful and may lead to undesirable side effects.

Progress check

1 Show the alkenes that are structural isomers of C$_4$H$_8$.
2 Show the *cis-trans* isomers of C$_4$H$_8$.
3 Show the optical isomers of C$_4$H$_9$OH.

[answers shown inverted at bottom of page]

1 CH$_3$CH$_2$CH=CH$_2$; CH$_3$CH=CHCH$_3$; (CH$_3$)$_2$C=CH$_2$

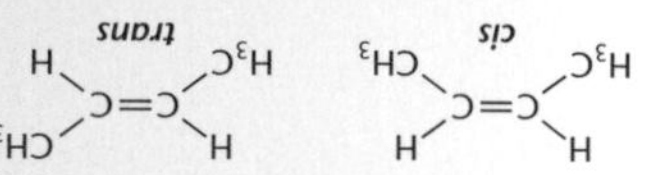

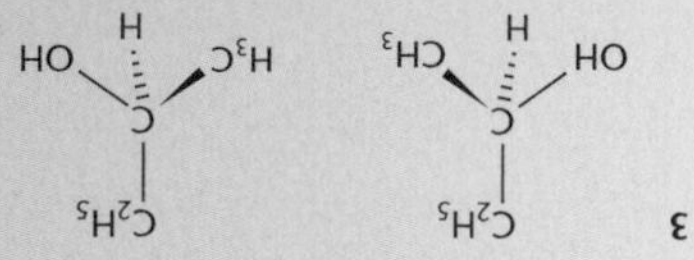

Common functional groups

AQA	M4	SALTERS	M4, M5
EDEXCEL	M4	WJEC	CH4
OCR	M4	NICCEA	M4
NUFFIELD	M4, M6		

It is essential that you can instantly identify a functional group within a molecule so that you can apply the relevant chemistry.

You must learn all of these!

The functional groups in the table below contain most of the functional groups you will meet in the A Level Chemistry course, including those met in AS Chemistry. It cannot be stressed too much just how important it is that you can instantly recognise a functional group. Without this, your progress in Organic Chemistry will be very limited.

name	functional group	examples structural formula		prefix or suffix (for naming)
alkane	C–H	$CH_3CH_2CH_3$ propane	$CH_3CH_2CH_3$	-ane
alkene	C=C	CH_3CHCH_2 propene		-ene
halogenoalkane	— Br	CH_3CH_2Br bromoethane	CH_3CH_2—Br	bromo-
alcohol	— OH	CH_3CH_2OH ethanol	CH_3CH_2—OH	-ol
aldehyde	—CHO	CH_3CHO ethanal		-al
ketone	—CO—	CH_3COCH_3 propanone		-one
carboxylic acid	—COOH	CH_3COOH ethanoic acid		-oic acid
ester	—COO—	CH_3COOCH_3 methyl ethanoate		-oate
acyl chloride	—COCl	CH_3COCl ethanoyl chloride		-oyl chloride
amine	— NH₂	$CH_3CH_2NH_2$ ethylamine	CH_3CH_2—NH₂	-amine
amide	—CONH₂	CH_3CONH_2 ethanamide		-amide
nitrile	—C≡N	CH_3CN ethanenitrile	H_3C—CN	-nitrile

Key points from AS

- **Functional groups** *Revise AS pages 115–116*

The nitrile carbon atom is included in the name. CH_3CN contains a longest carbon chain with **two** carbon atoms and its name is based upon ethane – hence ethanenitrile.

5.2 Aldehydes and ketones

After studying this section you should be able to:

- *understand the polarity and physical properties of carbonyl compounds*
- *describe nucleophilic addition to aldehydes and ketones*
- *describe a test to detect the presence of the carbonyl group*
- *describe tests to detect the presence of an aldehyde group*
- *describe the reduction of carbonyl compounds to form alcohols*

LEARNING SUMMARY

Carbonyl compounds General formula: $C_nH_{2n}O$

AQA	M4	SALTERS	M5
EDEXCEL	M4	WJEC	CH4
OCR	M4	NICCEA	M4
NUFFIELD	M4		

Key points from AS

- **Oxidation of alcohols**
  Revise AS pages 128–129

During AS Chemistry, you learnt about how alcohols can be oxidised to carbonyl compounds: aldehydes and ketones. For A2 Chemistry, you will learn about the reactions of aldehydes and ketones.

Types and naming of carbonyl compounds

The carbonyl group, C=O, is the functional group in aldehydes and ketones.

aldehyde, RCHO

ketone, RCOR'

ethanal
CH_3CHO

butanal
$CH_3CH_2CH_2CHO$

propanone
CH_3COCH_3

pentan-2-one
$CH_3COCH_2CH_2CH_3$

Uses of aldehydes and ketones

Propanone, $(CH_3)_2C=O$, is used in large quantities by industry as:
- an industrial solvent in paints, varnishes and nail-polish removers.

Methanal, $H_2C=O$ is used in large quantities in:
- preserving and embalming, as a germicide and insecticide
- for manufacturing plastic coatings such as *Bakelite*, *Formica* and melamine
- for manufacturing polymer adhesives such as those used to glue wood together.

Other aldehydes and ketones are mainly used as intermediates in the manufacture of plastics, dyes and pharmaceuticals, as solvents or as perfumes and flavouring agents.

Examples of aldehydes and ketones as flavours are:
- benzaldehyde, C_6H_5CHO (the flavour of fresh almonds)
- heptan-2-one, $C_5H_{11}COCH_3$ (smell of blue cheese).

Polarity of carbonyl compounds

Carbon and oxygen have different electronegativities, resulting in a polar C=O bond: carbonyl compounds have polar molecules.

The properties of aldehydes and ketones are dominated by the polar carbonyl group, C=O.

$$\overset{\delta+}{C}=\overset{\delta-}{O}$$

oxygen is more electronegative than carbon producing a dipole

Physical properties of carbonyl compounds

The polarity of the carbonyl group is less than that of the hydroxyl group in alcohols. Thus aldehydes and ketones have weaker dipole-dipole interactions and lower boiling points than alcohols of comparable molecular mass.

The polarity in propanone is such that it mixes with polar solvents such as water and also dissolves many organic compounds.

The low boiling point also makes it easy to remove by evaporation, a property exploited by its use in paints and varnishes.

$$\underset{\substack{\text{carbonyl group in}\\\text{aldehyde or ketone}}}{\overset{\delta+\quad\delta-}{\diagdown C=O}} \xrightarrow{\text{stronger dipole}} \underset{\substack{\text{hydroxyl group in}\\\text{alcohols}}}{-\overset{|}{\underset{|}{C}}-\overset{\delta-}{O}\diagdown \underset{H}{\overset{}{}}_{\delta+}}$$

Nucleophilic addition to carbonyl compounds

AQA	M4	SALTERS	M5
EDEXCEL	M4, M5	WJEC	CH4
OCR	M4	NICCEA	M4
NUFFIELD	M4		

The electron-deficient carbon atom of the polar $C^{\delta+}=O^{\delta}$ bond attracts **nucleophiles**. This allows an **addition** reaction to take place across the C=O double bond of aldehydes and ketones. This is called **nucleophilic addition**.

$$\overset{\delta+\quad\delta-}{\diagdown C=O} \quad \text{electron-rich nucleophile attracted to } \overset{\delta+}{C}$$

Nucleophilic addition of hydrogen cyanide, HCN

In the presence of cyanide ions, CN⁻, hydrogen cyanide, HCN, is added across the C=O bond in aldehydes and ketones.

HCN is added across the C=O double bond.

$$
\begin{array}{lll}
\text{aldehyde} & + & \text{HCN} \longrightarrow \text{hydroxynitrile} \\
CH_3CHO & + & \text{HCN} \longrightarrow CH_3CH(OH)CN
\end{array}
$$

$$H_3C-C\overset{\displaystyle O}{\underset{\displaystyle H}{\diagdown}} \xrightarrow{\text{KCN(aq) / HCN}} H_3C-\overset{\displaystyle OH}{\underset{\displaystyle H}{\overset{|}{\underset{|}{C}}}}-C\equiv N$$

This hydroxynitrile has optical isomers.

Hydrogen cyanide is a very poisonous gas and it is usually generated in solution as H⁺ and CN⁻ ions using:
- sodium cyanide, NaCN as a source of CN⁻
- dilute sulphuric acid as a source of H⁺.

Mechanism

The presence of cyanide ions is essential to provide the nucleophile for the first step of this mechanism.

$$H_3C-\overset{\delta+}{C}\overset{\overset{\displaystyle O}{\delta-}}{\underset{\displaystyle H}{\diagdown}} \quad :C\equiv N \longrightarrow H_3C-\overset{\overset{\displaystyle :\overset{..}{O}:^-}{}}{\underset{\displaystyle H}{\overset{|}{\underset{|}{C}}}}-C\equiv N \xrightarrow{\text{H}^+\text{ transfer}} H_3C-\overset{\overset{\displaystyle O-H}{}}{\underset{\displaystyle H}{\overset{|}{\underset{|}{C}}}}-C\equiv N$$

:CN nucleophile donates electron pair

'hydroxynitrile'

Increasing the carbon chain length

The nucleophilic addition of hydrogen cyanide is useful for increasing the length of a carbon chain. The nitrile product can then easily be reacted further in organic synthesis.
- Nitriles are easily *hydrolysed* by water in hot dilute acid to form a carboxylic acid.
- Nitriles are easily *reduced* by sodium in ethanol to form an amine.

The diagrams below show how 2-hydroxypropanenitrile, synthesised above, can be converted into a carboxylic acid and an amine.

nitrile → *carboxylic acid*

$$H_3C-\underset{\underset{H}{|}}{\overset{\overset{OH}{|}}{C}}-C\equiv N \xrightarrow[\text{hydrolysis}]{H^+ (aq) / H_2O, \text{ reflux}} H_3C-\underset{\underset{H}{|}}{\overset{\overset{OH}{|}}{C}}-C\overset{\overset{O}{\diagup\diagup}}{\diagdown_{OH}}$$

2-hydroxypropanoic acid
(lactic acid)

reduction | Na / ethanol

$$H_3C-\underset{\underset{H}{|}}{\overset{\overset{OH}{|}}{C}}-\underset{\underset{H}{|}}{\overset{\overset{H}{|}}{C}}-NH_2$$

amine

Testing for the carbonyl group

EDEXCEL	M4, M5	WJEC	CH4
OCR	M4	NICCEA	M4
NUFFIELD	M4		

Carbonyl compounds produce an orange-yellow crystalline solid with Brady's reagent

The carbonyl group can be detected using **Brady's reagent** – a solution of 2,4-dinitrophenylhydrazine (2,4-DNPH) in dilute acid.

- With 2,4-DNPH, both aldehydes and ketones produce bright **orange-yellow crystals** which identifies the carbonyl group, C=O.

2,4-DNPH   ethanal 2,4-dinitrophenylhydrazone

$$H_3C-C\overset{\overset{O}{\diagup\diagup}}{\diagdown_H} + \underset{\underset{H}{|}}{\overset{\overset{O_2N}{}}{N}}H_2N-\boxed{}-NO_2 \longrightarrow CH_3-\underset{\underset{H}{|}}{\overset{\overset{N}{\diagdown}}{C}}\quad N-\boxed{}-NO_2 + H_2O$$

orange crystalline precipitate

This is a **condensation reaction** – water is lost.

The carbonyl group can also be detected using i.r. spectroscopy (see pp. 132–133).

- The crystals have very sharp melting points which can be compared with known melting points from databases. Thus the actual carbonyl compound can be identified.

Testing for the aldehyde group

AQA	M4	NUFFIELD	M4
EDEXCEL	M4, M5	WJEC	CH4
OCR	M4	NICCEA	M4

Key points from AS

- **Oxidation of alcohols**
  Revise AS pages 128–129

An aldehyde can be distinguished from a ketone by using a combination of these two tests.

Both aldehyde and ketone produce a yellow-orange precipitate with 2,4-DNPH.

Only the aldehydes react in the three tests below.

You studied the oxidation of different types of alcohols during AS Chemistry. This is an extremely important reaction in organic chemistry, linking alcohols with aldehydes, ketones and carboxylic acids.

In terms of oxidation:

- aldehydes sit midway between primary alcohols and carboxylic acids.

$$RCH_2OH \underset{\text{reduction}}{\overset{\text{oxidation}}{\rightleftharpoons}} R-C\overset{\overset{O}{\diagup\diagup}}{\diagdown_H} \underset{\text{reduction}}{\overset{\text{oxidation}}{\rightleftharpoons}} R-C\overset{\overset{O}{\diagup\diagup}}{\diagdown_{OH}}$$

primary alcohol   aldehyde   carboxylic acid

The oxidation of aldehydes provides the basis of chemical tests used to identify this functional group.

In each test:
- the aldehyde **reduces** the reagents used in the test, producing a visible colour change
- the aldehyde is **oxidised** to a carboxylic acid.
  $$RCHO + [O] \longrightarrow RCOOH$$

Ketones cannot normally be oxidised and they **do not** react with Tollens' reagent, Fehling's solution or acidified dichromate(VI).

Heat aldehyde with Tollens' reagent

Tollens' reagent is a solution of silver nitrate in aqueous ammonia. The silver ions are reduced by an aldehyde producing a **silver mirror**.

Tollens' reagent $\longrightarrow$ **silver mirror**

$$Ag^+(aq) + e^- \longrightarrow Ag(s)$$

Heat aldehyde with Fehling's (or Benedict's) solution

Fehling's and Benedict's solutions contain Cu^{2+} ions dissolved in aqueous alkali. The Cu^{2+} ions are reduced to copper(I) ions, Cu^+, producing a **brick-red precipitate** of Cu_2O.

Key points from AS

- **Oxidation of primary alcohols**
  Revise AS pages 128–129

Benedict's / Fehling's solution $\longrightarrow$ **brick-red precipitate** of Cu_2O

$$2Cu^{2+}(aq) + 2e^- + 2OH^-(aq) \longrightarrow Cu_2O(s) + H_2O(l)$$

Acidified dichromate(VI) can also be used to oxidise alcohols.

Heat aldehyde with $H^+/Cr_2O_7^{2-}$

Concentrated sulphuric acid, H_2SO_4, is used as a source of H^+ ions and potassium dichromate(VI), $K_2Cr_2O_7$, as a source of $Cr_2O_7^{2-}$ ions. The **orange** $Cr_2O_7^{2-}$ ions are reduced to **green** chromium(III) ions, Cr^{3+}.

Reduction of carbonyl compounds

AQA	M4	SALTERS	M5
EDEXCEL	M4	WJEC	CH4
OCR	M4	NICCEA	M4
NUFFIELD	M4		

Key point from AS

- **Reduction of carbonyl compounds**
  Revise AS page 129

Aldehydes and ketones can be reduced to alcohols using a reducing agent containing the hydride ion, H^-.

Suitable reducing agents are:
- sodium tetrahydridoborate(III) (*sodium borohydride*), $NaBH_4$, in water (reflux)
- lithium tetrahydridoaluminate(III) (*lithium aluminium hydride*), $LiAlH_4$, in dry ether (at room temperature).

Aldehydes are reduced to primary alcohols.

For balanced equations, the reducing agent can be shown simply as [H].

$$H_3C-C{\overset{O}{\underset{H}{}}} + 2[H] \longrightarrow H_3C-CH_2OH$$

aldehyde   primary alcohol

$NaBH_4$ and $LiAlH_4$ both reduce the C=O double bond in aldehydes and ketones.

They do **not** reduce the C=C bond in alkenes.

Ketones are reduced to secondary alcohols.

$$\begin{array}{c} H_3C \\ C_2H_5 \end{array}C=O + 2[H] \longrightarrow \begin{array}{c} H_3C \\ C_2H_5 \end{array}CHOH$$

ketone   secondary alcohol

Mechanism (AQA only)

The reaction of carbonyl compounds with hydrogen cyanide (see p. 95) is also an example of nucleophilic addition.

The reduction of aldehydes and ketones to alcohols using $NaBH_4$ or $LiAlH_4$ is an example of **nucleophilic addition**. The reducing agent can be considered to release hydride ions, H^-.

$$NaBH_4 \longrightarrow H^- + NaBH_3^+$$

The hydride ion acts as a nucleophile in the first stage of the reaction.

$:H^-$ nucleophile
donates electron pair

protonation
with H_2O

primary alcohol

Test for the presence of the methyl carbonyl group

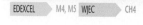

The presence of a methyl carbonyl group can be detected by heating a compound with iodine, I_2 in aqueous sodium hydroxide, OH^-(aq).

$$H_3C-\overset{\overset{\displaystyle O}{\|}}{C}-$$
methyl carbonyl

> Methyl carbonyl compounds produce pale-yellow crystals with I_2/OH^-.

- Methyl carbonyls produce **pale-yellow crystals** of triiodomethane (*iodoform*), CHI_3, with an antiseptic smell.
- $CH_3CHO + 3I_2 + 4OH^- \longrightarrow CHI_3 + HCOO^- + 3H_2O + 3I^-$

Progress check

1 (a) Draw the structure of:
   (i) 3-methylpentan-2-one
   (ii) 3-ethyl-2-methylpentanal.

2 **Three** isomers of C_4H_8O are carbonyl compounds.
   (a) (i) Show the formula for each isomer.
   (ii) Classify each isomer as an *aldehyde* or a *ketone*.
   (iii) Name each isomer.
   (b) Write the structural formulae of the **three** alcohols that could be formed by reduction of these isomers.

(b) $CH_3CH_2CH_2CH_2OH$; $CH_3CH_2CHOHCH_3$; $(CH_3)_2CHCH_2OH$

2 (a) (i)

$CH_3-CH_2-CH_2\underset{O}{\overset{H}{\diagdown}}C$
butanal
aldehyde

$CH_3-CH_2-\underset{\overset{\|}{O}}{C}-CH_3$
butanone
ketone

$CH_3-CH\underset{\overset{|}{CH_3}}{}-C\underset{O}{\overset{H}{\diagdown}}$
methylpropanal
aldehyde

1 (a) (i) $H_3C-CH_2-CH\underset{\overset{|}{CH_3}}{}-\underset{\overset{\|}{O}}{C}-CH_3$

(ii) $H_3C-C-CH_2\underset{\overset{|}{CH_2CH_3}}{}-CH\underset{\overset{|}{CH_3}}{}-C\underset{O}{\overset{H}{\diagdown}}$

5.3 Carboxylic acids

After studying this section you should be able to:

- *understand the polarity and physical properties of carboxylic acids*
- *describe acid reactions of carboxylic acids to form salts*
- *describe the esterification of carboxylic acids with alcohols*

Carboxylic acids   General formula: $C_nH_{2n+1}COOH$ RCOOH

AQA	M4	SALTERS	M4
EDEXCEL	M4	WJEC	CH4
OCR	M4	NICCEA	M4
NUFFIELD	M4		

The combination of carbonyl and hydroxyl groups in the carboxyl group modifies the chemistry of both groups.

Carboxylic acids have their own set of reactions and react differently from carbonyl compounds and alcohols.

The carboxyl group

The functional group in carboxylic acids is the **carboxyl** group, COOH. Although this combines both the **carb**onyl group and a hydr**oxyl** group its properties are very different from either.

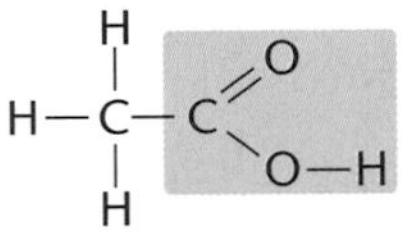

carboxyl group

Naming of carboxylic acids

Numbering starts from the carbon atom of the carboxyl group.

4-methylpentanoic acid
$(CH_3)_2CH_2CH_2CH_2COOH$

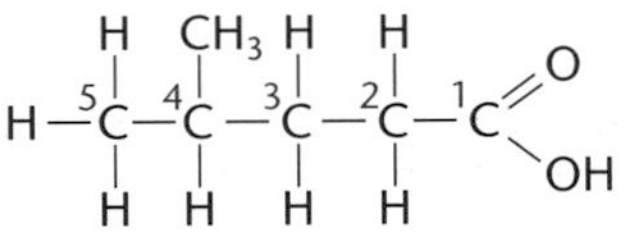

Natural carboxylic acids

Carboxylic acids are found commonly in nature. Their acidity is comparatively weak and their presence in food often gives a sour taste. Some examples of natural carboxylic acids are shown below.

structure	name	natural source
HCOOH	methanoic acid (*formic acid*)	ants, stinging nettles
CH_3COOH	ethanoic acid (*acetic acid*)	vinegar
COOH | COOH	ethanedioic acid (*oxalic acid*)	rhubarb
OH | $H_3C-CH-COOH$	2-hydroxypropanoic acid (*lactic acid*)	sour milk
CH_2-COOH | $HO-C-COOH$ | CH_2-COOH	2-hydroxypropane-1,2,3-tricarboxylic acid (*citric acid*)	oranges, lemons

Polarity

The carboxyl group is a combination of two polar groups: the hydroxyl –OH, **and** carbonyl C=O groups. This makes a carboxylic acid molecule more polar than a molecule of an alcohol or a carbonyl compound.

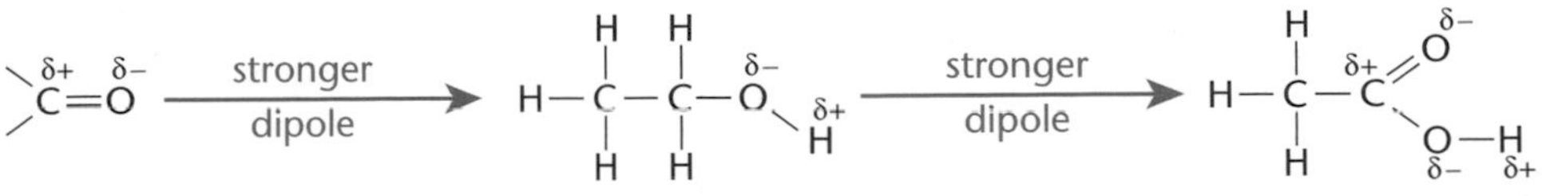

carbonyl group in aldehyde or ketone   →   stronger dipole   →   hydroxyl group in alcohols   →   stronger dipole   →   carboxyl group in carboxylic acids

The properties of carboxylic acids are dominated by the carboxyl group, COOH.

Physical properties of carboxylic acids

The COOH group dominates the physical properties of short-chain carboxylic acids. Hydrogen bonding takes place between carboxylic acid molecules, resulting in:

- higher melting and boiling points than alkanes of comparable M_r
- solubility in water.

The solubility of alcohols in water decreases with increasing carbon chain length as the non-polar contribution to the molecule becomes more important.

Carboxylic acids as 'acids'

Carboxylic acids are the 'organic acids'.

For more details of the dissociation of weak acids, see p. 36.

Carboxylic acids are only weak acids because they only partially dissociate in water.

$$CH_3COOH \rightleftharpoons CH_3COO^- + H^+$$

- Only 1 molecule in about 100 actually dissociates.
- Only a small proportion of the potential H^+ ions is released.

Carboxylates: salts of carboxylic acids

Carboxylic acid salts, 'carboxylates', are formed by neutralisation of a carboxylic acid by an alkali. In the example below, ethanoic acid produces ethanoate ions.

$$H-\overset{\overset{\displaystyle H}{|}}{\underset{\underset{\displaystyle H}{|}}{C}}-C\overset{\displaystyle O}{\underset{\displaystyle \underset{\delta-}{O}-\underset{\delta+}{H}}{}} + OH^- \longrightarrow H-\overset{\overset{\displaystyle H}{|}}{\underset{\underset{\displaystyle H}{|}}{C}}-C\overset{\displaystyle O}{\underset{\displaystyle O^-}{}} + H_2O$$

carboxylic acids exist in acidic conditions

carboxylates exist in alkaline conditions

On evaporation of water, a carboxylate salt crystallises out as an ionic compound. With aqueous sodium hydroxide as the alkali, sodium ethanoate, $CH_3COO^-Na^+$, is produced as the ionic salt.

Acid reactions of carboxylic acids

Key points from AS

- **Typical reactions of an acid**
  Revise AS page 110

Carboxylic acids react by the usual 'acid reactions' producing carboxylate salts.

Carboxylic acids take part in typical acid reactions. Note in the examples below that each salt formed is a carboxylate.

- They are **neutralised by alkalis**, forming a salt and water only.
  $$CH_3COOH + NaOH \longrightarrow CH_3COONa + H_2O$$
- They **react with carbonates** forming a salt, carbon dioxide and water.
  $$2CH_3COOH + CaCO_3 \longrightarrow (CH_3COO)_2Ca + CO_2 + H_2O$$
- They **react with** reactive **metals** forming a salt and hydrogen.
  $$2CH_3COOH + Mg \longrightarrow (CH_3COO)_2Mg + H_2$$

> Carboxylic acids are the only common organic group able to release carbon dioxide gas from carbonates. This provides a useful test to show the presence of a carboxyl group.
>
> **KEY POINT**

Properties of carboxylates

Carboxylates such as sodium ethanoate, $CH_3COO^-Na^+$, are ionic compounds. They have typical properties of an ionic compound.

- They are solids at room temperature with high melting and boiling points.
- They have a giant ionic lattice structure.
- They dissolve in water, totally dissociating into ions.

Reduction of carboxylic acids

EDEXCEL M4 NICCEA M4
WJEC CH4

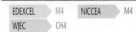

Just as primary alcohols can be oxidised to carboxylic acids, the reverse process can take place using a strong reducing agent such as lithium tetrahydridoalumuninate(III) (*lithium aluminium hydride*), $LiAlH_4$ in dry ether. (See also: Reduction of carbonyl compounds, page 97.)

$$RCOOH + 4[H] \longrightarrow RCH_2OH + H_2O$$

Esterification

AQA M4 SALTERS M4
EDEXCEL M4 WJEC CH4
OCR M4 NICCEA M4
NUFFIELD M4

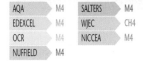

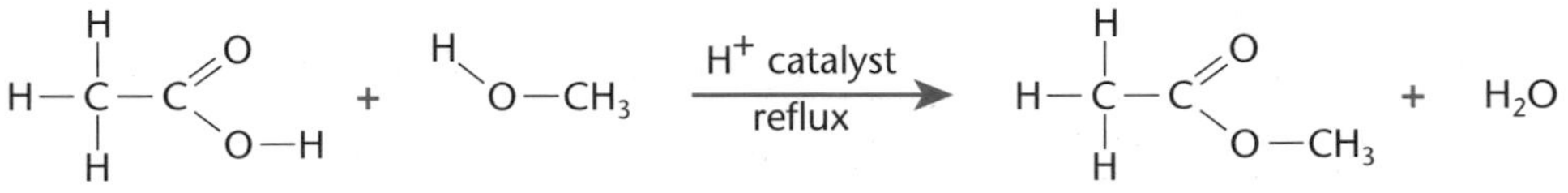

Esters are formed by reaction of a carboxylic acid with an alcohol.

Esterification is the formation of an ester by reaction of a **carboxylic acid** with an **alcohol** in the presence of an **acid catalyst** (e.g. concentrated sulphuric acid).

$$\text{carboxylic acid} + \text{alcohol} \longrightarrow \text{ester} + \text{water}$$

The esterification of ethanoic acid by methanol is shown below:

$$CH_3COOH + CH_3OH \longrightarrow CH_3COOCH_3 + H_2O$$

Conditions

- An acid catalyst (a few drops of conc. H_2SO_4) and reflux.
- The yield is usually poor due to incomplete reaction.

Progress check

1 Write down the structural formula of:
(a) propanoic acid
(b) the propanoate ion.

2 Explain why a carboxylic acid has a higher boiling point than the corresponding alcohol.

1 (a) CH_3CH_2COOH
(b) $CH_3CH_2COO^-$

2 Carboxylic acid has both polar carbonyl and hydroxyl groups. Alcohols have hydroxyl group only. Therefore, a carboxylic acid has greater intermolecular forces.

5.4 Esters

After studying this section you should be able to:

LEARNING SUMMARY

- understand the physical properties of esters
- describe the acid and base hydrolysis of esters
- describe the hydrolysis of fats and oils in soap making

Esters

General formula: $C_nH_{2n+1}COOC_mH_{2m+1}$ RCOOR'

The functional group in esters is the COOR' group, comprises an alkyl group in place of the acidic proton of a carboxylic acid.

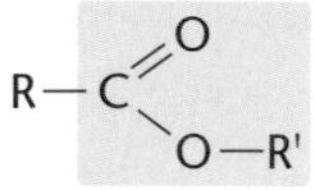

$$R-C \overset{\displaystyle O}{\underset{\displaystyle O-R'}{\big\langle}}$$

Naming of esters

The name of an ester is based on the carboxylic acid from which the ester is derived. The ester below, methyl butanoate, is derived from butanoic acid C_3H_7COOH with a methyl group in place of the acidic proton.

butanoate from
butanoic acid
C_3H_7COOH

$$C_3H_7-C \overset{\displaystyle O}{\underset{\displaystyle O-CH_3}{\big\langle}}$$

methyl from methanol
CH_3OH

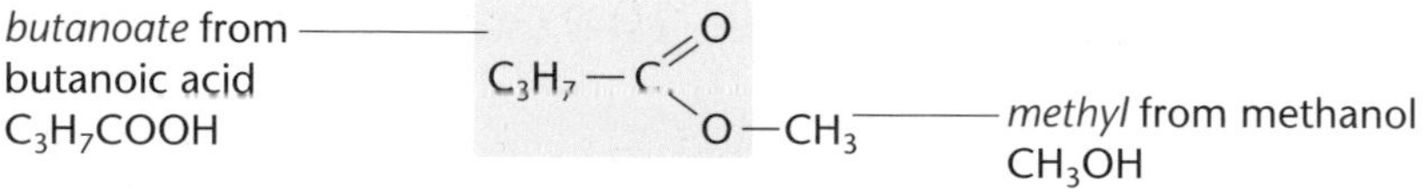

methyl butanoate

In the name of **methyl butanoate**,

- the alkyl group comes first as the ......*yl*: **methyl**
- the carboxylic acid part comes second as the ...*oate*: **butanoate**

Natural esters

Esters are found commonly in nature as fats and oils. They often have pleasant smells and contribute to the flavouring of many foods.

Some examples of natural esters are shown below.

Esters are common organic compounds present in fats and oils. They are also used in food flavourings and as perfumes.

structure	name	source
$HCOOCH_3$	methyl methanoate	raspberries
$C_3H_7COOC_4H_9$	butyl butanoate	pineapple
$CH_3COOCH_2CH_2CH(CH_3)_2$	3-methylbutyl ethanoate	pears

Physical properties of esters

Unlike carboxylic acids, esters are neutral. They are less polar than carboxylic acids and the absence of an –OH group means that esters cannot form hydrogen bonds and are generally insoluble in water.

Hydrolysis of esters

The hydrolysis of an ester is the reverse reaction to esterification (see page 101):

hydrolysis
ester + water ⇌ carboxylic acid + alcohol
esterification

Note that this reaction is reversible. The direction of reaction can be controlled by the reagents and reaction conditions used.

Isomerism, aldehydes, ketones and carboxylic acids

> **KEY POINT**
> Hydrolysis is the breaking down of a compound using **water** as the reagent.

Hydrolysis takes place by refluxing the ester with dilute aqueous acid or alkali.

- Acid hydrolysis $\longrightarrow$ alcohol + carboxylic acid.
- Alkaline hydrolysis $\longrightarrow$ alcohol + carboxylate.

Esterification **produces** water.

Hydrolysis **reacts** with water.

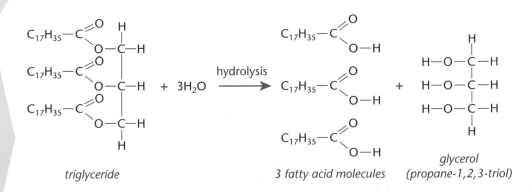

Hydrolysis of fats and oils

| AQA | M4 | SALTERS | M5 |
| OCR | M4 | NICCEA | M4 |

Fats are *triglyceryl esters* of fatty acids (long chain carboxylic acids) and propane-1,2,3-triol (*glycerol*).

The hydrolysis of fats and oils is an important reaction used in soap production. Acid hydrolysis of each molecule of the fat produces:

- **three** fatty acid molecules, 3RCOOH and
- **one** molecule of glycerol (a triol), $HOCH_2CHOHCH_2OH$.

Alkaline hydrolysis produces the carboxylate of the fatty acid.

Soaps are carboxylate salts, prepared by the alkaline hydrolysis of fats and oils. This process is called **saponification** from the Latin *sapo*: soap.

Alkaline hydrolysis with aqueous sodium hydroxide of the triglyceride above forms the sodium carboxylate salt $C_{17}H_{35}COO^-Na^+$.

Progress check

1. How could you prepare methyl propanoate from a named carboxylic acid and alcohol?

2. (a) Explain what is meant by the term *hydrolysis*.
   (b) Write down the products of the hydrolysis of ethyl butanoate in acidic and alkaline conditions.

1 Reflux propanoic acid with methanol in the presence of a few drops of concentrated sulphuric acid as catalyst.

2 (a) Breaking down a molecule using water.
   (b) Acid hydrolysis: ethanol and butanoic acid.
   Alkaline hydrolysis: ethanol and butanoate ions.

103

5.5 Acylation

After studying this section you should be able to:

- describe the formation of acyl chlorides from carboxylic acids
- describe nucleophilic addition-elimination reactions of acyl chlorides
- know the advantages of ethanoic anhydride for industrial acylations

LEARNING SUMMARY

Acyl chlorides

General formula: $C_nH_{2n+1}COCl$ $RCOCl$

AQA	M4	SALTERS	M4
EDEXCEL	M4	WJEC	CH4
NUFFIELD	M4	NICCEA	M4

The functional group in acyl chlorides is the COCl group,

$$-C \overset{\displaystyle O}{\underset{\displaystyle Cl}{=}}$$

Acyl chlorides are named from the parent carboxylic acid.

The examples below show that the suffix '*-oic acid*' is changed to '*-oyl chloride*' in the corresponding acyl chloride.

ethanoic acid   ethanoyl chloride   propanoic acid   propanoyl chloride

Properties

Acyl chlorides are the most reactive organic compounds that are commonly used. Unlike most other organic functional groups, acyl chlorides do **not** occur naturally because of their high reactivity with water.

> Acyl chlorides are the only common organic compounds that react violently with water.

$$R - C \overset{\delta-}{\underset{\underset{\delta-}{Cl}}{\overset{\overset{\delta+}{}}{=}} O}$$

The relatively large δ+ charge attracts the lone pair of a nucleophile

The electron-withdrawing effects of both the **chlorine** atom and carbonyl **oxygen** atom produces a relatively large δ+ charge on the carbonyl carbon atom. Nucleophiles are strongly attracted to the electron-deficient carbon atom, increasing the reactivity.

Acyl chlorides in organic synthesis

AQA	M4	WJEC	CH4
EDEXCEL	M4	NICCEA	M4
NUFFIELD	M4		

The high reactivity of acyl chlorides compared with carboxylic acids makes them particularly useful in the organic synthesis of related compounds.

Advantages of acyl chlorides over carboxylic acids.

- A good yield of product – reactions go to completion.
- Reactions often occur quickly and at lower temperatures.

Using acyl chlorides in synthesis

The high reactivity of an acyl chloride means that it is usually prepared from the parent carboxylic acid '*in situ*', i.e. when it is needed.

Acyl chlorides are prepared by reacting a carboxylic acid with:

- phosphorus pentachloride

$$RCOOH + PCl_5 \longrightarrow RCOCl + POCl_3 + HCl$$

> With acyl chlorides, anhydrous conditions are **essential** – acyl chlorides react with water.

- or sulphur dichloride oxide, $SOCl_2$

$$RCOOH + SOCl_2 \longrightarrow RCOCl + SO_2 + HCl$$

- The required nucleophile is then added to the acyl chloride.

Reactions of acyl chlorides with nucleophiles

AQA	M4	SALTERS	M4
EDEXCEL	M4, M6	WJEC	CH4
NUFFIELD	M4	NICCEA	M4

Nucleophiles of the type H–Y: react readily with acyl chlorides.

This is an **addition-elimination** reaction involving:
- addition of HY across the C=O double bond followed by
- elimination of HCl.

Addition-elimination reactions of acyl chlorides

Acyl chlorides can be reacted with different nucleophiles to produce a range of related functional groups. In each addition-elimination reaction, hydrogen chloride is eliminated as the second product.

The reaction scheme below shows the reactions of an acyl chloride RCOCl with the nucleophiles water, H_2O, methanol, CH_3OH, ammonia, NH_3 and methylamine, CH_3NH_2.

> The high reactivity of an acyl chloride means that these reactions take place at room temperature.

Ammonia and methylamine are bases – they react with the HCl eliminated:

$$NH_3 + HCl \longrightarrow NH_4^+Cl^-$$
$$CH_3NH_2 + HCl \longrightarrow CH_3NH_3^+Cl^-$$

Mechanism (AQA only)

The mechanism below shows the **nucleophilic addition-elimination** reaction of ethanoyl chloride with methanol to form an ester. Each nucleophile in the reaction scheme above will react in a similar manner.

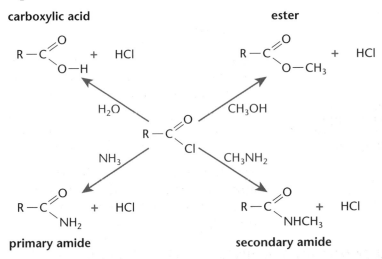

Acid anhydrides

AQA ▶ M4 · WJEC ▶ CH4

Acyl chlorides are ideal for small-scale preparations in the laboratory. However, they are too expensive and too reactive for large-scale preparations for which they are replaced by acid anhydrides.

The equation below shows the reaction between ethanoic anhydride and methanol.

> Notice the molecular structure of ethanoic anhydride – two ethanoic acid molecules have been condensed together with loss of H_2O. Hence the 'anhydride' from ethanoic acid.

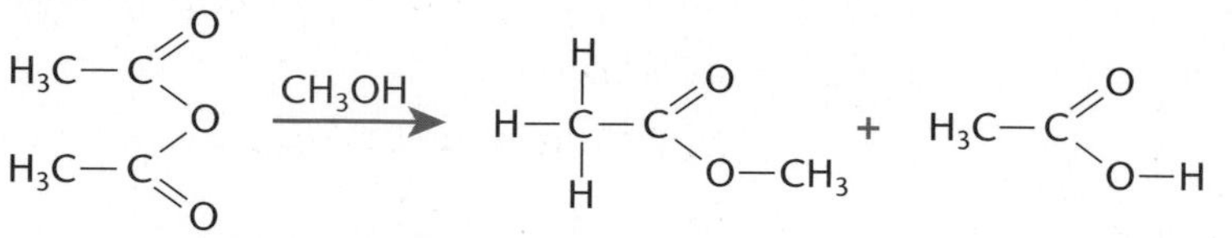

acid anhydride

carboxylic acid as second product

• The reaction is slower and easier to control on a large scale.
• This is the same essential reaction as with an acyl chloride but RCOOH forms instead of HCl as the second product.

Synthesis of aspirin

Aspirin can be synthesised by acylation of 2-hydroxybenzenecarboxylic acid (*salicylic acid*) using ethanoic anhydride.

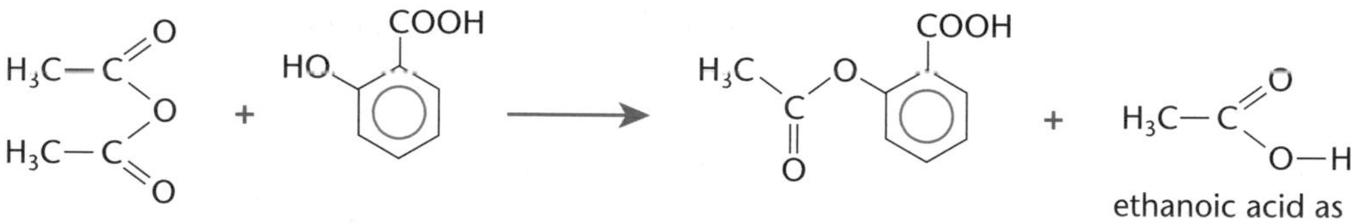

aspirin

ethanoic acid as second product

Progress check

1 Write down the structural formula of the organic product formed from the reaction of propanoyl chloride with:
(a) water
(b) ammonia.

2 What could you react together to make N-ethylbutanamide, $CH_3CH_2CH_2CONHCH_2CH_3$?

3 Write an equation for the preparation of aspirin using an acyl chloride.

3

2 $CH_3CH_2CH_2COCl$ and $CH_3CH_2NH_2$

1 (a) CH_3CH_2COOH
(b) $CH_3CH_2CONH_2$

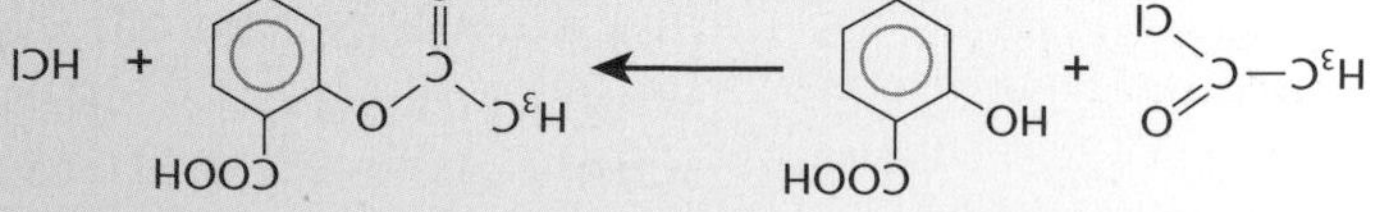

Sample question and model answer

(a) Outline the reaction of propanone with the following reagents. Give the equation for the reaction, the conditions, and the name of the organic product.

(i) Hydrogen cyanide

Equation:

$(CH_3)_2CO + HCN \longrightarrow (CH_3)_2C(OH)CN$ ✓

Conditions:

NaCN/dilute acid, heat ✓

Name of product:

ethanal hydroxynitrile ✓   [3]

In many questions, such as this, based on organic chemistry, you will be rewarded if you thoroughly learn the reactions, reagents and conditions.

(ii) Sodium tetrahydridoborate(III) (sodium borohydride).

Equation (you may represent $NaBH_4$ as [H]):

$(CH_3)_2CO + 2[H] \longrightarrow (CH_3)_2CHOH$ ✓

Conditions:

H_2O/reflux ✓

Name of product:

propan–2–ol ✓   [3]

(b) (i) Give a mechanism for the reaction in (a)(i).

[3]

Notice the care taken with the arrows.

Notice that the lone pair of the nucleophile is always shown.

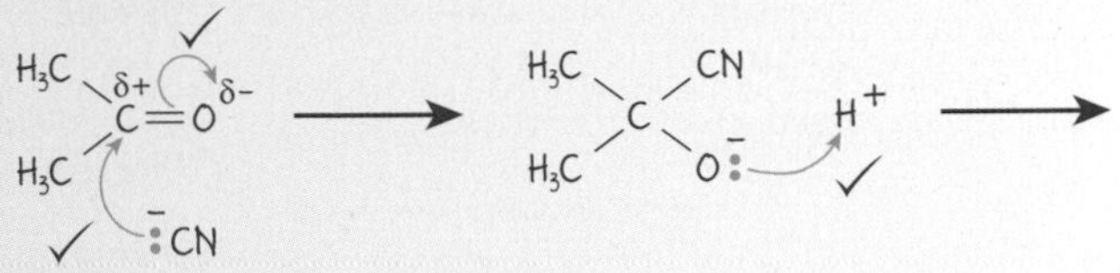

(ii) What type of mechanism is this?

nucleophilic addition ✓   [1]

'Nucleophilic' must be included for the mark.

(iii) What feature of the carbonyl group makes this type of mechanism possible? Explain how this feature arises.

dipole across C=O resulting in an electron-deficient carbon atom which can attract a nucleophile; ✓

oxygen is more electronegative than carbon. ✓   [2]

(iv) Explain briefly, by reference to its structure, why ethene would not react with HCN in a similar way.

The C=C double bond is between two identical atoms. ∴ there is no dipole and no electron-deficient carbon atom to attract a nucleophile. ✓   [1]

This is more difficult but tested often in exams.

*Alkenes react by **electrophilic** addition.*

*Carbonyl compounds react by **nucleophilic** addition.*

[Total: 13]

Edexcel Module Test 4 Q2 June 1998

Practice examination questions

1

(a) Two isomeric compounds, **A** and **B**, each have a relative molecular mass of 58 and the following percentage composition by mass: C, 62.1%; H, 10.3%; O, 27.6%. Calculate the molecular formula of **A** and of **B**.   [2]

(b) Draw the structural formula of **A** and of **B**.   [2]

(c) Name a reagent that reacts both with **A** and with **B**. State the observation.   [2]

(d) Name a reagent that reacts with only one of **A** and **B**. Identify which isomer reacts and state the observation.   [2]

[Total: 8]

Cambridge Chains and Rings Q1 Nov 1999

2

(a) Compound **G**, shown below, is a tri-ester.

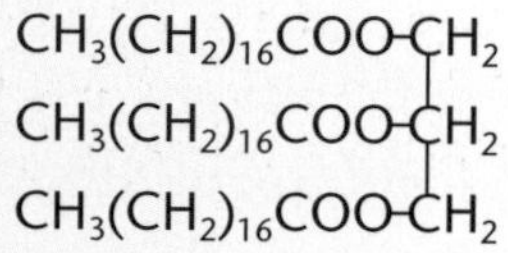

$$CH_3(CH_2)_{16}COO-CH_2$$
$$CH_3(CH_2)_{16}COO-CH_2$$
$$CH_3(CH_2)_{16}COO-CH_2$$

(i) Deduce the physical state of **G** at room temperature.

(ii) When completely hydrolysed by heating with aqueous sodium hydroxide, **G** forms an alcohol and the sodium salt of a carboxylic acid. Give the structural formula of the alcohol formed, write a formula for the sodium salt formed and state a use for this salt.   [4]

(b) (i) Complete and balance the equation below for the formation of di-ester **H**.

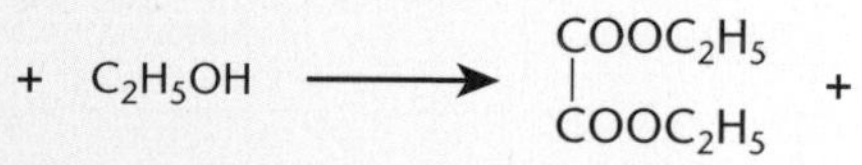

$$+ \ C_2H_5OH \longrightarrow \begin{array}{c} COOC_2H_5 \\ | \\ COOC_2H_5 \end{array} + $$
$$\mathbf{H}$$

(ii) Identify a substance which could catalyse this reaction to form **H**.

(iii) Draw a structural isomer of **H**, also a di-ester, which is formed by the reaction of ethane-1,2-diol with one other compound.   [5]

[Total: 9]

Assessment and Qualification Alliance Kinetics and Organic Chemistry Q5 March 1998

3

Consider the reaction scheme shown below.

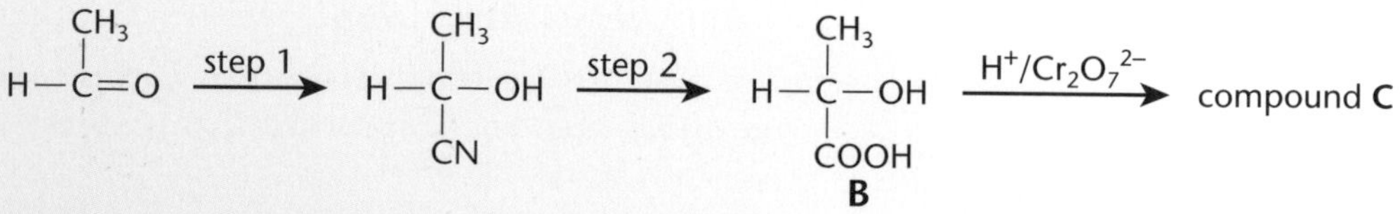

$$\begin{array}{c} CH_3 \\ | \\ H-C=O \end{array} \xrightarrow{\text{step 1}} \begin{array}{c} CH_3 \\ | \\ H-C-OH \\ | \\ CN \end{array} \xrightarrow{\text{step 2}} \begin{array}{c} CH_3 \\ | \\ H-C-OH \\ | \\ COOH \\ \mathbf{B} \end{array} \xrightarrow{H^+/Cr_2O_7^{2-}} \text{compound } \mathbf{C}$$

(a) State the reagent(s) that could be used for:
(i) step 1 (ii) step 2.   [2]

(b) Compound **B** has optical isomers. What structural feature in compound **B** results in optical isomerism?

(c) (i) Draw the structure of compound **C**.   [1]

(ii) Compound **C** can be reduced to propane-1,2-diol. Using [H] to represent the reducing agent, construct a balanced equation for this reduction.   [3]

[Total: 6]

Cambridge Chains and Rings Q5 June 2000

Chapter 6

Aromatics, amines, amino acids and polymers

The following topics are covered in this chapter:

- Arenes
- Reactions of arenes
- Phenols

- Amines
- Amino acids
- Polymers

6.1 Arenes

After studying this section you should be able to:

- apply rules for naming simple aromatic compounds
- understand the delocalised model of benzene
- explain the resistance to addition of benzene compared with alkenes

Aromatic organic compounds

AQA	M4	SALTERS	M5
EDEXCEL	M5	WJEC	CH4
OCR	M4	NICCEA	M5
NUFFIELD	M4		

Organic compounds with pleasant smells were originally classified as aromatic compounds. Many of these contain a benzene ring in their structure and nowadays an aromatic compound is one structurally derived from benzene, C_6H_6.

The diagrams below show different representations of a benzene molecule. It is usual practice to omit the carbon and hydrogen labels.

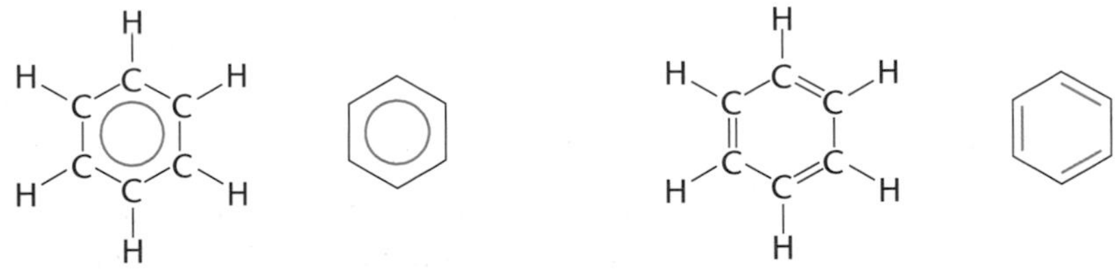

Arenes

An **arene** is an aromatic hydrocarbon containing a benzene ring. Benzene is the simplest arene and substituted arenes have alkyl groups attached to the benzene ring. Examples of arenes are shown below.

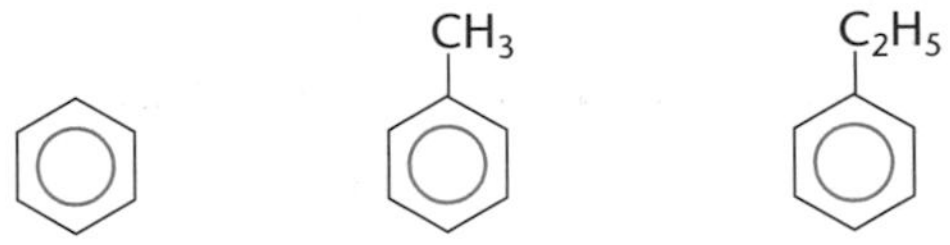

benzene   methylbenzene   ethylbenzene

Functional groups

Arenes can have a functional group next to an **aryl** group (a group containing a benzene ring):

 —X   or C_6H_5X

Aryl group   *Functional group*

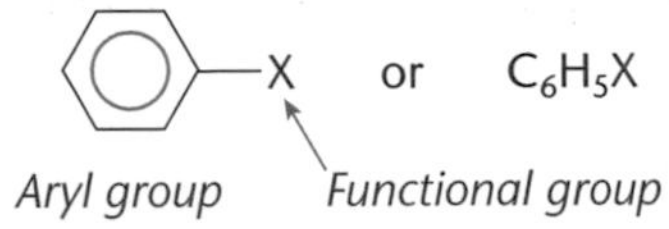

An aryl group contains a benzene ring.

- An aryl group is often represented simply as Ar—.
- The simplest aryl group is the **phenyl** group, C_6H_5, derived from benzene, C_6H_6.

Naming of aromatic organic compounds

The benzene ring of an aromatic compound is numbered from the carbon atom attached to a functional group or alkyl side-chain.

The names can be derived in two ways.

- Some compounds (e.g. hydrocarbons, chloroarenes and nitroarenes) are regarded as substituted benzene rings.

chlorobenzene   1,3-dimethylbenzene   2,4-dinitromethylbenzene

You should be able to suggest names for simple arenes.

- Other compounds (e.g. phenols, and amines) are considered as phenyl compounds of a functional group.

phenol
not hydroxybenzene

phenylamine
not aminobenzene

2,4,6-trichlorophenol

The stability of benzene

AQA	M4	SALTERS	M5
EDEXCEL	M5	WJEC	CH4
OCR	M4	NICCEA	M5
NUFFIELD	M4		

Two structures are used to represent benzene:

- the **Kekulé** structure developed between 1865 and 1872
- the modern **delocalised** structure developed in the 1930s.

The Kekulé structure of benzene

The Kekulé model of a benzene molecule has alternate double and single bonds making up the ring.

The original Kekulé structure showed benzene as a hexagonal molecule with alternate double and single bonds. Each carbon atom is attached to one hydrogen atom. This model was later modified to one with two isomers, rapidly interconverting into one another.

The Kekulé structure of benzene

The chemical name for the Kekulé structure of benzene is **cyclohexa-1,3,5-triene**, after the positions of the double bonds in the ring.

The delocalised structure of benzene

The delocalised model of a benzene molecule has identical carbon–carbon bonds making up the ring.

The delocalised structure of benzene shows a benzene molecule as a hybrid state between Kekulé's two isomers with no separate single and double bonds.

In this hybrid state:

- each carbon atom contributes one electron from its p-orbital to form π-bonds
- the π-bonds are spread out or **delocalised** over the whole ring.

Key point from AS

• **Alkenes**
  Revise AS page 122

This model helps to explain the low reactivity of benzene compared with alkenes (see also p. 112).

Although the Kekulé structure is used for some purposes, the delocalised structure is a better representation of benzene. You will find both representations in books.

The double bond in an alkene is a **localised** π-bond between two carbon atoms. In the delocalised structure of benzene, each bond is identical with electron density spread out to encompass and stabilise the whole ring. The diagram below shows one of the π-bonds formed by delocalisation of electrons in a benzene molecule.

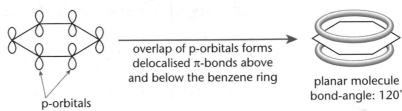

overlap of p-orbitals forms delocalised π-bonds above and below the benzene ring

p-orbitals

planar molecule
bond-angle: 120°

The delocalised structure of a benzene molecule is shown as a hexagon to represent the carbon skeleton and a circle to represent the six delocalised electrons.

delocalised electrons – all carbon-carbon bonds identical

The shape of a benzene molecule

Key points from AS

• **Electron-pair repulsion theory**
  Revise AS pages 50–51

Using electron-pair repulsion theory:

• there are **three** centres of electron density surrounding each carbon atom
• the shape around each carbon atom is trigonal planar with bond angles of 120°.

This results in a benzene molecule that is **planar**.

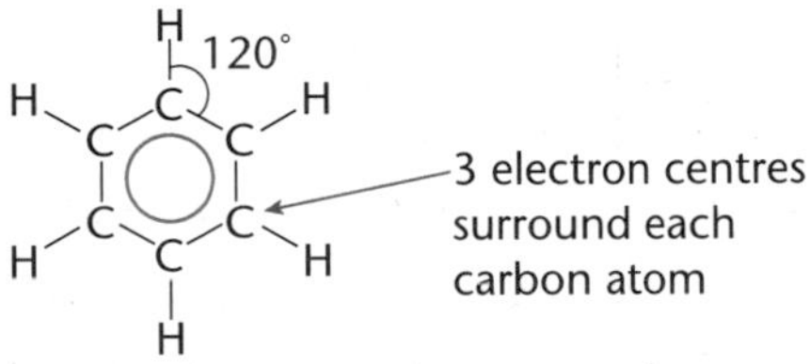

3 electron centres surround each carbon atom

Experimental evidence for delocalisation

AQA	M4	NUFFIELD	M4
EDEXCEL	M5	SALTERS	M5
OCR	M4	WJEC	CH4
		NICCEA	M5

Bond length data

The Kekulé structure of benzene as cyclohexa-1,3,5-triene suggests two bond lengths for the separate single and double bonds.

• C—C bond length = 0.154 nm
• C=C bond length = 0.134 nm

Experiment shows only one C—C bond length of 0.139 nm, between the bond lengths for single and double carbon-carbon bonds.

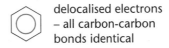

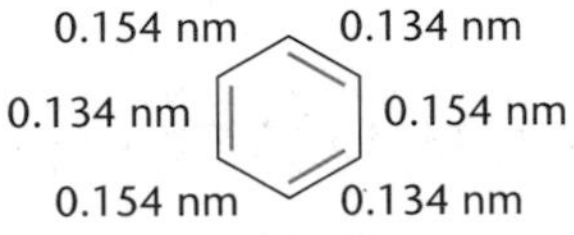

0.154 nm   0.134 nm
0.134 nm   0.154 nm
0.154 nm   0.134 nm

all C-C bonds are the same length: 0.139 nm

> This shows that each carbon-carbon bond in the benzene ring is intermediate between a single and a double bond.

KEY POINT

Thermochemical evidence

Hydrogenation of cyclohexene

Each molecule of cyclohexene has **one** C=C double bond. The enthalpy change for the reaction of cyclohexene with hydrogen is shown below.

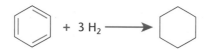 $+ \ H_2 \longrightarrow$ $\Delta H^{\ominus} = -120 \ kJ \ mol^{-1}$

Hydrogenation of benzene

The Kekulé structure of benzene as cyclohexa-1,3,5-triene has **three** double C=C bonds. It would be expected that the enthalpy change for the hydrogenation of this structure would be three times the enthalpy change for the **one** C=C bond in cyclohexene.

$+ \ 3 \ H_2 \longrightarrow$ predicted enthalpy change:
$\Delta H^{\ominus} = 3 \times -120 = -360 \ kJ \ mol^{-1}$

- When benzene is reacted with hydrogen, the enthalpy change obtained is far less exothermic, $\Delta H^{\ominus} = -208 \ kJ \ mol^{-1}$.

> - The difference between the thermochemical data for cyclohexa-1,3,5-triene and benzene suggests that benzene has **more stable bonding** than the Kekulé structure.
> - The delocalisation stability of benzene of $-152 \ kJ \ mol^{-1}$ is the difference between the two enthalpy changes above. This is extra energy that must be provided to break the delocalised benzene ring.

KEY POINT

The stability of benzene can also be demonstrated using thermochemical data for the reactions of bromine with cyclohexene and benzene.

Comparing electrophilic addition to arenes and alkenes

AQA	M4	SALTERS	M5
EDEXCEL	M5	WJEC	CH4
OCR	M4	NICCEA	M5
NUFFIELD	M4		

Key points from AS

- **Alkenes**
  Revise AS page 122
- **Addition reactions of alkenes**
  Revise AS pages 123–124

Arenes and alkenes both have π-bonds, but

- arenes have **delocalised** π-bonds
- alkenes have **localised** π-bonds
- the delocalised electron density in arenes is **less** than in alkenes.

Alkenes react with bromine in the dark at room temperature. Using the same conditions with benzene, there is no reaction. The stability of the delocalised system resists addition which would disrupt this stability.

Progress check

1. Name the **three** aromatic isomers of $C_6H_4Br_2$.
2. State two pieces of experimental evidence that support the delocalised structure of benzene.
3. Explain why alkenes, such as cyclohexene, are so much more reactive with electrophiles than arenes such as benzene.

3 Alkenes have localised π-bonds with a larger electron density than the delocalised π-bonds in benzene. The greater electron density is able to attract electrophiles more strongly. Also, the stability of the benzene ring must be disrupted if benzene is to react.
All the carbon-carbon bonds in the benzene ring are the same length.
2 The enthalpy change of hydrogenation of benzene is less exothermic than three times the enthalpy change of hydrogenation of cyclohexene.
1 1,2-dibromobenzene; 1,3-dibromobenzene; 1,4-dibromobenzene.

6.2 Reactions of arenes

After studying this section you should be able to:

LEARNING SUMMARY

- *describe electrophilic substitution of arenes: nitration, halogenation, alkylation, acylation and sulphonation*
- *describe the mechanism of electrophilic substitution in arenes*
- *understand the importance of reactions of arenes in the synthesis of commercially important materials*

Electrophilic substitution reactions of arenes

Many electrophiles react with alkenes by **addition**. However, electrophiles react with arenes by **substitution**, replacing a hydrogen atom on the ring. The difference in behaviour results from the high stability of the **delocalised** π-bonds in benzene compared with the **localised** π-bonds in alkenes (see pages 110–112).

> KEY POINT
> - Benzene reacts with only very reactive electrophiles.
> - The typical reaction of an arene is **electrophilic substitution**.

In this section, reactions of benzene are discussed to illustrate electrophilic substitution reactions of arenes.

Nitration of arenes

The nitration of arenes produces aromatic nitro compounds, important for the synthesis of many important products including explosives and dyes (see page 121).

Nitration of benzene

Benzene is nitrated by concentrated nitric acid at 55°C in the presence of concentrated sulphuric acid, which acts as a catalyst.

Although some heat is required (55°C), too much may give further nitration of the benzene ring forming 1,3-dinitrobenzene.

$$\text{C}_6\text{H}_6 + HNO_3 \xrightarrow[\text{55°C}]{H_2SO_4} \text{C}_6\text{H}_5NO_2 + H_2O$$

Mechanism

- The role of the concentrated sulphuric acid is to generate the **nitronium ion**, NO_2^+, as the active electrophile:

The nitronium ion is also called a **nitryl cation**.

$$HNO_3 + H_2SO_4 \longrightarrow H_2NO_3^+ + HSO_4^-$$
$$H_2NO_3^+ \longrightarrow NO_2^+ + H_2O$$

- The powerful NO_2^+ electrophile then reacts with benzene.

attack of NO_2^+ electrophile   proton loss

- The H^+ formed regenerates a molecule of H_2SO_4.

$$H^+ + HSO_4^- \longrightarrow H_2SO_4$$

The H_2SO_4 therefore acts as a **catalyst**.

- The H_2SO_4 molecule reacts with more nitric acid to form more nitronium ions.

Halogenation of arenes

Arenes react with halogens in the presence of a **halogen carrier**, which acts as a catalyst.

Suitable halogen carriers include:

*   iron
*   aluminium halides, e.g. $AlCl_3$ for chlorination.

Halogenation of benzene

The reaction takes place in warm conditions.

$$\text{benzene} + Br_2 \xrightarrow[\text{gentle heat}]{\text{halogen carrier e.g. Fe}} \text{bromobenzene} + HBr$$

Mechanism (Edexcel only)

*   Iron first reacts with bromine forming iron(III) bromide, $FeBr_3$.
* $FeBr_3$ acts as a halogen carrier polarising the Br—Br bond.

$$Br_2 + FeBr_3 \longrightarrow Br^{\delta+}\text{—}Br^{\delta-} \ FeBr_3$$

*   This electrophile reacts with benzene.

> This is a similar principle to the nitration of the benzene ring.

$$\text{attack of electrophile} \longrightarrow \longrightarrow + FeBr_4^- \longrightarrow \text{proton loss} + H^+$$

*   The H^+ formed generates $FeBr_3$.

$$H^+ + FeBr_4^- \longrightarrow FeBr_3 + HBr$$

> The $FeBr_3$ therefore acts as a **catalyst**.

* $FeBr_3$ can now polarise more bromine molecules.

Alkylation of arenes

Alkylation reactions of arenes are commonly known as *Friedel–Crafts* reactions. These are very important reactions industrially as they provide a means of introducing an alkyl group onto the benzene ring.

Alkylation of benzene

As with halogenation, a halogen carrier is required to generate a more reactive electrophile. With chloroethane, ethylbenzene is formed.

$$\text{benzene} \xrightarrow[\text{warm}]{\begin{array}{c} C_2H_5Cl \ / \\ \text{halogen carrier, e.g. } AlCl_3 \end{array}} \text{ethylbenzene } C_2H_5 + HCl$$

> This is a similar principle to the halogenation of the benzene ring.

Mechanism (AQA and Edexcel only)

Using a tertiary halogenoalkane, RCl, in the presence of $AlCl_3$ as a halogen carrier, a **carbonium ion** is generated.

$$RCl + AlCl_3 \longrightarrow R^+ + AlCl_4^-$$

- This carbonium ion acts as a powerful electrophile which reacts with benzene.

attack of R^+ electrophile   proton loss

- The H^+ formed generates more $AlCl_3$.

$$H^+ + AlCl_4^- \longrightarrow AlCl_3 + HCl$$

- The $AlCl_3$ reacts with more of the tertiary halogenoalkane RCl to generate more carbonium ions. The $AlCl_3$ acts as a **catalyst**.

Formation of ethylbenzene in industry

An important example of alkylation is the production of ethylbenzene, used to synthesise poly(phenylethene) (*polystyrene*). See page 126.

In industry, benzene is alkylated with ethene in the presence of HCl and $AlCl_3$ which both act as catalysts. Ethene is far cheaper and more readily available than chloroethane.

Chloroethane is first produced: $CH_2{=}CH_2 + HCl \longrightarrow CH_3CH_2Cl$

A complex forms: $CH_3CH_2Cl + AlCl_3 \longrightarrow CH_3CH_2^+ \, AlCl_4^-$

Ethylbenzene forms: $C_6H_6 + CH_3CH_2^+AlCl_4^- \longrightarrow C_6H_5CH_2CH_3 + AlCl_3 + HCl.$

Note that $AlCl_3$ and HCl are regenerated and can react again.

Acylation of arenes

Acylation introduces an acyl group such as $CH_3C{=}O$ onto the benzene ring. An acyl chloride is used in the presence of a halogen carrier.

acylium ion

The acylium ion is then able to react with the benzene ring, introducing an acyl group onto the ring.

Acylation of benzene

Using ethanoyl chloride, CH_3COCl, the ethanoyl group $CH_3C{=}O$ can be introduced onto the benzene ring.

This reaction takes place by a similar mechanism to the alkylation reaction above (Edexcel only).

Sulphonation of arenes

NUFFIELD M4   SALTERS M5

Sulphonation of arenes produces sulphonic acids. Sodium salts of sulphonic acids are used for detergents and fabric conditioners.

Sulphonation of benzene

Benzene reacts with fuming sulphuric acid (concentrated sulphuric acid saturated with sulphur trioxide) forming a sulphonic acid.

The reaction takes place between benzene and sulphur trioxide.

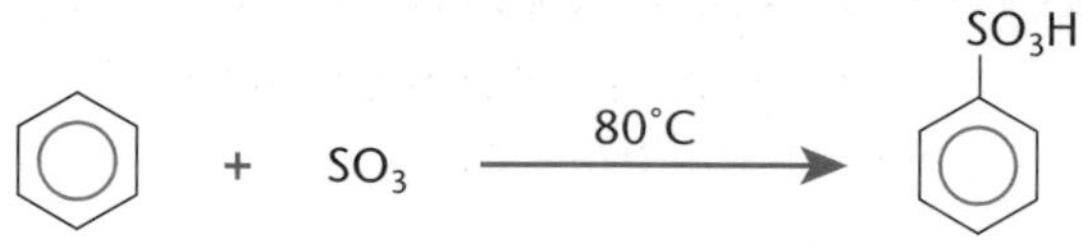

$$+ \quad SO_3 \quad \xrightarrow{80°C} \quad$$

benzenesulphonic acid

Oxidation of substituted arenes

EDEXCEL M5   WJEC CH4

The side chains of substituted arenes can be readily oxidised by prolonged reflux with a strong oxidising agent.

Oxidation of methylbenzene

Using alkaline potassium manganate(VII), followed by addition of dilute aqueous acid, methylbenzene is oxidised to benzenecarboxylic acid (**benzoic acid**).

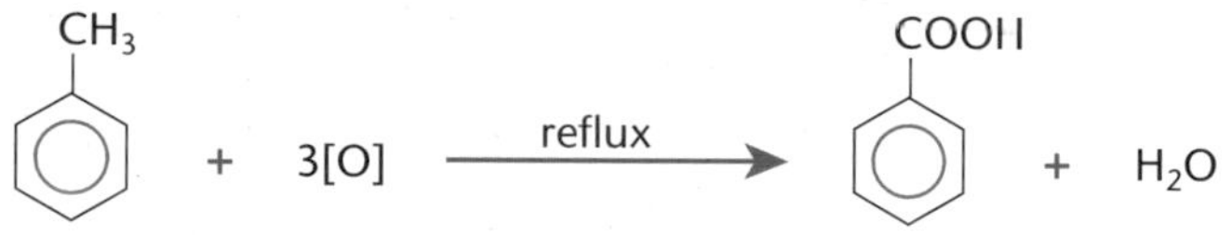

$$+ \quad 3[O] \quad \xrightarrow{reflux} \quad + \quad H_2O$$

benzenecarboxylic acid

Progress check

1 What is the common type of reaction of arenes?

2 Benzene reacts with bromine, nitric acid and chloroethane.
   (a) Using C_6H_6 to represent benzene and C_6H_5 to represent the phenyl group, write balanced equations for each of these reactions.
   (b) State the essential conditions that are needed for each of these reactions.

With chloroethane, a halogen carrier such as Fe.
With nitric acid, concentrated sulphuric acid at 55°C.
(b) With bromine, a halogen carrier such as Fe.
$C_6H_6 + C_2H_5Br \longrightarrow C_6H_5C_2H_5 + HBr$
$C_6H_6 + HNO_3 \longrightarrow C_6H_5NO_2 + H_2O$
2 (a) $C_6H_6 + Br_2 \longrightarrow C_6H_5Br + HBr$
1 Electrophilic substitution.

6.3 Phenols

After studying this section you should be able to:

LEARNING SUMMARY

- state the uses of phenols in antiseptics and disinfectants
- describe the reactions of phenol with sodium and bases
- explain the ease of bromination of phenol compared with benzene

Phenols

General structure: ArOH (Ar is an aromatic ring)

Phenols have an aromatic ring bonded directly to a hydroxyl (–OH) group. Some phenols are shown below.

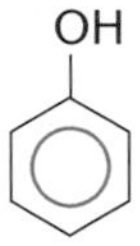

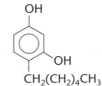

| phenol, C₆H₅OH | 2,4,6-trichlorophenol (TCP) | 4-hexyl-3-hydroxyphenol (in cough drops) | thymol (in thyme) |

phenol, C_6H_5OH

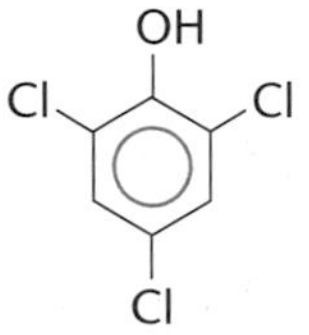

2,4,6-trichlorophenol (TCP)

4-hexyl-3-hydroxyphenol (in cough drops)

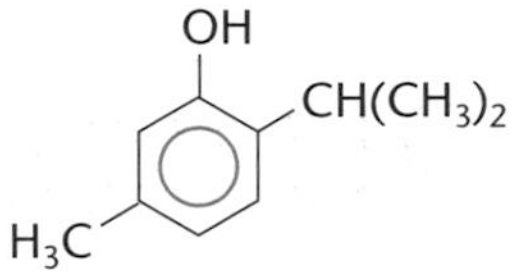

thymol (in thyme)

Uses

Dilute solutions of phenol are used as disinfectants and phenol was probably the first antiseptic. However, phenol is toxic and can cause burns to the skin. Less toxic phenols are used nowadays in antiseptics.

Phenols are also used in the production of plastics, such as the widely used phenol-formaldehyde resins. More complex phenols, such as thymol shown above, can be used as flavourings and aromas and these are obtained from essential oils of plants.

Properties

Like alcohols, a phenol has a **hydroxyl group** which is able to form intermolecular **hydrogen bonds**. The resulting properties include:

- higher melting and boiling points than hydrocarbons with similar relative molecular masses
- some solubility in water.

Reactions of phenol as an acid

Unlike the neutrality shown by alcohols, phenols are very weak acids. The hydroxyl group donates a proton, H^+, by breaking the O–H bond.

$$C_6H_5OH \rightleftharpoons H^+ + C_6H_5O^-$$

Phenols takes part in some typical acid reactions, reacting with sodium and bases to form ionic salts called **phenoxides**. However, the acidity is too weak to release carbon dioxide from carbonates (see carboxylic acids, page 100).

Key points from AS

- **Tests for alcohols**
  Revise AS page 131

Reaction with sodium

As with the hydroxyl groups in water and ethanol, phenol reacts with sodium, releasing hydrogen gas and forming a solution of sodium phenoxide, $C_6H_5O^-Na^+$.

$$2C_6H_5OH + 2Na \longrightarrow 2C_6H_5ONa + H_2$$

Compare these reactions with those of carboxylic acids p. 100.

Reaction with an alkali

Phenol is a weak acid and is neutralised by aqueous alkalis. For example, aqueous sodium hydroxide forms the salt sodium phenoxide.

$$C_6H_5OH + NaOH \longrightarrow C_6H_5ONa + H_2O$$

Electrophilic substitution of the aryl ring of phenol

| EDEXCEL | MS | NUFFIELD | M4 |
| OCR | M4 | WJEC | CH4 |

The aryl ring of phenol has a greater electron density than benzene because the adjacent oxygen reinforces the electron density in the ring.

This results in:

- a greater electron density in the aryl ring
- greater reactivity of the ring towards electrophiles.

Bromination of phenol

- Phenol reacts directly with bromine. However Benzene reacts with bromine only in the presence of a halogen carrier (see page 114).
- Phenol undergoes multiple substitution with bromine. Benzene is monosubstituted only.

- The organic product, 2,4,6-tribromophenol separates as a white solid.
- The bromine is decolourised.

Formation of esters

| EDEXCEL | MS | WJEC | CH4 |

Phenols form esters with carboxylic acids **only** in the presence of aqueous alkali.

The negative inductive effect from the aryl ring reduces the electron density of the phenolic –OH group and its effectiveness as a nucleophile. An alkali increases the electron density by generating the phenoxide ion:

$$OH^- + C_6H_5OH \longrightarrow C_6H_5O^- + H_2O$$
$$CH_3COOH + C_6H_5O^- \longrightarrow CH_3COOC_6H_5 + OH^-$$

Phenols readily form esters with acyl chlorides.

E.g. $CH_3COCl + C_6H_5OH \longrightarrow CH_3COOC_6H_5 + HCl$

Progress check

Phenols are easily nitrated with dilute nitric acid in the cold. As with bromine, multiple substitution takes place.
Suggest an equation for the reaction that takes place.

6.4 Amines

After studying this section you should be able to:

- explain the relative basicities of ethylamine and phenylamine
- describe the reactions of primary amines with acids to form salts
- describe the formation of phenylamine by reduction of nitrobenzene
- describe the synthesis of an azo-dye from phenylamine
- describe the reactions of amines with halogenoalkanes and acyl chlorides

Aliphatic and aromatic amines

AQA	M4	SALTERS	M4
EDEXCEL	M4	WJEC	CH4
OCR	M4	NICCEA	M5
NUFFIELD	M6		

Amines are organic compounds containing nitrogen derived from ammonia, NH_3. Amines are classified as primary, secondary or tertiary amines depending on how many of the hydrogen atoms in ammonia have been replaced by organic groups. The diagram below shows examples of aliphatic and aromatic amines.

Aliphatic amines			**Aromatic amine**
1 hydrogen replaced	2 hydrogens replaced	3 hydrogens replaced	
primary amine	*secondary amine*	*tertiary amine*	
$H_3C—NH_2$	$\begin{array}{c} H_3C \\ \\ H_3C \end{array}\!\!>NH$	$\begin{array}{c} H_3C \\ H_3C—N \\ H_3C \end{array}$	⬡—NH_2
NH_3			
ammonia	methylamine	dimethylamine	trimethylamine	phenylamine

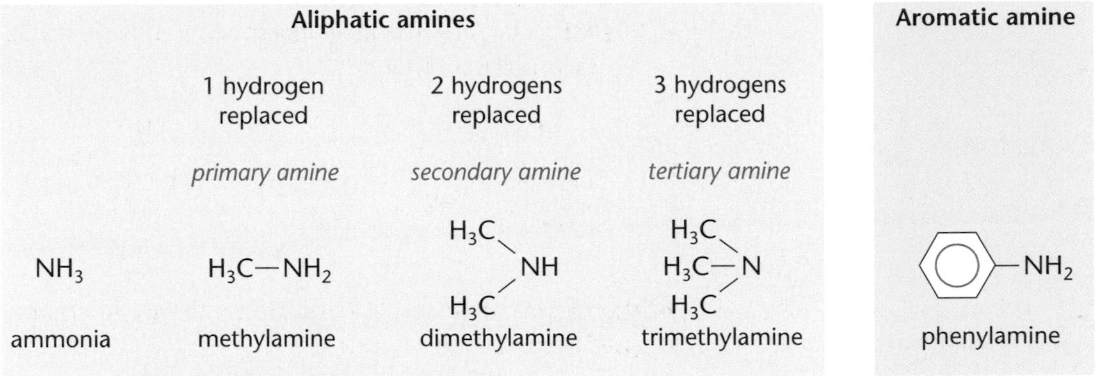

Amines in nature

Amines are found commonly in nature. They are weak bases and the shortest chain amines, e.g. ethylamine $C_2H_5NH_2$, smell of fish. Some diamines, such as putrescine $H_2N(CH_2)_4NH_2$, and cadaverine $H_2N(CH_2)_5NH_2$, are found in decaying flesh.

Polarity

> Notice the similarity with ammonia.

The presence of an electronegative nitrogen atom in amines results in polar molecules. The diagram shows the polarity of the primary amine methylamine.

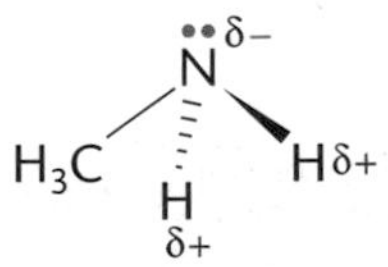

$$H_3C \begin{array}{c} \overset{\bullet\bullet\,\delta-}{N} \\ \overset{|}{H} \\ \overset{}{\delta+} \end{array} H\delta+$$

Physical properties of amines

> As with other polar functional groups, the solubility of amines in water decreases with increasing carbon chain length as the non-polar contribution to the molecule becomes more important.

The amino group dominates the physical properties of short-chain amines.

Hydrogen bonding takes place between amine molecules, resulting in:

- higher melting and boiling points than alkanes of comparable relative molecular mass
- solubility in water – amines with fewer than six carbons mix with water in all proportions.

Amines as bases

AQA	M4	SALTERS	M4
EDEXCEL	M4	WJEC	CH4
OCR	M4	NICCEA	M5
NUFFIELD	M6		

> Amines are the organic bases.

Amines are weak bases because they only partially associate with protons:

$$RNH_2 + H^+ \rightleftharpoons RNH_3^+$$

The basic strength of an amine measures its ability to **accept** a proton, H^+. This depends upon:

- the size of the $\delta-$ charge on the amino nitrogen atom
- the availability of the nitrogen lone pair.

The basicity of aliphatic and aromatic amines

Aliphatic amines are stronger bases than ammonia; aromatic amines are substantially weaker.

> **KEY POINT**
>
> - Electron-donating groups, e.g. alkyl groups, **increase** the basic strength.
> - Electron-withdrawing groups, e.g. C_6H_5, **decrease** the basic strength.

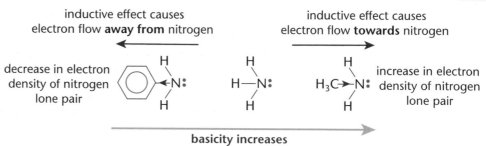

| inductive effect causes electron flow **away from** nitrogen | | inductive effect causes electron flow **towards** nitrogen |

decrease in electron density of nitrogen lone pair

increase in electron density of nitrogen lone pair

basicity increases

> Amines react by the usual 'base reactions' producing organic ammonium salts.

Organic ammonium salts

Amines are **neutralised by acids** forming salts.

In the example below, methylamine is neutralised by hydrochloric acid forming a primary **ammonium salt**.

> A proton, H^+, is added to the amino nitrogen atom.

$$CH_3NH_2 + HCl \longrightarrow CH_3NH_3^+Cl^-$$
methylammonium chloride

On evaporation of water, the primary ammonium salt crystallises out as an ionic compound.

Preparation of amines

AQA	M4	WJEC	CH4
EDEXCEL	M4, M5	NICCEA	M5
OCR	M4		

> Other reducing agents can also be used:
> - H_2/Ni catalyst
> - $LiAlH_4$ in dry ether
>
> $LiAlH_4$ is an almost universal reducing agent and works in most examples of organic reduction.

Preparation of aliphatic amines from nitriles

Aliphatic amines can be prepared by **reducing** a nitrile with a suitable reducing agent (e.g. Na in ethanol; $LiAlH_4$ in dry ether). (See also page 96.)

$$CH_3CH_2C\equiv N + 4[H] \longrightarrow CH_3CH_2CH_2NH_2$$

Preparation of aromatic amines from nitroarenes

Aromatic amines can be prepared by **reducing** a nitroarene. For example nitrobenzene, is reduced by Sn in concentrated HCl to form phenylamine.

phenylamine

- Using tin and hydrochloric acid, the salt $C_6H_5NH_3^+Cl^-$ is formed.
- The amine is obtained from this salt by adding aqueous alkali.

This process is often called the *Hofmann degradation* and can be used to move down a homologous series.

See also: Increasing the carbon chain length p. 95.

Preparation of amines from amides (Edexcel only)

Primary amines can be prepared by heating a primary amide with bromine in alkali. The amine formed has **one less carbon** atom than the starting amide.

$$RCONH_2 + Br_2 + 4NaOH \longrightarrow RNH_2 + Na_2CO_3 + 2NaBr + 2H_2O$$

The preparation of dyes from aromatic amines

Aromatic amines, such as phenylamine, are important industrially for the production of dyes. Modern dyes are formed in a two-stage synthesis:

* the aromatic amine is converted into a **diazonium salt**
* the diazonium salt is **coupled** with an aromatic compound such as phenol, forming an **azo dye**.

Formation of diazonium salts

An aromatic amine, such as phenylamine, forms a diazonium salt in the presence of nitrous acid, HNO_2, and hydrochloric acid.

* HNO_2 is unstable and is prepared *in situ* from $NaNO_2$ and HCl(aq).

$$NaNO_2 + HCl \longrightarrow HNO_2 + NaCl$$

* This mixture is then reacted with phenylamine. It is important to keep the temperature **below 10°C** because diazonium salts decompose above this temperature.

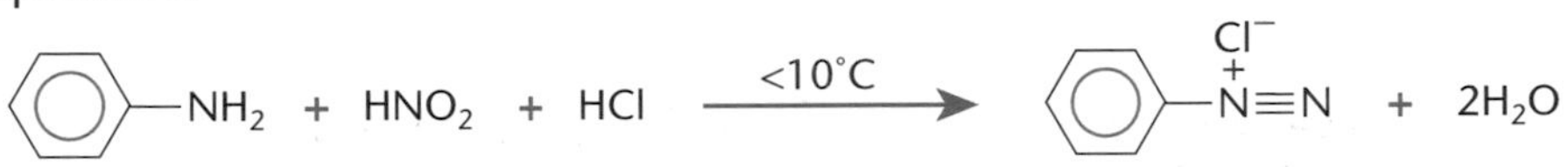

benzenediazonium chloride
(diazonium salt)

$$C_6H_5NH_2 + HNO_2 + HCl \longrightarrow C_6H_5N_2^+Cl^- + 2H_2O$$

Formation of azo dyes by coupling

The diazonium salt is coupled with a suitable aromatic compound **below 10°C** in **aqueous alkali** to produce an azo dye.

E.g. coupling of benzenediazonium chloride with phenol:

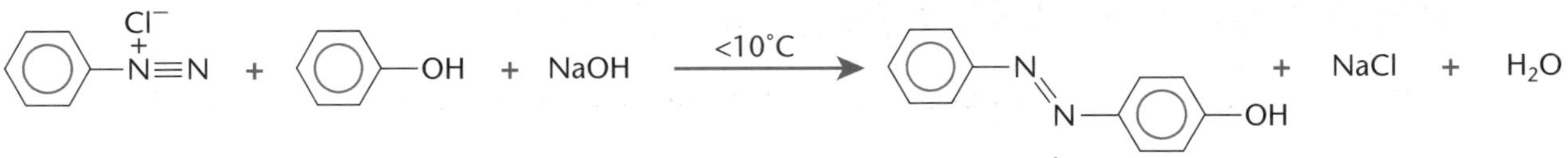

azo dye

* The azo dye is coloured.
* Coupling of a diazonium salt with different aromatic compounds forms dyes with different colours.

Using benzene for the synthesis of dyes

Benzene is used as the raw material for the synthesis of dyes.
A four-stage synthesis is shown below:

* benzene is nitrated to nitrobenzene (see page 113)
* nitrobenzene is reduced to phenylamine (see above)
* phenylamine is converted into a **diazonium salt**
* the diazonium salt is **coupled** with an aromatic compound such as phenol, forming an **azo dye**.

Reactions of amines with halogenoalkanes

AQA	M4	WJEC	CH4
NUFFIELD	M6	NICCEA	M5

Formation of a primary amine by nucleophilic substitution

Key points from AS

- Nucleophilic substitution reactions of halogenoalkanes
  Revise AS pages 132–133

Ammonia and amines react as **nucleophiles** with halogenoalkanes in substitution reactions.

- **Excess ammonia** reacts with halogenoalkanes in hot ethanol, forming a primary ammonium salt.

$$RX + NH_3 \longrightarrow RNH_3^+X^-$$

- Proton transfer with ammonia forms a **primary amine** RNH_2

$$RNH_3^+X^- + NH_3 \longrightarrow RNH_2 + NH_4X$$

Further substitution (AQA only)

At each stage, a hydrogen atom is replaced by an alkyl group.

- With an **excess** of the **halogenoalkane**, further substitution takes place which continues until a quaternary ammonium salt is obtained.

$$RNH_2 \xrightarrow{RX} R_2NH_2^+X^- \xrightarrow{RX} R_3NH^+X^- \xrightarrow{RX} R_4N^+X^-$$

secondary   *tertiary*   *quaternary*
ammonium salt   *ammonium salt*   *ammonium salt*

Each reaction stage proceeds by **nucleophilic substitution**. The mechanism for the final stage in the formation of the quaternary salt tetramethylammonium bromide is shown below.

Quaternary ammonium salts are used as cationic surfactants in fabric softeners.

$$(CH_3)_3\ddot{N} \qquad \overset{\delta+}{C}\!-\!\overset{\delta-}{Br} \longrightarrow (CH_3)_4N^+Br^-$$

Reactions of amines with acyl chlorides

AQA	M4	SALTERS	M4
EDEXCEL	M4	WJEC	CH4
NUFFIELD	M6	NICCEA	M5

Amines react with acyl chorides forming amides.

E.g.: $RCOCl + CH_3NH_2 \longrightarrow RCONHCH_3 + HCl$

$$CH_3NH_2 + HCl \longrightarrow CH_3NH_3^+Cl^-$$

For details, see also acyl chlorides p.104.

Progress check

1 Explain why ethylamine is a stronger base than phenylamine.

2 (a) Write equations for the conversion of:
   (i) benzene to nitrobenzene
   (ii) nitrobenzene to phenylamine
   (iii) phenylamine to benzenediazonium chloride.

3 (a) Write down the formula of the organic product formed between:
   (i) bromoethane and excess ammonia
   (ii) ammonia and excess bromoethane.

1 The ethyl group has a positive inductive effect which reinforces the electron density of the nitrogen atom of the amine. This lone pair on the nitrogen will attract protons more strongly.
The phenyl group has a negative inductive effect which reduces the electron density on the nitrogen atom of the amine. This lone pair on the nitrogen will attract protons less strongly.

2 (a) (i) $C_6H_6 + HNO_3 \longrightarrow C_6H_5NO_2 + H_2O$
   (ii) $C_6H_5NO_2 + 6[H] \longrightarrow C_6H_5NH_2 + 2H_2O$
   (iii) $C_6H_5NH_2 + HNO_2 + HCl \longrightarrow C_6H_5N_2^+Cl^- + 2H_2O$

3 (a) (i) $C_2H_5NH_2$ (ii) $(C_2H_5)_4N^+Br^-$

6.5 Amino acids

After studying this section you should be able to:

- describe the acid–base properties of amino acids and the formation of zwitterions
- explain the formation of polypeptides and proteins as condensation polymers of amino acids
- describe the acid hydrolysis of proteins and peptides

Amino acids

AQA	M4	SALTERS	M4
EDEXCEL	M4	WJEC	CH4
OCR	M4	NICCEA	M5
NUFFIELD	M6		

There are 22 naturally occurring amino acids.

A typical amino acid has organic molecules with both acidic and basic properties comprising:

- a basic amino group, $-NH_2$
- an acidic carboxyl group, $-COOH$.

Each amino acid has a unique organic R group or side chain. The formula of a typical amino acid is shown below.

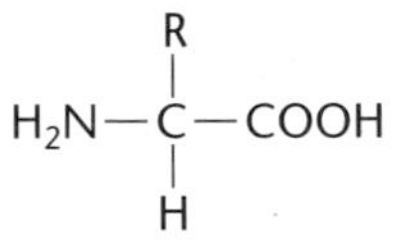

$$H_2N-\underset{\underset{H}{|}}{\overset{\overset{R}{|}}{C}}-COOH$$

Examples of some amino acids are shown below:

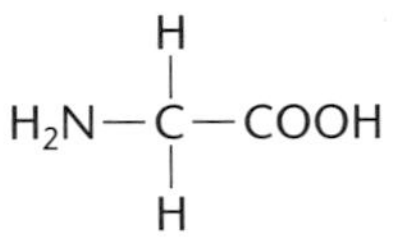

$$H_2N-\underset{\underset{H}{|}}{\overset{\overset{H}{|}}{C}}-COOH \qquad H_2N-\underset{\underset{H}{|}}{\overset{\overset{CH_3}{|}}{C}}-COOH \qquad H_2N-\underset{\underset{H}{|}}{\overset{\overset{CH(CH_3)_2}{|}}{C}}-COOH$$

R = H	R = CH_3	R = $CH(CH_3)_2$
glycine	*alanine*	*valine*

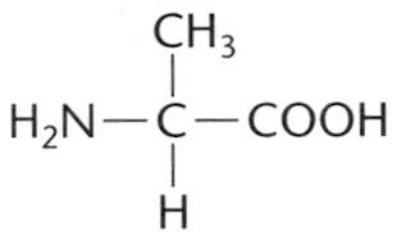

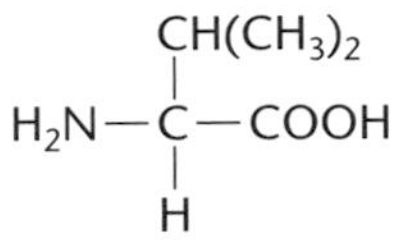

Acid–base properties of amino acids

AQA	M4	SALTERS	M4
EDEXCEL	M4	WJEC	CH4
OCR	M4	NICCEA	M5
NUFFIELD	M6		

Isoelectric points and zwitterions

Each amino acid has a particular pH called the **isoelectric point** at which the overall charge on an amino acid molecule is zero.

Examples of isoelectric points

amino acid	aspartic acid	glycine	histidine	arginine
isoelectric point	3.0	6.1	7.6	10.8

At the isoelectric point, an amino acid exists as a zwitterion:

- the **carboxyl** group **has donated** a proton to the **amino** group which form a positive NH_3^+ ion.

A **zwitterion** is a dipolar ion with both positive and negative charges in different parts of the molecule.

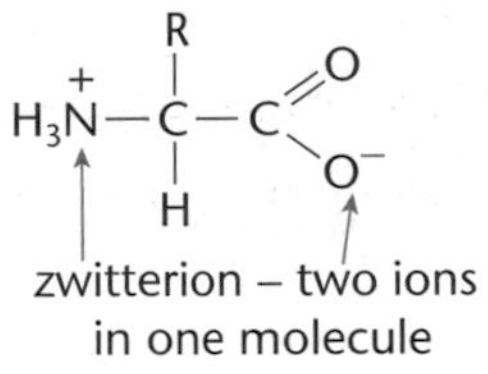

At the isoelectric point, the amino acids exist in equilibrium with its zwitterion form.

$$H_3\overset{+}{N}-\underset{\underset{H}{|}}{\overset{\overset{R}{|}}{C}}-C\overset{\displaystyle O}{\underset{\displaystyle O^-}{}}$$

zwitterion – two ions in one molecule

Amino acids as bases

In strongly **acidic** conditions a **positive ion** forms:

- an amino acid behaves as a **base**
- the COO^- ion gains a proton.

$$H_3\overset{+}{N}-\underset{H}{\overset{R}{C}}-C\overset{O}{\underset{O^-}{}} + H^+ \longrightarrow H_3\overset{+}{N}-\underset{H}{\overset{R}{C}}-C\overset{O}{\underset{OH}{}}$$

positive ion

> Because of their reactions with strong acids and strong bases, amino acids act as buffers and help to stabilise the pH of living systems.

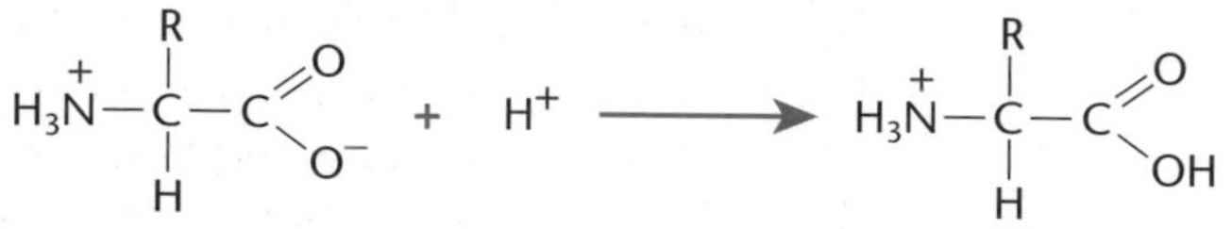

Amino acids as acids

In strongly **alkaline** conditions a **negative ion** forms:

- an amino acid behaves as an **acid**
- the NH_3^+ ion loses a proton.

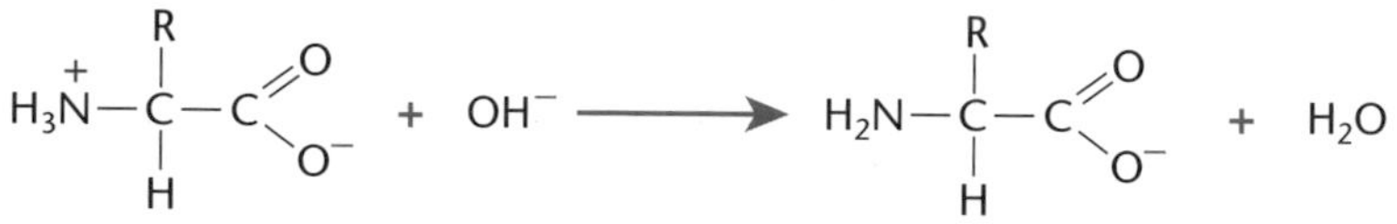

$$H_3\overset{+}{N}-\underset{H}{\overset{R}{C}}-C\overset{O}{\underset{O^-}{}} + OH^- \longrightarrow H_2N-\underset{H}{\overset{R}{C}}-C\overset{O}{\underset{O^-}{}} + H_2O$$

negative ion

> **KEY POINT**
>
> - At the **isoelectric point**, the amino acid is **neutral**.
> - At a pH more **acidic** than the isoelectric point, the amino acid forms a **positive ion**.
> - At a pH more **alkaline** than the isoelectric point, the amino acid forms a **negative ion**.

Physical properties of amino acids

AQA	M4	SALTERS	M4
EDEXCEL	M4	WJEC	CH4
OCR	M4	NICCEA	M5
NUFFIELD	M6		

Solid amino acids

Solid amino acids have higher melting points than expected from their molecular masses and structure. This suggests that amino acids crystallise in a giant lattice with strong electrostatic forces between the **zwitterions**.

Optical isomers

With the exception of aminoethanoic acid (*glycine*), H_2NCH_2COOH, all amino acids have a chiral centre and are optically active (see page 92).

Polypeptides and proteins

AQA	M4	SALTERS	M4
OCR	M4	WJEC	CH4
NUFFIELD	M6	NICCEA	M5

In nature, individual amino acids are linked together in chains as **polypeptides** and **proteins** (see page 127).

The diagram below shows the condensation of the amino acids glycine and alanine to form a **dipeptide**. The amino acids are bonded together by a **peptide link**.

> A protein is formed by condensation polymerisation of amino acids. See p. 127.

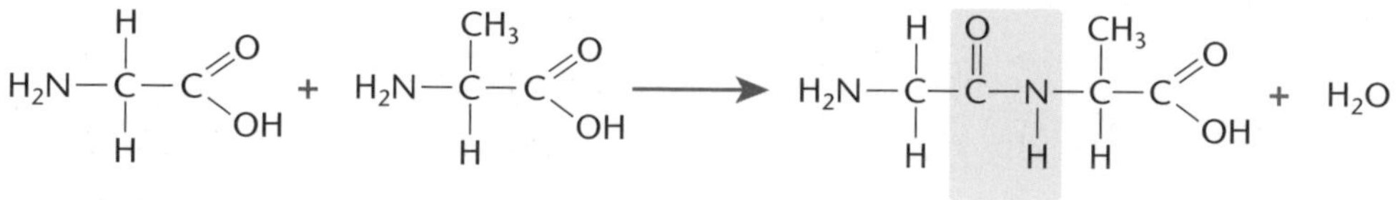

$$H_2N-\underset{H}{\overset{H}{C}}-C\overset{O}{\underset{OH}{}} + H_2N-\underset{H}{\overset{CH_3}{C}}-C\overset{O}{\underset{OH}{}} \longrightarrow H_2N-\underset{H}{\overset{H}{C}}-\overset{O}{\overset{\|}{C}}-\underset{H}{\overset{}{N}}-\underset{H}{\overset{CH_3}{C}}-C\overset{O}{\underset{OH}{}} + H_2O$$

glycine　　　　*alanine*　　　　　　*peptide link*

A polypeptide is the name given to a short chain of amino acids linked by peptide bonds.

A protein is simply the name given to a long-chain polypeptide.

Each peptide link forms:

- between the **carboxyl group** of glycine and the **amino group** of alanine
- with loss of a water molecule in a **condensation reaction**.

Further condensation reactions between amino acids build up a **polypeptide** or **protein**.

- For each amino acid added to a protein chain one water molecule is lost.
- Most common proteins contain more than 100 amino acids.
- Each protein has a unique sequence of amino acids and a complex three-dimensional shape held together by intermolecular bonds including hydrogen bonds.

Hydrolysis of proteins

Hydrolysis breaks down a protein into its separate amino acids.

- The protein is refluxed with 6 mol dm^{-3} HCl(aq) for 24 hours.
- The resulting solution is neutralised.

The equation below shows the hydrolysis of a dipeptide.

$$H_2N-\overset{\underset{|}{H}}{\underset{\underset{H}{|}}{C}}-\overset{\overset{O}{\|}}{C}-\overset{\underset{H}{|}}{N}-\overset{\underset{\underset{H}{|}}{CH_3}}{C}-C\overset{O}{\underset{OH}{}} + H_2O \xrightarrow[\text{2 neutralise}]{\text{1 6M HCl, reflux}} H_2N-\overset{\underset{H}{|}}{\underset{\underset{H}{|}}{C}}-C\overset{O}{\underset{OH}{}} + H_2N-\overset{\underset{H}{|}}{\underset{\underset{H}{|}}{C}}\overset{CH_3}{}-C\overset{O}{\underset{OH}{}}$$

peptide link   glycine   alanine

- Hydrolysis of a protein is the reverse process to the condensation polymerisation that forms proteins.
- Biological systems use enzymes to catalyse this hydrolysis, which takes place at body temperature without the need for acid or alkali.

Identifying the amino acids in a protein

To determine the amino acids present in a protein, the protein is first boiled with 6 mol dm^{-3} hydrochloric acid. The amino acids formed are separated using paper chromatography and made visible by spraying the paper with ninhydrin.

Each amino acid moves a different distance on the chromatography paper, making for easy identification of the amino acids in the protein.

Progress check

1 The isoelectric point of serine (R = –CH$_2$OH) is 5.7. Draw the form of the molecule in aqueous solutions of pH 3.0, pH 5.7 and pH 10.0.

2 Draw the structure of the tripeptide with the sequence alanine–serine–aspartic acid. (alanine: R = –CH$_3$; aspartic acid: R = –CH$_2$COOH).

[answers shown inverted at bottom of page]

2.

1.   pH = 3.0 / pH = 5.7 / pH = 10.0

125

6.6 Polymers

After studying this section you should be able to:

- *describe the characteristics of addition polymerisation*
- *describe the characteristics of condensation polymerisation in polypeptides and proteins, polyamides and polyesters*
- *discuss the disposal of polymers*

LEARNING SUMMARY

During the study of alkenes in AS Chemistry, you learnt about addition polymers. For A2 Chemistry, addition polymerisation is reviewed and compared with condensation polymerisation.

Monomers and polymers

A **polymer** is a compound comprising very large molecules that are multiples of simpler chemical units called **monomers**.

Monomers are small molecules which can combine together to form a single large molecule, called a **polymer**.

> **Two processes lead to formation of a polymer.**
> - **Addition polymerisation** – monomers react together forming the polymer **only**. There are no by-products.
> - **Condensation polymerisation** – monomers react together forming the polymer **and** a simple compound, usually water.
>
> **KEY POINT**

Addition polymerisation

AQA	M4	NUFFIELD	M6
EDEXCEL	M5	SALTERS	M4
OCR	M4	WJEC	CH4

In addition polymerisation:

- the monomer is an **unsaturated** molecule with a double **C=C** bond
- the double bond is **lost** as the **saturated** polymer forms.

Many different addition polymers can be formed using different monomer units based upon alkenes.

Key points from AS

- **Addition polymerisation of alkenes**
  Revise AS page 125

Notice that

n monomer molecules produce **one** polymer molecule with **n** repeat units.

You should be able to draw a short section of a polymer given the monomer units (and vice versa).

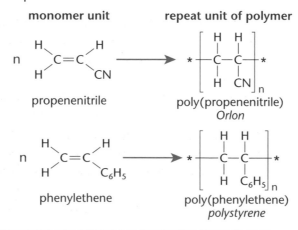

monomer unit — repeat unit of polymer

propenenitrile → poly(propenenitrile) *Orlon*

phenylethene → poly(phenylethene) *polystyrene*

Properties of monomers and polymers

The monomers are volatile liquids or gases. Polymers are solids.
This difference can be explained in terms of van der Waals' forces.

- The van der Waals' forces acting between the large polymer molecules are much stronger than those acting between the much smaller monomer molecules.

Condensation polymerisation

AQA	M4	SALTERS	M4
EDEXCEL	M4, M5	WJEC	CH4
OCR	M4	NICCEA	M4, M5
NUFFIELD	M6		

In condensation polymerisation, the formation of a bond between monomer units also produces a small molecule such as H_2O or HCl.

Condensation polymers can be divided into natural polymers and man-made (synthetic) polymers.

- Natural polymers in living organisms include proteins, cellulose, rayon and DNA.
- Man-made polymers include synthetic fibres such as polyamides (e.g. *nylon*) and polyesters (e.g. *terylene*).

Proteins and polypeptides

A protein is a chain of many amino acids linked together with 'peptide bonds' formed by **condensation polymerisation**.

Each amino acid molecule has both **amino** and **carboxyl** groups.
A **peptide** bond forms, with loss of a water molecule, between:

- the **amino** group of one amino acid molecule and
- the **carboxyl** group of another amino acid molecule.

The diagram below shows how amino acid molecules are linked together during condensation polymerisation.

> The formation of proteins from amino acids is also discussed in more detail on pp. 124–125.

> In the **repeat unit** shown here the R group may differ.

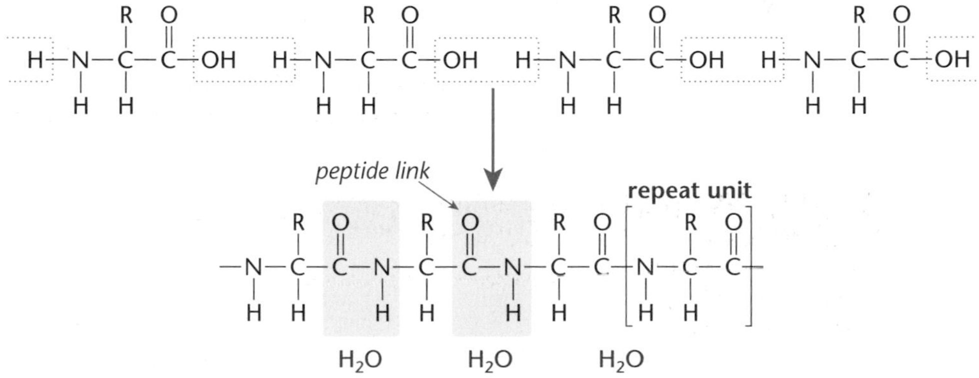

Polyamides

> Nylon, proteins and polypeptides are all polyamides.

Proteins and polypeptides are natural condensation polymers. The link between the amino acid monomer units is usually described as a **peptide link** but chemically this is identical to an **amide** group. Hence polypeptides and proteins are **natural polyamides**.

Nylon-6,6 was the first man-made condensation polymer and was synthesised as an artificial alternative to natural protein fibres such as wool and silk.

The principle used was to mimic the natural polymerisation process above but, instead of using an amino acid monomer with two different functional groups, **two** chemically different monomers are usually used:

> Compare the use of two monomers in the production of synthetic nylon-6,6 with the natural condensation polymerisation using amino acids only.

- a dicarboxylic acid **A** $HO-C(=O)-\square-C(=O)-OH$ $\square$ carbon chain or ring structure $\bigcirc$

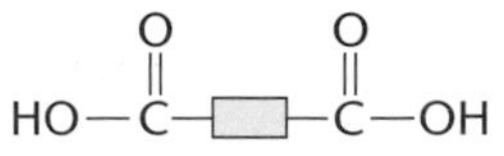

- a diamine **B** $H_2N-\bigcirc-NH_2$

> Many different polyamides can be made using different carbon chains or rings which bridge the double functional group.

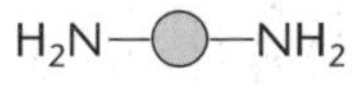

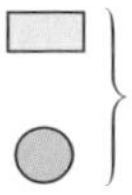

Each diamine molecule bonds to a dicarboxylic acid molecule with loss of water molecule.

- the two monomers **A** and **B** join alternately: –A–B–A–B–A–B–A–B–

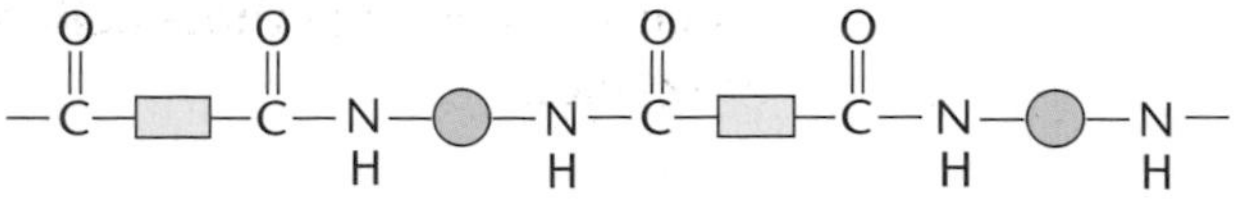

127

Nylon-6,6 gets its name from the number of carbon atoms in each monomer (diamine first).

The diamine has 6 carbon atoms.

The dicarboxylic acid has 6 carbon atoms.

Hence: nylon-6,6.

Other polyamides include Kevlar, one of the hardest materials known. Kevlar is used for bulletproof vests, belts for radial tyres, cables and reinforced panels in aircraft and boats.

The diacyl chloride for preparing nylon-6,6 would be ClOC(CH$_2$)$_4$COCl.

The diagrams below show how *nylon-6,6* is formed from its monomers.

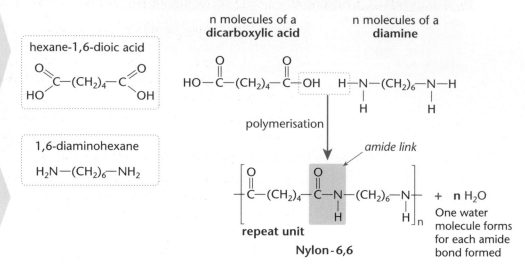

hexane-1,6-dioic acid

1,6-diaminohexane

$H_2N-(CH_2)_6-NH_2$

n molecules of a **dicarboxylic acid**

n molecules of a **diamine**

polymerisation

amide link

repeat unit

Nylon-6,6

One water molecule forms for each amide bond formed

Polyamides can also be made using a diacyl chloride instead of a dicarboxylic acid.

- The greater reactivity of an acyl chloride results in easier polymerisation.
- Hydrogen chloride is lost instead of water.

Polyesters

Polyesters are polymers made by a condensation reaction between monomers with formation of an ester group as the linkage between the molecules.

The first man-made polyester produced was *Terylene*. As with polyamides, polyesters are used as fibres for clothing.

Learn the principle behind condensation polymerisation.

In exams, you may be required to predict structures from unfamiliar monomers. However, the principle is the same.

As with artificial polyamides, man-made polyesters are usually made from **two** different monomers:

- a dicarboxylic acid
- a diol

carbon chain or ring structure

Each diol molecule bonds to a dicarboxylic acid molecule with loss of a water molecule.

Many different polyesters can be made by using different carbon chains or rings bridging the double functional group.

The diagrams below show how *Terylene* is formed from its monomers.

This is the same basic principle as for polyamides.

n molecules of a **dicarboxylic acid**

n molecules of a **diol**

benzene-1,4-dicarboxylic acid

monomer A

ethane-1,2-diol

monomer B

polymerisation

ester link

repeat unit

Terylene

One water molecule forms for each ester bond formed

Disposal of polymers

AQA	M4	SALTERS	M4
EDEXCEL	M5	WJEC	CH4
OCR	M4	NICCEA	M4
NUFFIELD	M6		

Problems with addition polymers

Addition polymers are non-polar and chemically inert.
This creates potential environmental problems during disposal of polymers.
Disposal by landfill causes long-term problems.

- Addition polymers are **non-biodegradable** and take many years to break down.

Disposal by burning can produce toxic fumes.

- Depolymerisation produces poisonous monomers.
- Disposal of poly(chloroethene) (*pvc*) by incineration can lead to the formation of very toxic dioxins if the temperature is too low.

Condensation polymers

Condensation polymers are polar.

- Condensation polymers are broken down naturally by **hydrolysis** into their monomer units. They are therefore biodegradable and easier to dispose off than addition polymers.

Progress check

1 (a) Draw the structure of the monomer needed to make poly(tetrafluoroethene).
(b) Draw the repeat unit of the polymer perspex, made from the monomer methyl 2-methylpropenoate, shown below.

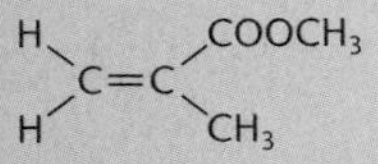

$$\begin{array}{c} H \\ \diagdown \\ H \end{array} C = C \begin{array}{c} COOCH_3 \\ \diagup \\ \diagdown CH_3 \end{array}$$

2 (a) Show the structures of the two monomers needed to make nylon-4,6.
(b) Draw a short section of nylon 4,6 and show its repeat unit.

3 Draw the structures of the two monomers needed to make the polyester shown below.

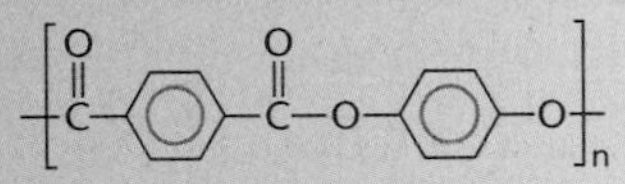

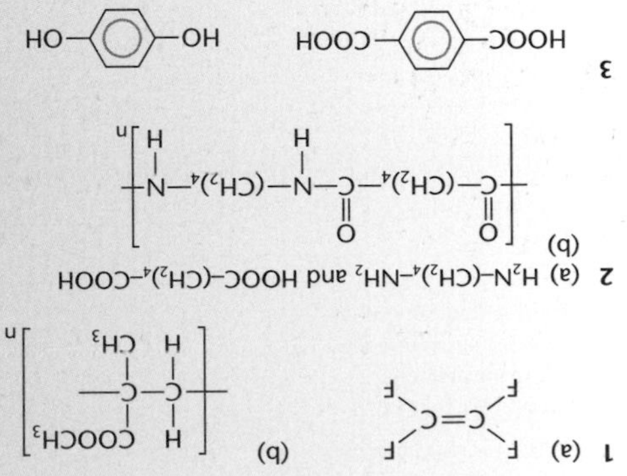

$$\left[\begin{array}{c} O \\ \parallel \\ C \end{array} - \bigcirc - \begin{array}{c} O \\ \parallel \\ C \end{array} - O - \bigcirc - O \right]_n$$

(answers, printed upside-down at foot of page):

3   HO—⬡—OH   HOOC—⬡—COOH

2 (a) $H_2N-(CH_2)_4-NH_2$ and $HOOC-(CH_2)_4-COOH$

(b) $\left[\begin{array}{c} H \\ | \\ N \end{array} -(CH_2)_4-\begin{array}{c} H \\ | \\ N \end{array} -\begin{array}{c} C \\ \parallel \\ O \end{array} -(CH_2)_4-\begin{array}{c} C \\ \parallel \\ O \end{array} \right]_n$

1 (a) $\begin{array}{c} F \\ \diagdown \\ F \end{array} C = C \begin{array}{c} F \\ \diagup \\ \diagdown F \end{array}$

(b) $\left[\begin{array}{c} H \\ | \\ -C- \\ | \\ H \end{array} \begin{array}{c} CH_3 \\ | \\ -C- \\ | \\ COOCH_3 \end{array} \right]_n$

Sample question and model answer

The characteristic reaction of benzene is **electrophilic substitution**.

(a) Select a reaction of benzene which illustrates this type of reaction. Give the reagents, the equation, the conditions under which it occurs and the name of the organic product for the reaction you have chosen.

Reagent(s):   concentrated HNO_3/H_2SO_4 ✓

Equation: $C_6H_6 + HNO_3 \longrightarrow C_6H_5NO_2 + H_2O$ ✓

Conditions: warm to 55°C ✓

Name of organic product: nitrobenzene ✓   [4]

This is standard bookwork that must be learnt.

Other reactions could have been chosen (e.g. bromination and alkylation).

Remember that 'reagents' are the chemicals 'out of the bottle' that are reacted with the organic compound.

You must give the full name or formula for any 'reagent'.

(b) For the reaction selected in (a):

(i) identify the electrophile

NO_2^+ ✓

(ii) give an equation to show its formation;

$HNO_3 + H_2SO_4 \longrightarrow H_2NO_3^+ + HSO_4^-$ ✓
$H_2NO_3^+ \longrightarrow NO_2^+ + H_2O$ ✓

More standard bookwork. An alternative response that would get both marks here is:
$HNO_3 + 2H_2SO_4 \rightarrow NO_2^+ + 2HSO_4^-$

(iii) give the mechanism for the substitution reaction.

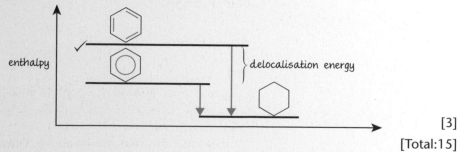

[6]

Notice how precise you need to be to score all three marks:
• arrow to electrophile ✓
• correct intermediate ✓
• arrow on C–H for loss of H+ ✓.

(c) Give two **specific** safety precautions you would need to take in **carrying out** the reaction in (a).

Use of gloves to prevent reagents coming in contact with skin ✓

Fume cupboard to prevent breathing in of poisonous fumes ✓   [2]

Notice that the answer given here justifies each safety precaution given.

(d) The enthalpy of hydrogenation of a single C=C bond is in the order of –120 kJ mol^{-1}.

(i) Assuming that benzene consists of a ring with three separate double bonds, predict the enthalpy change for the reaction.

3×-120 kJ mol^{-1} = -360 kJ mol^{-1} ✓

An easy mark.
Don't forget the sign!

(ii) The enthalpy of hydrogenation of a benzene is actually –205 kJ mol^{-1}. What can you deduce from this and your answer to part (i) about the stability of the benzene ring?
Use an enthalpy level diagram to illustrate your answer.

Benzene has extra stability arising from delocalisation of π-electrons ✓

An enthalpy diagram is a good way of showing the delocalisation energy of delocalised benzene.

1 mark here is awarded for the relative positions of the delocalised and Kekulé structures of benzene.

[3]

[Total:15]

Edexcel Specimen paper Unit Test 4 Q3 January 2000 modified

Practice examination questions

1

A hydrocarbon is known to contain a benzene ring. It has a relative molecular mass of 106 and has the following composition by mass: C, 90.56%; H, 9.44%.

(a) (i) Use the data above to show that the empirical formula is C_4H_5.

(ii) Deduce the molecular formula.

(iii) Draw structures for all possible isomers of this hydrocarbon that contain a benzene ring. [7]

(b) In the presence of a catalyst (such as aluminium chloride), one of these isomers, **A** reacts with chlorine to give only one monochloro-product, **B**.

(i) Deduce which of the isomers in **(a)(iii)** is **A**.

(ii) Draw the structure of **B**. [2]

[Total: 9]

Cambridge Chains and Rings Q2 (a)–(b) Jun 1997

2

(a) Explain why ethylamine is a Brønsted–Lowry base. [2]

(b) Why is phenylamine a weaker base than ethylamine? [2]

(c) Ethylamine can be prepared from the reaction between bromoethane and ammonia.

(i) Name the type of reaction taking place. [1]

(ii) Give the structures of three other organic substitution products which can be obtained from the reaction between bromoethane and ammonia. [3]

(d) Write an equation for the conversion of ethanenitrile into ethylamine and give one reason why this method of synthesis is superior to that in part (c). [2]

[Total: 10]

Assessment and Qualifications Alliance Specimen Test Unit 4 Q6 2000

3

2-Aminopropanoic acid (alanine), $CH_3CH(NH_2)CO_2H$, has a chiral centre and hence can exist as two optical isomers.

(a) (i) State what is meant by a *chiral centre*.

(ii) Explain how a chiral centre gives rise to optical isomerism.

(iii) Draw diagrams to show the relationship between the two optical isomers. State the bond angle around the chiral centre. [5]

(b) In aqueous solution, 2-aminopropanoic acid exists in different forms at different pH values. The zwitterion predominates between pH values of 2.3 and 9.7. Draw the displayed formula of the predominant form of 2-aminopropanoic acid at pH values of 2.0, 6.0 and 10.0. [3]

(c) 2-Aminopropanoic acid can react with an amino acid, **K**, to form the dipeptide below.

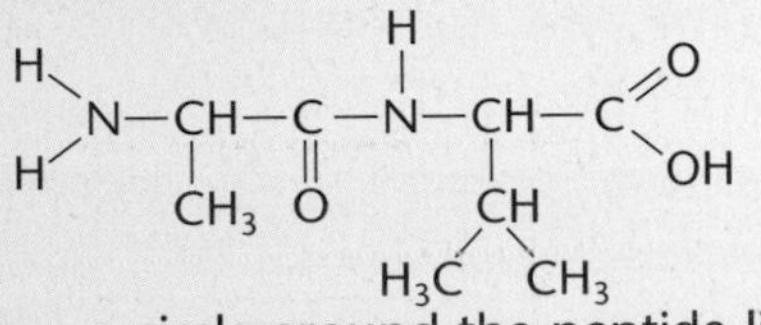

(i) Draw a circle around the peptide linkage.

(ii) Draw the displayed formula of the amino acid, **K**.

(iii) 2-Aminopropanoic acid can react with the amino acid, **K**, to form a different dipeptide from that shown above. Draw the structural formula of this other dipeptide. [3]

[Total: 11]

Cambridge Chains and Rings Q5 June 1998

Chapter 7

Analysis and synthesis

The following topics are covered in this chapter:

- Infra-red spectroscopy
- Mass spectrometry
- N.m.r. spectrosopy
- Organic synthetic routes

7.1 Infra-red spectroscopy

After studying this section you should be able to:

- know that i.r. spectroscopy can be used to identify functional groups
- interpret a simple infra-red spectrum

Using infra-red spectroscopy in analysis

AQA	M4	SALTERS	M5
EDEXCEL	M5	WJEC	CH4
OCR	M4	NICCEA	M5
NUFFIELD	M6		

The frequency of i.r. absorption is measured in wavenumbers, units: cm⁻¹.

Basic principles

Molecules are able to convert energy from infra-red radiation into energy to vibrate their bonds. Different bonds absorb different frequencies of infra-red radiation. An infra-red spectrum is obtained by passing a range of infra-red frequencies through a compound producing **absorption peaks**. The frequencies of the absorption peaks can be matched to those of known bonds to identify structural features in an unknown compound.

> Infra-red spectroscopy is useful for identifying the functional groups in a molecule.

KEY POINT

Important i.r. absorptions

You don't need to learn the absorption frequencies – the data is provided.

bond	functional group	wavenumber/cm⁻¹
O–H	hydrogen bonded in alcohols	3230 – 3550
N–H	amines	3100 – 3500
O–H	hydrogen bonded in carboxylic acids	2500 – 3300 (broad)
C=O	aldehydes, ketones, carboxylic acids, esters	1680 – 1750
C–O	alcohols, esters	1000 – 1300

An infra-red spectrum is particularly useful for identifying:
- an **alcohol** from absorption of the O–H bond
- a **carbonyl** compound from absorption of the C=O bond
- a **carboxylic** acid from absorption of the C=O bond **and** broad absorption of the O–H bond.

Interpreting infra-red spectra

AQA	M4	SALTERS	M4
EDEXCEL	M5	WJEC	CH4
OCR	M4	NICCEA	M5
NUFFIELD	M6		

Carbonyl compounds (aldehydes and ketones)

Butanone,
$CH_3COCH_2CH_3$

- C=O absorption
  1680 to 1750 cm⁻¹

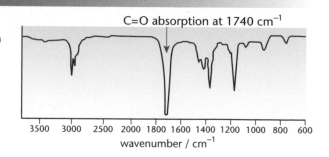

C=O absorption at 1740 cm⁻¹

wavenumber / cm⁻¹

Infra-red spectroscopy is most useful for identifying C=O and O–H bonds.

Look for the distinctive patterns.

Alcohols

Ethanol,
C_2H_5OH

- O–H absorption
  3230 to 3500 cm^{-1}
- C–O absorption
  1000 to 1300 cm^{-1}

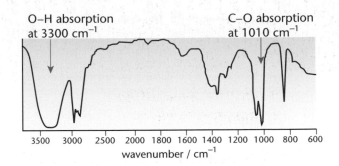

O–H absorption at 3300 cm^{-1}

C–O absorption at 1010 cm^{-1}

wavenumber / cm^{-1}

Note that all these molecules contain C–H bonds which absorb in the range 2840 to 3045 cm^{-1}.

Carboxylic acids

Propanoic acid,
C_2H_5COOH

- Very broad O–H
  absorption
  2500 to 3500 cm^{-1}
- C=O absorption
  1680 to 1750 cm^{-1}

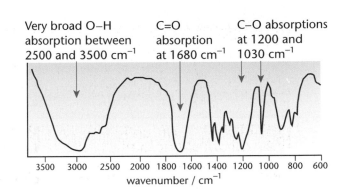

Very broad O–H absorption between 2500 and 3500 cm^{-1}

C=O absorption at 1680 cm^{-1}

C–O absorptions at 1200 and 1030 cm^{-1}

wavenumber / cm^{-1}

There are other organic groups (e.g. N–H, C=C) that absorb i.r. radiation but the principle of linking the group to the absorption wavenumber is the same.

Ester

Ethyl ethanoate,
$CH_3COOC_2H_5$

- C=O absorption
  1680 to 1750 cm^{-1}
- C–O absorption
  1000 to 1300 cm^{-1}

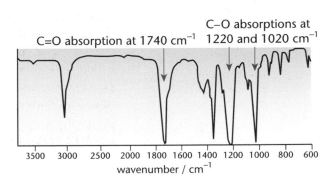

C=O absorption at 1740 cm^{-1}

C–O absorptions at 1220 and 1020 cm^{-1}

wavenumber / cm^{-1}

The fingerprint region is unique for a particular compound.

Fingerprint region

- Between 1000 and 1550 cm^{-1}

Many spectra show a complex pattern of absorption in this range.

- This pattern can allow the compound to be identified by comparing its spectrum with spectra of known compounds.

Progress check

1 Ethanol, CH_3CH_2OH was oxidised to ethanal, CH_3CHO and then to ethanoic acid, CH_3COOH. How could you use i.r. spectroscopy to follow the course of this reaction?

2 Two compounds **A** and **B** both have the molecular formula $C_3H_6O_2$. Compound **A** absorbs i.r. at 1720 cm^{-1} and 1030 cm^{-1}. Compound **B** absorbs i.r. at 1700 cm^{-1} and shows broad absorption between 2600 and 3300 cm^{-1}. Identify possible structures for **A** and **B**.

2 **A**: $HCOOC_2H_5$ or CH_3COOCH_3; **B**: CH_3CH_2COOH.

1 Ethanol absorbs at about 3300 cm^{-1} (OH); ethanal absorbs at about 1700 cm^{-1}; ethanoic acid absorbs at about 1700 cm^{-1} and 2500–3500 cm^{-1} (broad)

7.2 Mass spectrometry

After studying this section you should be able to:

- *use a mass spectrum to determine the relative molecular mass of an organic molecule*
- *use the fragmentation pattern of a mass spectrum to deduce a likely structure for an unknown compound*

LEARNING SUMMARY

Key points from AS
• **Measuring relative atomic masses** *Revise AS page 31*

For AS Chemistry, you learnt that mass spectrometry can be used to measure relative atomic masses from a mass spectrum.

For A2 Chemistry, you will study how a mass spectrometer can be used to measure relative molecular masses and identify the molecular structure of organic compounds.

Formation of molecular ions

AQA	M4	SALTERS	M5
EDEXCEL	M5	WJEC	CH4
OCR	M4	NICCEA	M5
NUFFIELD	M6		

Organic molecules can be analysed using mass spectrometry.

In the mass spectrometer, organic molecules are bombarded with electrons. This can lead to the formation of a **molecular ion**.

The equation below shows the formation of a molecular ion from butanone, $CH_3COCH_2CH_3$.

> The molecular ion peak is usually give the symbol M.
>
> m/e refers to the mass/charge ratio.

$$H_3C-\overset{\overset{\displaystyle O}{\|}}{C}-CH_2CH_3 \ + \ e^- \longrightarrow \left[H_3C-\overset{\overset{\displaystyle O}{\|}}{C}-CH_2CH_3\right]^+ + \ 2e^-$$

molecular ion, m/e: 72

> **KEY POINT**
>
> The molecular ion peak, *M*, provides the relative molecular mass of the compound.

Formation of fragment ions

AQA	M4	SALTERS	M5
EDEXCEL	M5	WJEC	CH4
OCR	M5.4	NICCEA	M5
NUFFIELD	M6		

In the conditions within the mass spectrometer, some molecular ions are fragmented by bond fission.

- The bond fission that takes place is a fairly random process:
  - different bonds are broken
  - a **mixture** of **fragment ions** is obtained.
- The mass spectrum contains both the molecular ion and the mixture of fragment ions.

> The fragmentation pattern provides clues about the molecular structure of the compound.

> **KEY POINT**
>
> Mass spectrometry of organic compounds is useful for identifying:
> - the relative molecular mass
> - parts of the structural formula.

Fragmentation of butane, C_4H_{10}

Bond fission forms a fragment ion and a free radical. The diagram below shows that fission of the same C–C bond in a butane molecule can form two different fragment ions.

> The mass spectrum of butane would contain peaks for these four ions:
>
> $C_4H_{10}^+$: m/e = 58
> $C_3H_7^+$: m/e = 43
> $C_2H_5^+$: m/e = 29
> CH_3^+: m/e = 15

$$H_3C-CH_2-CH_2\overset{\vdots}{}CH_3{}^+$$
m/e = 58

loss of 15 mass units → $H_3C-CH_2-CH_2{}^+ \ + \ {}^\bullet CH_3$
m/e = 43

loss of 43 mass units → $H_3C-CH_2-\overset{\bullet}{C}H_2 \ + \ CH_3{}^+$
m/e = 15

- Fragmentation of the molecular ion produces a fragment ion by loss of a free radical.
- The mass spectrometer only detects the fragment ion.
- The fragment ion may itself fragment into another ion and free radical.

The mass spectrum of butanone

The mass spectrum of butanone, $CH_3CH_2COCH_3$, is shown below. The m/e values of the main peaks have been labelled.

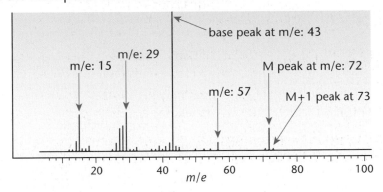

The table below shows the identity of the main peaks.

m/e	ion	fragment lost
72	$CH_3COCH_2CH_3^+$	-
57	$CH_3CH_2CO^+$	$CH_3\bullet$
43	CH_3CO^+	$CH_3CH_2\bullet$
29	$CH_3CH_2^+$	$CH_3CO\bullet$
15	CH_3^+	$CH_3CH_2CO\bullet$

The most abundant peak in the mass spectrum is the base peak.

- The base peak in the mass spectrum of butanone has m/e: 43, formed by the loss of a $CH_3CH_2\bullet$ free radical:

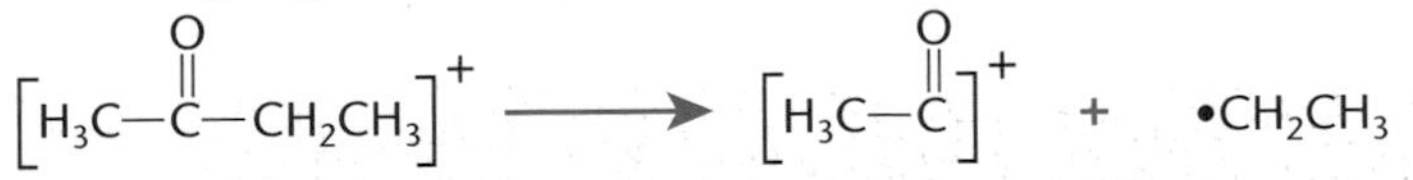

$$\left[H_3C-\overset{\overset{O}{\|}}{C}-CH_2CH_3 \right]^+ \longrightarrow \left[H_3C-\overset{\overset{O}{\|}}{C} \right]^+ + \bullet CH_2CH_3$$

molecular ion   *fragment ion*   *free radical*
m/e: 72   m/e: 43   M – 29

Common patterns in mass spectra

Different fragmentations are possible depending on the structure of the molecular ion.

- The table above shows common fragment ions and fragmented free radicals.
- The skill in interpreting a mass spectrum is in searching for known patterns, evaluating all the evidence and reassembling all the information into a molecular structure.
- You should also look for a peak at m/e 77 or loss of 77 units. This is a giveaway for the presence of a phenyl group, C_6H_5, in a molecule.

Sidebar notes:

Note that only ions are detected in the mass spectrum.

Uncharged species such as the free radical cannot be deflected in the mass spectrometer.

The M+1 peak is a small peak 1 unit higher than the molecular ion peak.

The origin of the M+1 peak is the small proportion of carbon–13 in the carbon atoms of organic molecules.

A common mistake in exams is to show both the fragmentation products as ions.

The base peak is given a relative abundance of 100. Other peaks are compared with this base peak.

Progress check

1 Write an equation for the formation of the fragment ion at m/e=29 in the mass spectrum of butanone above.

2 What peaks would you expect in the mass spectrum of pentane?

2 m/e: 72, $C_5H_{12}^+$; m/e: 57, $C_4H_9^+$; m/e: 43, $C_3H_7^+$; m/e: 29, $C_2H_5^+$; m/e CH_3^+, 15.

1 $CH_3CH_2COCH_3^+ \longrightarrow CH_3CH_2^+ + CH_3CO\bullet$

135

7.3 N.m.r. spectroscopy

After studying this section you should be able to:

- predict the different types of proton present in a molecule from chemical shift values
- predict the relative numbers of each type of proton present from an integration trace
- predict the number of protons adjacent to a given proton from the spin-spin coupling pattern
- predict possible structures for a molecule
- predict the chemical shifts and splitting patterns of the protons in a given molecule
- describe the use of D_2O in n.m.r. spectroscopy

LEARNING SUMMARY

Nuclear magnetic resonance

AQA	M4	SALTERS	M4, 5
EDEXCEL	M5	WJEC	CH4
OCR	M4	NICCEA	M5
NUFFIELD	M6		

Key principles

The nucleus of an atom of hydrogen (i.e. a proton) has a magnetic spin. When placed in a strong electromagnetic field:

- the nucleus can absorb energy from the **radio-frequency** region of the spectrum to move to a higher energy state
- **nuclear magnetic resonance** occurs as protons resonate between their spin energy states.

The different nuclear spin states in an applied magnetic field

> Only nuclei with an odd number of nucleons (neutrons + protons) possess a magnetic spin: e.g. ^1H, ^{13}C. Proton n.m.r. spectroscopy is the most useful general purpose technique.

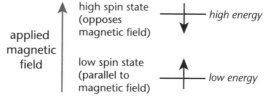

> The energy gap is equal to that provided by radio waves.

An n.m.r. spectrum shows **absorption peaks** corresponding to the radio-frequency absorbed.

Chemical shift, δ

Electrons around the nucleus **shield** the nucleus from the applied field.

- The magnetic field at the nucleus of a particular proton is different from the applied magnetic field.

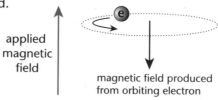

> Protons in different environments absorb at different chemical shifts.

- Different radio-frequencies are absorbed, depending on the **environment** of the proton.
- **Chemical shift** is a measure of the magnetic field experienced by protons in different environments resulting from nuclear shielding.

> Chemical shift, δ, is measured relative to a standard: tetramethylsilane (TMS), $Si(CH_3)_4$.
> - The chemical shift of TMS is defined as δ = 0.
>
> **KEY POINT**

The frequencies of the absorption peaks in an n.m.r. spectrum can be matched to those of known types of protons. This allows structural features of an unknown compound to be identified.

Typical chemical shifts

Chemical shifts indicate the **types of protons** and **functional groups** present. The table below shows typical chemical shift values for protons in different chemical environments.

The presence of an electronegative atom or group causes chemical shift 'downfield'. This is called 'deshielding'.

Notice the chemical shift caused by the carbonyl group, oxygen, halogens and a benzene ring.

You don't need to learn these chemical shifts – the data is provided on exam papers.

type of proton	chemical shift, δ
R-CH$_3$	0.7–1.6
R-CH$_2$-R	1.2–1.4
H$_3$C—C$\diagdown^O$ R—CH$_2$—C$\diagdown^O$	2.0–2.9
⬡—CH$_3$ ⬡—CH$_2$—R	2.3–2.7
X–CH$_3$ X–CH$_2$–R (X = halogen)	3.2–3.7
–O–CH$_3$ –O–CH$_2$–R	3.3–4.3
R–O–H	3.5–5.5
⬡—H	7
H$_3$C—C$\diagdown^O_H$	9.5–10
H$_3$C—C$\diagdown^O_{OH}$	11.0–11.7

The only reliable means of identifying O–H in n.m.r. is to use D$_2$O: see Identifying O–H protons, p. 139.

- The actual chemical shift may be slightly different depending upon the actual environment of the proton.
- The chemical shift for O–H can vary considerably and depends upon concentration, solvent and other factors.

Interpreting low resolution n.m.r. spectra

AQA	M4	SALTERS	M4, M5
EDEXCEL	M5	WJEC	CH4
OCR	M4	NICCEA	M5
NUFFIELD	M6		

Low resolution n.m.r. spectrum of ethanol

A low resolution n.m.r. spectrum of ethanol shows absorptions at **three** chemical shifts showing the **three different types of proton**:

The relative areas of each peak are usually measured by running a second n.m.r. spectrum as an *integration trace*.

An n.m.r. spectrum is obtained in solution. The solvent must be proton-free, usually CCl$_4$ or CDCl$_3$ being used.

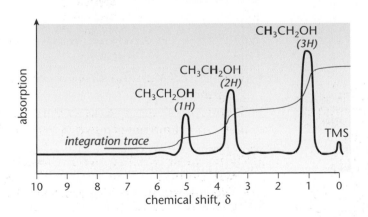

- The **area** under each peak is in direct proportion to the **number of protons** responsible for the absorption.
- The three chemical shifts can be matched to the table of chemical shifts above so that the protons responsible for each absorption can be identified.

chemical shift	number of protons	environment	type of proton
$\delta = 1.0$	3H	CH_3CH_2OH	CH_3 adjacent to a carbon chain
$\delta = 3.5$	2H	CH_3CH_2OH	CH_2 adjacent to $-O$
$\delta = 4.9$	1H	CH_3CH_2OH	OH

> **KEY POINT**
>
> A low resolution n.m.r. spectrum is useful for identifying:
> - the **number** of different types of proton from the number of peaks
> - the **type** of environment of each proton from the chemical shift
> - **how many** protons of each type from the integration trace.

Interpreting high resolution n.m.r spectra

| AQA | M4 | WJEC | CH4 |
| OCR | M4 | NICCEA | M5 |

Different types of proton have different chemical shifts.

A high resolution n.m.r. spectrum shows splitting of peaks into a pattern of sub-peaks.

Spin–spin coupling patterns:
- arise from interactions between protons on **adjacent** carbon atoms which have **different chemical shifts**
- indicate the number of **adjacent** protons.

Equivalent protons (i.e. protons with the same chemical shift) will **not** couple with one another.

The **spin-spin coupling** pattern shows as a **multiplet** – a doublet, triplet, quartet, etc.

A singlet is next to C.
A doublet is next to CH.
A triplet is next to CH$_2$.
A quadruplet is next to CH$_3$.

> **KEY POINT**
>
> We can interpret the spin-spin coupling pattern using the **n+1 rule**.
> For **n adjacent protons**, the number of peaks in a multiplet = **n+1**.

The high resolution n.m.r. spectrum of ethanol

The high resolution n.m.r. spectrum of ethanol shows spin-spin coupling patterns – some of the signals have been split into multiplets:

The multiplicity (doublet, triplet, quartet) does not indicate the number of protons on that carbon. The number of protons is given by the integration trace.

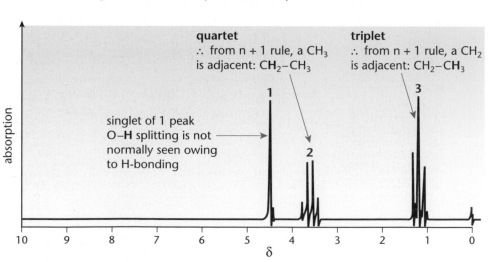

Notice the different sizes of each sub-peak within the splitting pattern:
- *doublet 1:1*
- *triplet 1:2:1*
- *quartet 1:4:4:1.*

The **chemical shift** identifies the **type of protons** responsible for each peak.

The **n+1 rule** can be used to identify the **number of adjacent protons**.

chemical shift	environment	number of adjacent protons (n)	splitting pattern (n+1)
$\delta = 1.2$	CH_3CH_2OH	2H	2+1 = 3: triplet
$\delta = 3.6$	CH_3CH_2OH	3H	3+1 = 4: quartet
$\delta = 4.5$	CH_3CH_2OH	–	singlet

- Note that O–H splitting is not normally seen owing to H-bonding or exchange with the solvent used.

Equivalent protons do not interact with each other.
- The three equivalent CH_3 protons in ethanol cause splitting of the adjacent CH_2 protons, but not amongst themselves.

> A high resolution n.m.r. spectrum is useful for identifying
> - the **number** of **adjacent** protons from the spin-spin coupling pattern.

KEY POINT

Identifying O–H protons

> The OH absorption peak at 4.5 δ is absent in D_2O.

O–H protons can absorb at different chemical shifts depending on the solvent used and the concentration. The signals are often broad and usually show no splitting pattern.

These factors can make it difficult to identify O–H protons. However, they can be identified by using deuterium oxide, D_2O.

- A second n.m.r. spectrum is run with a small quantity of D_2O added to the solvent.
- The D_2O exchanges with O–H protons and any such peak will disappear.

Solvents

Deuterated solvents such as $CDCl_3$ are used in n.m.r. spectroscopy. Any protons in a solvent such as $CHCl_3$ would produce a large proton absorption peak. By using $CDCl_3$, this absorption is absent and the spectrum is that of the organic compound alone.

Progress check

1 For each structure, predict the number of peaks in its low resolution n.m.r. spectrum corresponding to the different types of proton and also the number of each different type of proton
(e.g. CH_3CH_2OH has 3 peaks in ratio 3:2:1).
(a) CH_3OH (b) CH_3CH_2CHO (c) CH_3COCH_3 (d) $(CH_3)_2CHOH$.

2 The n.m.r. spectrum of compound **X** (C_2H_4O) has a doublet at δ 2.1 and a quartet at δ 9.8.
(a) Identify compound **X**. (Use the table of chemical shifts on page 137).
(b) How many protons are responsible for each multiplet?

2 (a) ethanal, CH_3CHO.
(b) doublet at δ 2.1, 3H; quartet at δ 9.8, 1H.

1 (a) 2 peaks, 3:1 (b) 3 peaks 3:2:1 (c) 1 peak
(d) 3 peaks, 6:1:1

7.4 Organic synthetic routes

There are many variations possible and the schemes below could not include all reactions without appearing more complicated.

Organic chemists are frequently required to synthesise an organic compound in a multistage process. This is fundamental to the design of new organic compounds such as those needed for modern drugs to combat disease, a new dye or a new fibre. With so many reactions to consider, it is essential to see how different functional groups can be interconverted and summary charts are useful for showing these links. The flow-charts on the next pages show reactions drawn from both the AS and A2 parts of A Level Chemistry.

Construct a scheme of your own using only those reactions in your course.

In revision it is useful to draw your own schemes from memory.

Aliphatic synthetic routes

AQA	M4	SALTERS	M5	
EDEXCEL	M5	WJEC	CH4	
OCR	M4	NICCEA	M4, M5	
NUFFIELD	M6			

The scheme below is based upon two main sets of reactions:
- a set based around bromoalkanes
- a set based around carboxylic acids and their derivatives.

The two sets of reactions are linked via the oxidation of primary alcohols.
- A reaction scheme based upon a secondary alcohol would result in oxidation to a ketone only.

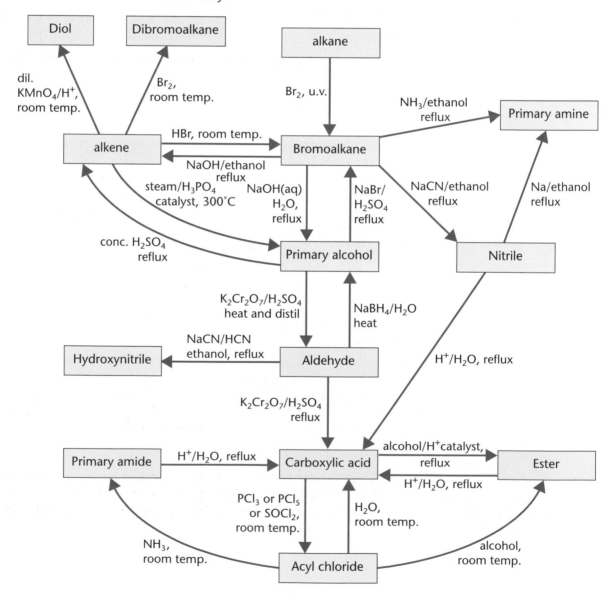

Increasing the carbon chain length

AQA	M4	SALTERS	M4, M5
EDEXCEL	M4, M5	WJEC	CH4
OCR	M4	NICCEA	M4, M5
NUFFIELD	M4, M6		

Hydrogen cyanide is useful for increasing the length of a carbon chain. The nitrile product can easily be reacted further in organic synthesis.

- A nitrile is easily **hydrolysed** by water in hot dilute acid to form a carboxylic acid.
- A nitrile is easily **reduced** by sodium in ethanol to form an amine.

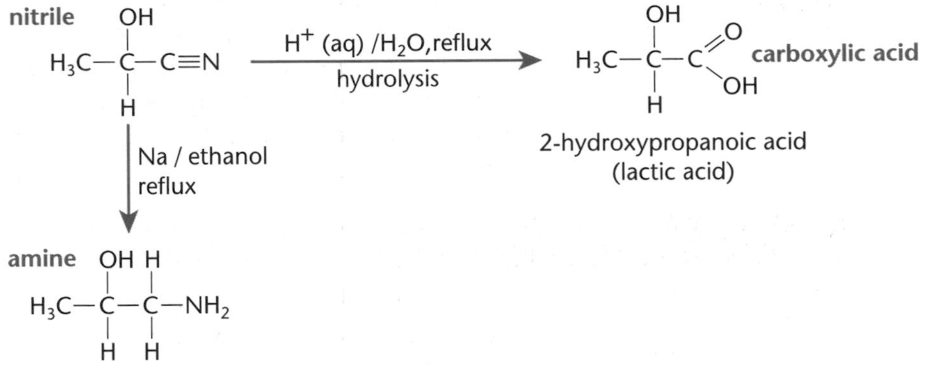

> For more details of these reactions, see 'Carboxylic acids' p. 99 and 'Amines' p. 119.

nitrile

$$H_3C-\underset{\underset{H}{|}}{\overset{\overset{OH}{|}}{C}}-C\equiv N \xrightarrow[\text{hydrolysis}]{H^+ (aq) /H_2O, reflux} H_3C-\underset{\underset{H}{|}}{\overset{\overset{OH}{|}}{C}}-C\overset{O}{\underset{OH}{}} \quad \textbf{carboxylic acid}$$

2-hydroxypropanoic acid
(lactic acid)

Na / ethanol
reflux

amine

$$H_3C-\underset{\underset{H}{|}}{\overset{\overset{OH}{|}}{C}}-\underset{\underset{H}{|}}{\overset{\overset{H}{|}}{C}}-NH_2$$

Aromatic synthetic routes

AQA	M4	SALTERS	M4, M5
EDEXCEL	M5	WJEC	CH4
OCR	M4	NICCEA	M5
NUFFIELD	M4, M6		

Compared with aliphatic organic chemistry at A Level, there are comparatively few aromatic reactions.

Benzene

- Reactions involving the benzene ring are mainly electrophilic substitution.

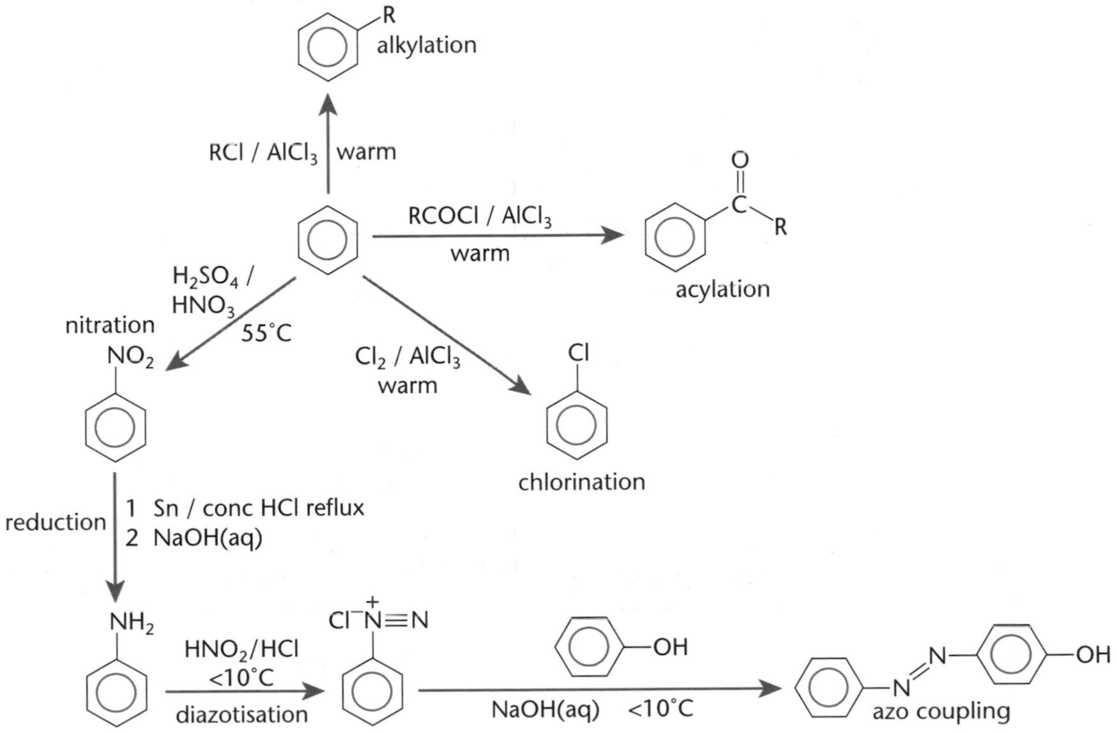

alkylation

RCl / AlCl₃ warm

RCOCl / AlCl₃
warm

acylation

H₂SO₄ / HNO₃

nitration   55°C

NO₂

Cl₂ / AlCl₃
warm

chlorination

reduction   1 Sn / conc HCl reflux
2 NaOH(aq)

NH₂

HNO₂/HCl
<10°C
diazotisation

Cl⁻N≡N (⁺)

—OH

NaOH(aq)   <10°C

N=N   —OH
azo coupling

> It is essential, if you are to answer such questions, that you thoroughly learn suitable reagents and conditions for all the reactions in your syllabus.

- Questions are often set in exams asking for reagents, conditions or products.
- More searching problems may expect a synthetic route from a starting material to a final product. The synthetic route may include several stages.

Sample question and model answer

An ester **A** of molecular formula $C_4H_8O_2$ was hydrolysed to form a carboxylic acid **B** and an alcohol **C**.

(a) The structures of the four esters of molecular formula $C_4H_8O_2$ are shown below.

$$CH_3CH_2COOCH_3 \quad CH_3COOCH_2CH_3 \quad HCOOCH_2CH_2CH_3 \quad HCOOCH(CH_3)_2$$
$$\text{I} \qquad\qquad\quad \text{II} \qquad\qquad\quad \text{III} \qquad\qquad\qquad \text{IV}$$

For each structure predict the number of peaks in its low resolution n.m.r. spectrum corresponding to the different types of hydrogen and also the number of each different type of hydrogen.

Compound I, $CH_3CH_2COOCH_3$, has 3 peaks ✓ in the ratio 3:2:3 ✓
Compound II, $CH_3COOCH_2CH_3$, has 3 peaks ✓ in the ratio 3:2:3 ✓
Compound III, $HCOOCH_2CH_2CH_3$, has 4 peaks ✓ in the ratio 1:2:2:3 ✓
Compound IV, $HCOOCH(CH_3)_2$, has 3 peaks ✓ in the ratio 1:1:6 ✓   [8]

> When we are predicting the number of peaks, we are identifying the number of different types of proton.
>
> In compound IV, $HCOOCH(CH_3)_2$, both the methyl groups (6H) are equivalent and will have the same chemical shift.

(b) The low resolution n.m.r. spectra of **A** and **B** are shown below with the number of hydrogens for each peak indicated. Use these spectra to decide which of the four esters is the correct one for **A** and explain your answer.

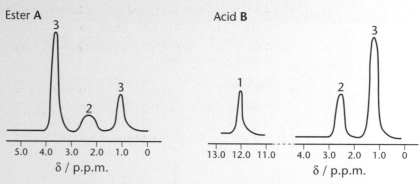

Ester A has peaks in the ratio 3:2:3 and must be either I, $CH_3CH_2COOCH_3$, or II, $CH_3COOCH_2CH_3$. ✓
The acid obtained by hydrolysis of I would be CH_3CH_2COOCH – 3 peaks.
The acid obtained by hydrolysis of II would be CH_3COOH – 2 peaks. ✓
Acid B has 3 peaks. ∴ ester A must be I, $CH_3CH_2COOCH_3$. ✓   [3]

> In this question, no chemical shifts were supplied but they were not needed. The problem can be solved using logic and intuition.

(c) Predict, with reasons, what you would expect to see in a high-resolution n.m.r. spectrum of ester **A**.

The methyl CH_3 group will be split into a triplet ✓ (2 adjacent protons) ✓
The CH_2 group will be split into a quartet ✓ (3 adjacent protons) ✓
The OCH_3 group will appear as a singlet ✓ (no adjacent protons) ✓
On hydrolysis, the OCH_3 group has been lost. Comparing the spectra of ester A and acid B, this must be the peak at δ 3.7. ✓
∴ the n.m.r. of ester A will show
    a singlet at δ 3.7
    a quartet at δ 2.2
    a triplet at δ 1.0 ✓   [8]

[Total: 19]

> The n+1 rule is applied here. We have solved this by logic without chemical shift data.

NEAB Kinetics and Organic Chemistry Q6(a)-(b) Feb 1995

Part (c) added

Practice examination questions

1

The infra-red spectra shown below were obtained from two isomeric compounds **J** and **K**, of general formula $C_pH_qO_r$.

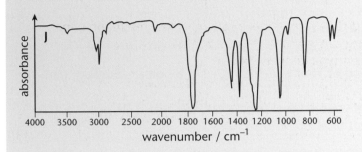
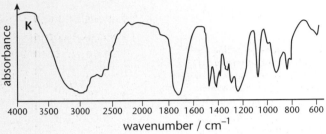

(a) Identify characteristic absorptions and hence name the functional groups present in each of **J** and **K**. (Use the table of i.r. absorptions on page 132.)

(b) Analysis of **J** showed that it contained 54.5% by mass of carbon, and 9.10% by mass of hydrogen. Use these data to determine the empirical formula of **J**.

(c) The M_r of **J** is 88. Show a possible structure for **J**.   [10]

[Total: 10]

Cambridge Methods of Analysis and Detection Q6 (d) modified March 2000

2

The proton n.m.r. spectrum of an alcohol, **A**, $C_5H_{12}O$, is shown below.

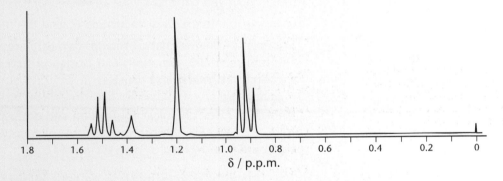

The measured integration trace gives the ratio 0.90 to 0.45 to 2.70 to 1.35 for the peaks at δ 1.52, 1.39, 1.21 and 0.93, respectively.

(a) What compound is responsible for the signal at δ 0?   [1]

(b) How many different types of proton are present in compound **A**?   [1]

(c) What is the ratio of the numbers of each type of proton?   [1]

(d) The peaks at δ 1.52 and δ 0.93 arise from the presence of a single alkyl group. Identify this group and explain the splitting pattern.   [5]

(e) What can be deduced from the single peak at δ 1.21 and its integration value?   [1]

(f) Give the structure of compound **A**.   [1]

[Total: 10]

Assessment and Qualifications Alliance Further Organic Chemistry Q5 March 1999

Chapter 8
Synoptic assessment

What is synoptic assessment?

Part of your chemistry course is assessed using **synoptic questions**. These are written so that you can **draw together** knowledge, understanding and skills learned in **different parts** of AS and A2 Chemistry.

Synoptic assessment:

- takes place at the **end of your course**, mainly in module exams but part may be assessed in coursework
- accounts for **20% of your A Level course** but, because it takes place at the end of the course, it makes up **40% of A2**.

> Synoptic assessment emphasises your understanding and your application of the principles included in your Chemistry course.

What type of questions will be asked?

You will need to answer **two** main types of synoptic question.

1 **You make links** and **use connections** between different areas of chemistry. Some examples are given below.

> You make the links between different areas of chemistry yourself. You choose and use the context for your answer.

   Using examples drawn from different parts of your chemistry course:
   - discuss the role of a lone pair in chemistry…
   - discuss the chemistry of water…
   - compare the common types of chemical bonding…

2 **You use ideas and skills** which **permeate chemistry**.

This type of question will be more structured and you are unlikely to have as much choice in how you construct your answer.

Your task is to interpret any information using the 'big ideas' of chemistry.

> The context has been made for you. Often this will be a situation or will involve data that you will not have seen before.

   Examples of themes that permeate chemistry are shown below:
   - formulae, moles, equations and oxidation states
   - chemical bonding and structure
   - periodicity
   - reaction rates, chemical equilibrium and enthalpy changes.

How will you gain synoptic skills?

Luckily chemistry is very much a synoptic subject. Throughout your study of chemistry, you apply many of the ideas and skills learnt during AS chemistry or your GCSE course. In studying A2 Chemistry, you will have been using synoptic skills naturally, probably without realising it. You certainly can make little real progress in chemistry without a sound grasp of concepts such as equations, the mole, structure and bonding!

> A synoptic question may present you with a new piece of information.
>
> You may be expected to calculate formulae from data, write correct formulae, balance equations and perform quantitative calculations using the Mole concept, use knowledge of the Periodic Table to predict reactions of unfamiliar elements or compounds…

How are synoptic questions different?

Different Exam Boards have used different strategies for their synoptic assessment and some of these are listed below.

- Structured questions: short and extended answers based on important chemical themes drawn from AS and A2 Chemistry.
- An essay-type question.
- A problem set in a practical context.
- Objective questions: multiple-choice and either matching pairs or multiple completion items.

Worked synoptic questions and model answers

Here some of the big ideas of chemistry are being tested: formulae, moles and equations. These will always appear on an exam paper with synoptic questions.

You have nothing to fear from this type of question provided that your chemistry is sound.

Structured type question

1

Nitrogen oxides such as nitrogen dioxide, NO_2, are pollutants and cause considerable harm to the environment.

(a) Nitrogen dioxide and dinitrogen tetroxide exist in the following equilibrium:

$$2NO_2(g) \rightleftharpoons N_2O_4(g)$$

When 11.04 g of nitrogen dioxide were placed in a vessel at a fixed temperature, 5.52 g of dinitrogen tetroxide were produced at equilibrium under a pressure of 100 kPa.

Equilibrium masses:
NO_2 = 5.52 g;
N_2O_4 = 11.04 – 5.52 = 5.52 g.

moles = $\dfrac{\text{mass}}{\text{molar mass}}$.

(i) Calculate the mole fractions of NO_2 and N_2O_4 present in the equilibrium mixture.

amount NO_2 = $\dfrac{5.52}{46.0}$ = 0.120 mol ✓; amount N_2O_4 = $\dfrac{5.52}{92.0}$ = 0.060 mol ✓

When working with mole fractions, remember the total number of moles
0.120 + 0.060 = 0.180 mol.

mole fraction (NO_2) = $\dfrac{\text{moles}(NO_2)}{\text{total number of moles}}$ = $\dfrac{0.120}{0.180}$ = 0.667 ✓

mole fraction (N_2O_4) = $\dfrac{0.060}{0.180}$ = 0.333 ✓

[4]

(ii) Calculate the value of the equilibrium constant, K_p, and state its units.

For K_p, don't use square brackets – these mean concentrations!

units: $\dfrac{\text{(kPa)}}{\text{(kPa)}^2}$ = kPa^{-1}

partial pressure, p = mole fraction × total pressure, P

p_{NO_2} = 0.667 × 100 = 66.7 kPa; $p_{N_2O_4}$ = 0.333 × 100 = 33.3 kPa ✓

$K_p = \dfrac{p_{N_2O_4}}{p_{NO_2}^2}$ ✓ = $\dfrac{(33.3)}{(66.7)^2}$ = 7.49 × 10^{-3} ✓ kPa^{-1} ✓

[4]

(iii) State and explain the effect on the mole fraction of NO_2 when the pressure is increased at constant temperature.

Use le Chatelier's process. There are 3 key points here for the 3 marks.

From the equation: $2NO_2(g) \rightleftharpoons N_2O_4(g)$,
the system will oppose an increase in pressure ✓ by moving towards the side of the equation with fewer gas moles. ✓
∴ the $[NO_2]$ decreases and mole fraction of NO_2 decreases. ✓   [3]

(b) State **two** environmental consequences of nitrogen oxides.

nitrogen oxides cause acid (rain) ✓, greenhouse effect ✓, photochemical smog/ozone build-up ✓   [2 max]

(c) Not all nitrogen compounds are harmful: some, such as nitrogen fertilisers, are beneficial to man.

A nitrogen fertiliser, **D**, has the composition by mass Na, 27.1%; N, 16.5%; O, 56.4%. On heating, 3.40 g of **D** was broken down into sodium nitrite, $NaNO_2$, and oxygen gas.

Suggest an identity for the fertiliser, **D**, and calculate the volume of oxygen that was formed. [Assume that 1 mole of gas molecules occupy 24 dm^3.]

D Na: N: O = 27.1/23 : 16.5/14.0 : 56.4/16.0 ✓ = $NaNO_3$ ✓
$2NaNO_3 \longrightarrow 2NaNO_2 + O_2$
∴ 0.04 mol $\longrightarrow$ 0.02 mol ✓
volume of $O_2(g)$ = 24 × 0.0200 dm^3 = 0.48 dm^3 ✓   [4]

[Total: 17]

Worked synoptic questions and model answers (continued)

Essay type question

2

Write an essay on the chemistry of ammonia.

In your answer you should include a discussion of the bonding in, and structure of the molecule, consider its industrial preparation and you should bring together its reactions in both inorganic and organic chemistry.

Ammonia, NH_3 is bonded by 3 covalent bonds ✓. There are 4 centres of electron density around the central N atom (3 covalent bonds + 1 lone pair ✓ of electrons). This gives a pyramidal shape ✓ with a H–N–H bond angle of 107° ✓:

[4]

Ammonia is prepared in the Haber process by the reaction of nitrogen with hydrogen, catalysed by iron✓:

$$N_2(g) + 3H_2(g) \rightleftharpoons 2NH_3(g) \checkmark$$

The optimum equilibrium yield is a high pressure and low temperature ✓. However, the reaction proceeds too slowly at low temperatures and a compromise is necessary between a high enough temperature to overcome activation energy and as low a temperature as possible to secure a reasonable equilibrium yield. ✓

[4]

Ammonia is a Brønsted–Lowry base ✓

e.g: ammonia forms salts with inorganic acids:
$NH_3 + HCl \longrightarrow NH_4^+Cl^-$ ✓
ammonia forms salts with organic acids:
$NH_3 + CH_3COOH \longrightarrow CH_3COO^-NH_4^+$ ✓

Ammonia reacts with complex aqua–ions,
    as a base forming a metal hydroxide precipitate: ✓
$[Cu(H_2O)_6]^{2+}(aq) + 2NH_3(aq) \longrightarrow Cu(OH)_2(H_2O)_4(s) + 2NH_4^+(aq)$ ✓
as a ligand, forming complex ions by ligand substitution with metal aqua–ions: ✓
$[Cr(H_2O)_6]^{3+} + 6NH_3 \longrightarrow [Cr(NH_3)_6]^{3+} + 6H_2O$ ✓

Ammonia can behave as a nucleophile, ✓
    reacting with halogenoalkanes in a nucleophilic substitution reaction:✓
$RBr + 2NH_3 \longrightarrow RNH_2 + NH_4^+Br^-$ ✓

[7 max]

[Total: 17]

Assessment and Qualifications Alliance Specimen Unit Test 5 Q8 2000

You have been given a little help about what needs to go into your essay.

You will always score more marks by answering 'a bit about everything' than 'a lot about one part'.

Bonding and structure of NH_3.

Industrial preparation of NH_3.

The choice of reaction is yours – you make the links.

To gain full marks in this part you must include reactions from both inorganic and organic reactions of NH_3.

In the marking scheme, there will be a 'ceiling' of marks for each branch of chemistry and for each 'type of reaction'.

A description of ten ligand substitution reactions would only answer from one area of chemistry and, even if correct, would score poorly.

Practice examination questions

Structured type questions

1

A solution of cobalt(II) chloride was reacted with ammonia and ammonium chloride while a current of air was blown through the mixture. A red compound **X** was produced which contained a complex ion of cobalt. The compound had the following composition.

	% by mass
Co	23.6
N	27.9
H	6.0
Cl	42.5

(a) Use the data to calculate the empirical formula of the compound **X**. [3]

(b) In the reaction in which **X** is formed from cobalt(II) chloride, explain the role of:
   (i) the air
   (ii) the ammonia and ammonium chloride. [3]

(c) A solution of cobalt(II) chloride reacts with concentrated hydrochloric acid to form a stable complex ion. A solution of calcium chloride does not form a corresponding complex. Use your knowledge of atomic structure to explain this difference, and suggest other differences you would expect between the chemistry of cobalt and calcium. [4]

[Total: 10]

Nuffield Specimen Unit Test 6 Q3 2000

2

The structure of the silkworm moth sex attractant, bombykol, is:

$$H_3C-(CH_2)_2-CH=CH-CH=CH-(CH_2)_8-CH_2OH$$

(a) Predict some of the properties you would expect for bombykol.
   You should comment on:

   (i) the likely solubility of bombykol in water

   (ii) the number of possible geometric isomers

   (iii) its likely reaction with four reactants of your choice.
      Write equations or reaction schemes for the reactions you choose, showing the structures of the organic products clearly. [8]

(b) When bombykol is treated with ozone in the presence of water the molecule splits into fragments wherever there is a C=C double bond.

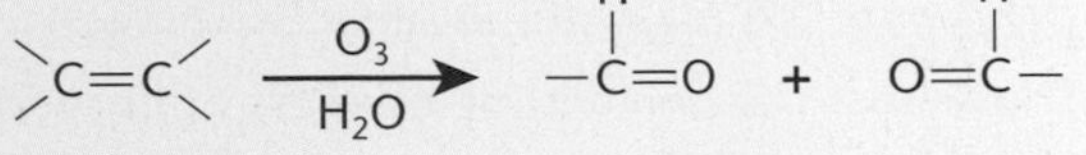

The aldehyde groups are converted to carboxylic acids in oxidizing conditions.

Predict the oxidation products formed when bombykol is reacted with ozone in oxidising conditions. [2]

[Total: 10]

Nuffield Specimen Unit Test 6 Q2 2000

Practice examination questions (continued)

Essay type question

3

Redox reactions are an important type of reaction in chemistry.

Explain what is meant by a redox reaction. Illustrate your answer with **two** examples drawn from inorganic chemistry (one of which should involve a transition element) and **two** examples from organic chemistry. [Total: 14]

Cambridge Trend and Patterns Specimen Unit Test Q4 2000

Interpretation of data

4

Using knowledge, principles and concepts from different areas of chemistry, explain and interpret, as fully as you can, the data given in the table below. In order to gain full credit, you will need to consider each type of information separately and also to link this information together.

compound	boiling point /K	properties of a 0.1 mol dm^{-3} solution	
		electrical conductivity	[H$^+$]/mol dm^{-3}
NaCl	1686	good	1.0×10^{-7}
CH$_3$COOH	391	slight	1.3×10^{-3}
CH$_3$CH$_2$OH	352	poor	1.0×10^{-7}
AlCl$_3$	451	good	3.0×10^{-1}

[Total: 14]

Cambridge Unifying concepts Specimen Unit Test Q4 2000

Practical-based problem

5

You are required to plan an experiment to determine the percentage by mass of bromine in a bromoalkane. The bromoalkane, which boils at 75°C, can be hydrolysed completely by heating with an appropriate amount of boiling, aqueous sodium hydroxide for about 40 minutes. The bromide ion released can be estimated by converting it into a silver bromide precipitate which is subsequently weighed.

(a) Write equations for the reactions which occur.

(b) Describe how you would carry out the hydrolysis, giving details of the apparatus and the conditions which you would use.

(c) Describe, giving details of the apparatus and reagents, how you would obtain a silver bromide precipitate from the hydrolysis solution and how you would determine the mass of the silver bromide.

(d) Show how the percentage by mass of bromide ion, in the original haloalkane, can be calculated.

[Total: 15]

Assessment and Qualifications Alliance Specimen Unit Test 5 Q9 2000

Practice examination answers

1 Reaction rates

1 (a) The colour changes from colourless to orange-brown as iodine forms. ✓
This can be monitored using a colorimeter. ✓ [2]

  (b) (i) 1st order with respect to H_2O_2 ✓
Double concentration of H_2O_2, rate doubles ✓
1st order with respect to I^- ✓
Quadruple concentration of I^-, rate x4 ✓
Zero order with respect to H^+ ✓
Double concentration of H^+, rate stays constant ✓ [6]

    (ii) rate = $k[H_2O_2][I^-]$ ✓ [1]

    (iii) 2.8×10^{-2} dm^3 mol^{-1} s^{-1} ✓✓ (from experiment 1) [2]

  (c) (i) The slowest step of a multi-step process ✓ [1]

    (ii) $H_2O_2 + I^- \longrightarrow H_2O + IO^-$ ✓✓ [2]

                                                  [Total: 14]

2 (a) (i) rate = $k[X][Y]^2$ ✓✓ [2]

    (ii) 3 ✓ [1]

    (iii) 8 ✓ [1]

  (b) (i) Overall order of reaction = 2 ✓
Concentrations of both A and B doubled, rate $\times$ 4 ✓ [2]

    (ii) Order with respect to B = 0 ✓
When [A] doubled and [B] is constant (experiments 2 and 3), rate $\times$ 4.
∴ [B] has no effect ✓ [2]

    (iii) rate = $k[A]^2$ ✓ [1]

    (iv) $k = 8.75 \times 10^{-3}$ dm^3 mol^{-1} s^{-1} ✓✓ [2]

                                                  [Total: 11]

2 Chemical equilibrium

1 (a) (i) $K_c = \dfrac{[CH_3COOCH_2CH_3]\,[H_2O]}{[CH_3COOH]\,[C_2H_5OH]}$ ✓ [1]

    (ii) 3.86 ✓✓✓ [3]

    (iii) Equilibrium moves to the right ✓ to counteract the added CH_3COOH ✓ [2]

  (b) No effect ✓ The catalyst is not in the expression for K_c ✓ [2]

                                                  [Total: 8]

2 (a) Yield of SO_3 is less. ✓ Forward reaction is exothermic. ✓ Equilibrium position
moves to the left to counteract the increase in temperature. ✓ [3]

  (b) Partial pressure of O_2 = 48 kPa. ✓ Mole fraction of O_2 = 48/120 = 0.4 ✓✓ [3]

149

(c) (i) $K_p = \dfrac{p_{SO_3}^2}{p_{SO_2}^2 \times p_{O_2}}$ ✓ [1]

    (ii) 2.91×10^{-2} kPa^{-1} ✓✓✓ [3]

[Total: 10]

3 (a) The mixture becomes more concentrated ✓ [1]

(b) Increase in pressure. ✓ Equilibrium position moves to the left (the side with least gas molecules) to relieve the increase in pressure. ✓ [2]

(c) $K_p = \dfrac{p_{NO_2}^2}{p_{N_2O_4}}$ ✓ [1]

(d) 2.68 atm ✓✓✓ [3]

[Total: 7]

4 (a) Proton donor ✓ [1]

(b) $HCl(g) + H_2O(l) \rightleftharpoons H_3O^+(aq) + Cl^-(aq)$ ✓
H_2O is a base because it accepts a proton. ✓ [2]

(c) $NH_3(g) + H_2O(l) \rightleftharpoons NH_4^+(aq) + OH^-(aq)$ ✓
H_2O is an acid because it donates a proton. ✓ [2]

(d) $H_2SO_4 + HNO_3 \rightleftharpoons H_2NO_3^+ + HSO_4^-$ ✓
$H_2NO_3^+ \longrightarrow NO_2^+ + H_2O$ ✓
HNO_3 is a base because it accepts a proton. ✓ [3]

(e) (i) Species that only partially dissociates to donate protons ✓ [1]

    (ii) $K_a = \dfrac{[H^+(aq)]\,[A^-(aq)]}{[HA(aq)]}$ ✓ [1]

    (iii) Concentration of undissociated HX is very small ✓ [1]

    (iv) Strong acid ✓ because [HX] is much smaller than [H$^+$] and [Cl$^-$]. ✓ [2]

[Total: 13]

5 (a) (i) $pH = -\log[H^+(aq)]$ ✓ [1]

    (ii) An acid that is partially ionised ✓ [1]
        $HCOOH + H_2O \rightleftharpoons HCOO^- + H_3O^+$ ✓ [1]

(b) (i) $pH = 0.82$ ✓✓ [2]

    (ii) $pH = 13.9$ ✓✓✓ [3]

    (iii) $pH = 2.28$ ✓✓✓✓ [4]

(c) (i) It resists changes in pH ✓ with small amounts of acid or base. ✓ [2]

    (ii) $pH = 5.26$ ✓✓✓ [3]

[Total: 17]

3 Energy changes in chemistry

1 (a)

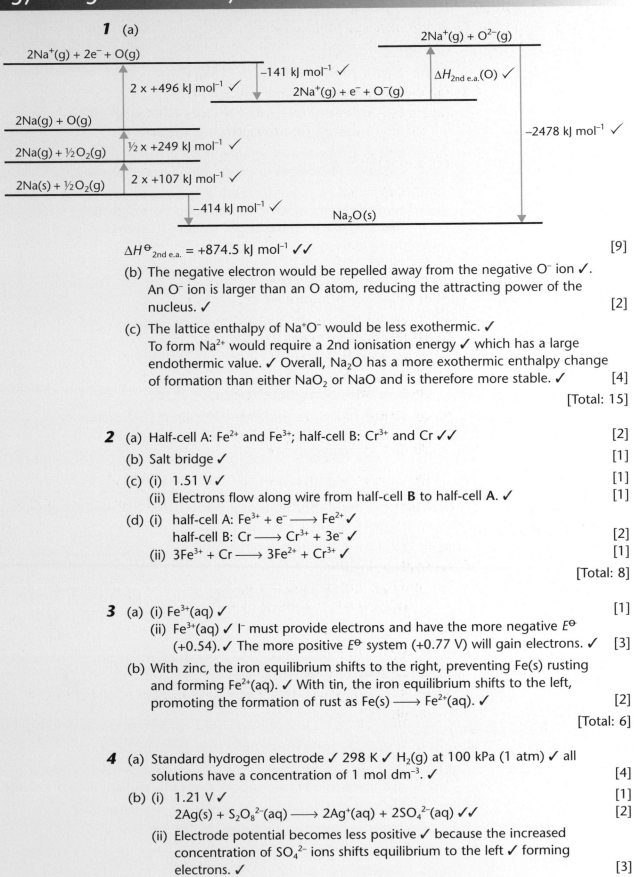

$\Delta H^{\ominus}_{\text{2nd e.a.}}$ = +874.5 kJ mol^{-1} ✓✓ [9]

(b) The negative electron would be repelled away from the negative O$^-$ ion ✓.
An O$^-$ ion is larger than an O atom, reducing the attracting power of the
nucleus. ✓ [2]

(c) The lattice enthalpy of Na$^+$O$^-$ would be less exothermic. ✓
To form Na^{2+} would require a 2nd ionisation energy ✓ which has a large
endothermic value. ✓ Overall, Na$_2$O has a more exothermic enthalpy change
of formation than either NaO$_2$ or NaO and is therefore more stable. ✓ [4]

[Total: 15]

2 (a) Half-cell A: Fe^{2+} and Fe^{3+}; half-cell B: Cr^{3+} and Cr ✓✓ [2]

(b) Salt bridge ✓ [1]

(c) (i) 1.51 V ✓ [1]
  (ii) Electrons flow along wire from half-cell **B** to half-cell **A**. ✓ [1]

(d) (i) half-cell A: Fe^{3+} + e$^-$ $\longrightarrow$ Fe^{2+} ✓
    half-cell B: Cr $\longrightarrow$ Cr^{3+} + 3e$^-$ ✓ [2]
  (ii) 3Fe^{3+} + Cr $\longrightarrow$ 3Fe^{2+} + Cr^{3+} ✓ [1]

[Total: 8]

3 (a) (i) Fe^{3+}(aq) ✓ [1]
  (ii) Fe^{3+}(aq) ✓ I$^-$ must provide electrons and have the more negative $E^{\ominus}$
    (+0.54). ✓ The more positive $E^{\ominus}$ system (+0.77 V) will gain electrons. ✓ [3]

(b) With zinc, the iron equilibrium shifts to the right, preventing Fe(s) rusting
and forming Fe^{2+}(aq). ✓ With tin, the iron equilibrium shifts to the left,
promoting the formation of rust as Fe(s) $\longrightarrow$ Fe^{2+}(aq). ✓ [2]

[Total: 6]

4 (a) Standard hydrogen electrode ✓ 298 K ✓ H$_2$(g) at 100 kPa (1 atm) ✓ all
solutions have a concentration of 1 mol dm^{-3}. ✓ [4]

(b) (i) 1.21 V ✓ [1]
    2Ag(s) + S$_2$O$_8$$^{2-}$(aq) $\longrightarrow$ 2Ag$^+$(aq) + 2SO$_4$$^{2-}$(aq) ✓✓ [2]

  (ii) Electrode potential becomes less positive ✓ because the increased
    concentration of SO$_4$$^{2-}$ ions shifts equilibrium to the left ✓ forming
    electrons. ✓ [3]

[Total: 10]

4 The Periodic Table

1 (a)

	Na	Mg	Al	Si
Formula of anhydrous chloride	NaCl	$MgCl_2$	Al_2Cl_6	$SiCl_4$

✓✓ [2]

(b) (i) $NaCl(s) + aq \longrightarrow Na^+(aq) + Cl^-(aq)$ ✓

$MgCl_2(s) + aq \longrightarrow Mg^{2+}(aq) + 2Cl^-(aq)$ ✓

$Al_2Cl_6(s) + 3H_2O(l) \longrightarrow 2Al(OH)_3(s) + 6HCl(aq)$ ✓

$SiCl_4(l) + 2H_2O(l) \longrightarrow SiO_2(s) + 4HCl(aq)$ ✓ [4]

(ii) Ionic chlorides dissolve in water ✓

Covalent chlorides are hydrolysed and react with water ✓ [2]

(iii) Carbon has no vacant d orbitals ✓ to accept a lone pair ✓ from the oxygen atom in water. ✓ The four chlorine atoms sterically hinder approach to the central carbon atom by water molecules. ✓ [4]

(c) $Al(OH)_3(s) + 3H^+(aq) \longrightarrow Al^{3+}(aq) + 3H_2O(l)$ ✓

$Al(OH)_3(s) + OH^-(aq) \longrightarrow Al(OH)_4^-(aq)$ ✓ [2]

[Total: 14]

2 (a) (i) $Na_2O(s) + H_2O(l) \longrightarrow 2NaOH(aq)$ ✓ ; pH = 14 ✓

(ii) $SO_2(g) + H_2O(l) \longrightarrow H_2SO_3(aq)$ ✓ ; pH = 3 ✓ [4]

(b) Ionic oxides form alkaline solutions ✓

Covalent oxides form acidic solutions ✓ [2]

(c) (i) $MgCl_2(s) + aq \longrightarrow Mg^{?+}(aq) + 2Cl^-(aq)$ ✓ ; pH = 6 ✓

(ii) $SiCl_4(l) + 2H_2O(l) \longrightarrow SiO_2(s) + 4HCl(aq)$ ✓ ; pH = 1 ✓ [4]

[Total: 10]

3 (a) A dative covalent bond forms ✓ between a lone pair on a ligand and the central metal ion. ✓ [2]

(b) (i) 1,2-diaminoethane ✓ [1]

(ii) Bidentate ✓ [1]

(iii) +3 ✓ [1]

(iv) 6 ✓ [1]

(v) Octahedral ✓ [1]

(c) To $CoCl_2$, add 1,2-diaminoethane ✓ and concentrated hydrochloric acid ✓

Bubble O_2 through the solution ✓. [3]

[Total: 10]

4 (a) Co: $1s^2 2s^2 2s^6 3s^2 3p^6 3d^7 4s^2$ ✓; Co^{2+}: $1s^2 2s^2 2s^6 3s^2 3p^6 3d^7$ ✓ [2]

(b) A d-block element has its highest energy electron in a d sub-shell. ✓

A transition element has at least one ion with a partially filled d sub-shell. ✓ [2]

(c) (i) Lone pair of electrons on electronegative oxygen atom ✓ [1]

(ii) Covalent bonding and ✓ dative covalent bonding ✓ [2]

(iii) Octahedral ✓ [1]

(d) (i) Cobalt(II) hydroxide ✓

$[Co(H_2O)_6]^{2+}(aq) + 2OH^-(aq) \longrightarrow Co(OH)_2(H_2O)_4(s) + 2H_2O(l)$ ✓ [2]

(ii) Precipitation ✓ [1]

      (iii) $[Co(OH)_6]^{4-}$ ✓   [1]

      (iv) the hexaamminecobalt(II) ion ✓   [1]

      (v) $[Co(H_2O)_6]^{2+}(aq) + 4Cl^-(aq) \longrightarrow [CoCl_4]^{2-}(aq) + 6H_2O(l)$ ✓

              ligand substitution ✓   [2]

                                                        [Total: 15]

5 (a) Reducing agent ✓   [1]

   (b) Dilute sulphuric acid ✓   [1]

   (c) Colourless $\longrightarrow$ pink ✓   [1]

   (d) 0.0362 mol dm^{-3} ✓✓✓✓   [4]

                                                        [Total: 7]

5 Isomerism, aldehydes, ketones and carboxylic acids

1 (a) C_3H_6O ✓✓   [2]

   (b)

      $H_3C-CH_2-C\overset{O}{\underset{H}{\diagdown}}$ ✓ $H_3C-\overset{O}{\overset{\|}{C}}-CH_3$ ✓

                                                         [2]

   (c) 2,4-dinitrophenylhydrazine ✓   [2]

      Observation: yellow/orange precipitate ✓

   (d) Tollens' reagent ✓ reacts with propanal forming a silver mirror. ✓   [2]

                                                       [Total: 8]

2 (a) (i) Solid ✓   [1]

      (ii) Alcohol: $CH_2OHCHOHCH_2OH$; ✓ sodium salt: $CH_3(CH_2)_{16}COO^-Na^+$ ✓;

         use: soap ✓   [3]

   (b) (i)

      ✓ $\underset{COOH}{\overset{COOH}{|}}$ $+ \; 2\,C_2H_5OH \longrightarrow \underset{COOC_2H_5}{\overset{COOC_2H_5}{|}}$ ✓ $+ \; 2\,H_2O$ ✓

                                                        [3]

      (ii) Concentrated H_2SO_4 ✓   [1]

      (iii) $\underset{CH_3COO-CH_2}{\overset{CH_3COO-CH_2}{|}}$ ✓   [1]

                                                      [Total: 9]

3 (a) (i) NaCN/dilute acid ✓ (ii) HCl(aq)/H_2O ✓   [2]

   (b) Chiral or asymmetric carbon atom ✓   [1]

   (c) (i) $CH_3COCOOH$ ✓   [1]

      (ii) $CH_3COCOOH + 6[H] \longrightarrow CH_3CHOHCH_2OH + H_2O$ ✓✓   [2]

                                                        [Total: 6]

6 Aromatics, amines, amino acids and polymers

1 (a) (i) molar ratio C:H = 90.56/12 : 9.44/1 ✓ = 1:1.25 ✓ = 4:5   [2]

     (ii) C_8H_{10} ✓   [1]

     (iii)   [4]

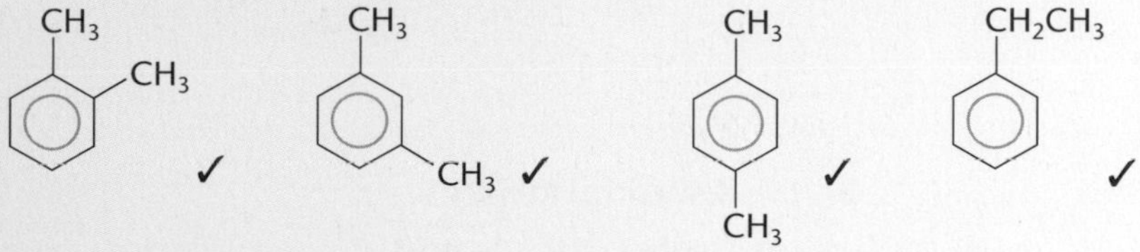

   (b) (i) 1,4-dimethylbenzene ✓   [1]

     (ii)   [1]

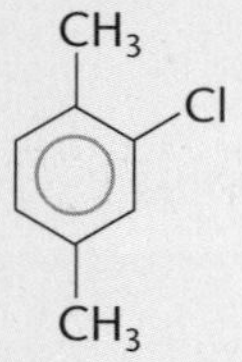

                                      [Total: 9]

2 (a) A lone pair of electrons on the nitrogen atom ✓ accepts a proton. ✓   [2]

   (b) The lone pair has less electron density ✓ due to the negative inductive effect from the aryl ring. ✓   [2]

   (c) (i) Nucleophilic substitution ✓   [1]

     (ii) $(CH_3CH_2)_2NH$ ✓ $(CH_3CH_2)_3N$ ✓ $(CH_3CH_2)_4N^+Br^-$ ✓   [3]

   (d) $CH_3CN + 4[H] \longrightarrow CH_3CH_2NH_2$ ✓

      Only one product is formed ✓   [2]

                                        [Total: 10]

3 (a) (i) A chiral carbon atom has **four** different groups attached to it. ✓   [1]

     (ii) It provides two possible forms that are non-superimposable mirror images of one another. ✓   [1]

     (iii) Bond angle = 109.5° ✓

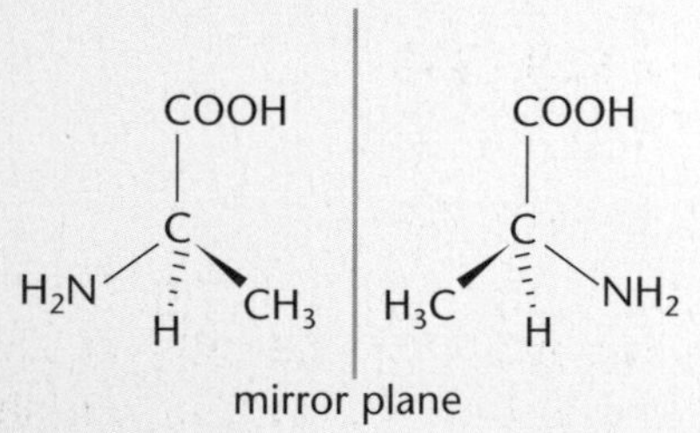

mirror plane   [3]

   (b)

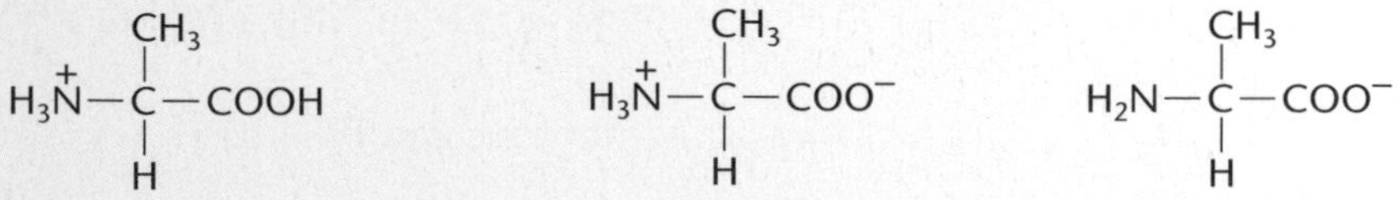

       pH = 2.0 ✓   pH = 6.0 ✓   pH = 10.0 ✓

                                           [3]

(c) (i)

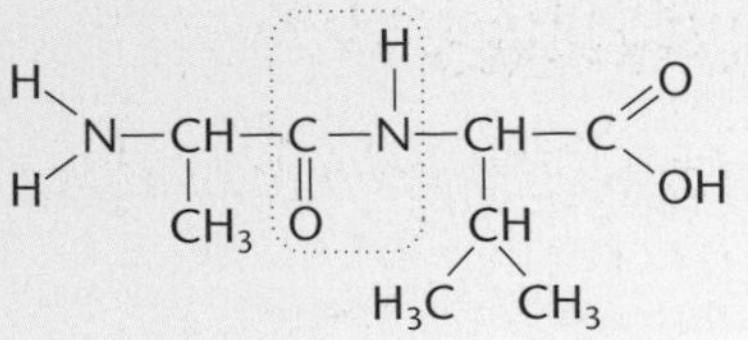

[1]

(ii)

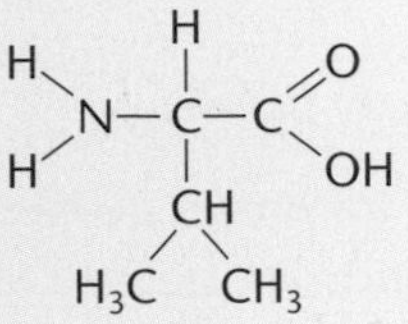

[1]

(iii)

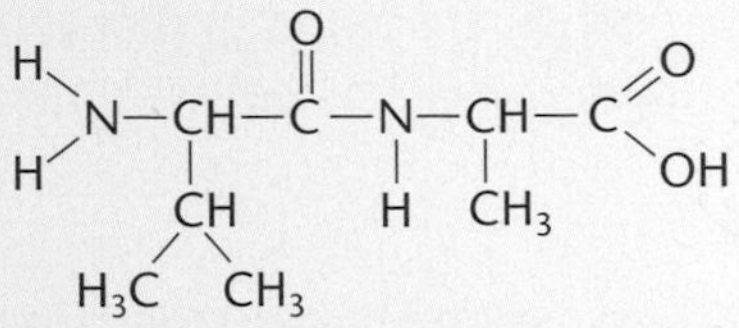

[1]

[Total: 11]

7 Analysis and synthesis

1 (a) **J** 1750 cm^{-1} C=O ✓ ; 1250 cm^{-1} and 1050 cm^{-1} C–O ✓
Evidence suggests an ester ✓
K broad 2500–3300 cm^{-1} O–H (carboxylic acid) ✓
1720 cm^{-1} C=O; 1240 cm^{-1} C–O ✓
Evidence suggests a carboxylic acid ✓ [6]

(b) Molar ratio C:H:O = 54.5/12 : 9.1/1 : 36.4/16 ✓ = 2:4:1 ∴ C_2H_4O ✓ [2]

(c) **J** = $C_4H_8O_2$ ✓ 4 possible structures, e.g. $CH_3CH_2COOCH_3$ ✓ [2]

[Total: 10]

2 (a) Trimethylsilane (TMS) ✓ [1]

(b) 4 ✓ [1]

(c) From left to right, 2:1:6:3 ✓ [1]

(d) C_2H_5 ✓
The CH_2 group at δ 1.53 is split into a quartet ✓ (3 adjacent protons) ✓
The CH_3 group at δ 0.93 is split into a triplet ✓ (2 adjacent protons) ✓ [5]

(e) 6 equivalent protons (2 equivalent CH_3 groups) ✓ [1]

(f) $CH_3CH_2COH(CH_3)_2$ ✓ [1]

[Total: 10]

155

8 Synoptic assessment

1 (a) Co : N : H : Cl = 23.6/59 : 27.9/14 : 6.0/1 : 42.5/35.5 ✓
Ratio Co : N : H : Cl = 0.4 : 1.99 : 6.0 : 1.2 = 1 : 5 : 15 : 3 ✓
Empirical formula is $CoN_5H_{15}Cl_3$ ✓ [3]

(b) (i) Air is an oxidising agent, ✓ (oxidising Co^{2+} to Co^{3+}) [1]

(ii) NH_3 is ligand ✓ [1]
NH_4^+ is buffer ✓ (and provides Cl^-) [1]

(c) Cobalt uses 3d orbitals for dative covalent bond formation; calcium does not. ✓
Cobalt and its compounds are better catalysts than calcium. ✓
Cobalt compounds coloured, calcium are colourless ✓
Cobalt compounds are found in more than one oxidation state; calcium only
has compounds with +2 oxidation state. ✓ [4]

[Total: 10]

2 (a) (i) Long hydrocarbon chain suggests insolubility in water ✓ [1]
OH groups suggest solubility in water ✓ [1]

(ii) 4 geometric isomers ✓
cis:cis cis:trans trans:cis and *trans:trans* ✓ [2]

(iii) with Na:
$H_3C-(CH_2)_2-CH=CH-CH=CH-(CH_2)_8-CH_2O^-Na^+$ ✓
with $H_2SO_4/K_2Cr_2O_7$:
$H_3C-(CH_2)_2-CH=CH-CH=CH-(CH_2)_8-CHO$ ✓
with H_2SO_4/KBr:
$H_3C-(CH_2)_2-CH=CH-CH=CH-(CH_2)_8-CH_2Br$ ✓
with Br_2:

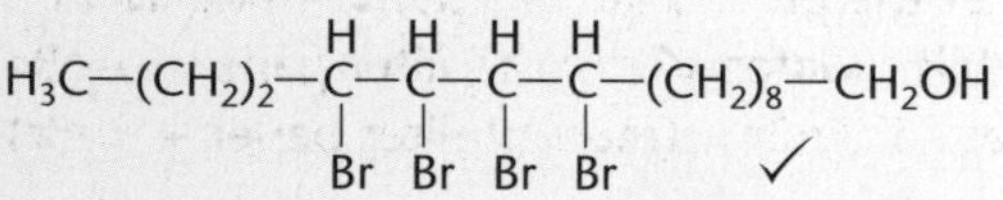

$$H_3C-(CH_2)_2-\overset{H}{\underset{Br}{C}}-\overset{H}{\underset{Br}{C}}-\overset{H}{\underset{Br}{C}}-\overset{H}{\underset{Br}{C}}-(CH_2)_8-CH_2OH \checkmark$$
[4]

(b) $CH_3(CH_2)_2COOH$; $(COOH)_2$ and $HOOC(CH_2)_8COOH$ ✓✓ [2]

[Total: 10]

3 Reduction is gain of electrons ✓ decrease in oxidation number ✓
Oxidation is loss of electrons ✓ increase in oxidation number ✓
Inorganic:
$Na + Cl_2 \longrightarrow 2NaCl$ ✓✓
Na oxidised; Cl reduced ✓
$5Fe^{2+}(aq) + MnO_4^-(aq) + 8H^+(aq) \longrightarrow Mn^{2+}(aq) + 4H_2O(l)$ ✓✓
Fe oxidised; Mn reduced ✓

Organic:
Oxidation of alcohol with $H_2SO_4/K_2Cr_2O_7$ ✓
$CH_3CH_2OH + [O] \longrightarrow CH_3CHO + H_2O$ ✓✓
Reduction of nitrobenzene with Sn/HCl ✓
$C_6H_5NO_2 + 6[H] \longrightarrow C_6H_5NH_2 + 2H_2O$ ✓✓
Clear, well-organised, using specialist terms ✓

[Total: 14 max]

4 NaCl has a giant lattice ✓ with strong forces between ions giving a high boiling point. ✓ In solution, the ions are free to move ✓ and conduct electricity ✓
CH_3COOH, CH_3CH_2OH and $AlCl_3$ have simple molecular structures ✓ with weak forces between molecules giving a low boiling point. ✓

In solution, CH_3CH_2OH has mobile uncharged molecules ✓ which cannot conduct electricity. ✓
CH_3COOH partially dissociates in solution giving a small proportion of mobile ions and slight conductivity. ✓
$CH_3COOH \rightleftharpoons CH_3COO^- + H^+$ ✓
$AlCl_3$ conducts by reacting with water:
$AlCl_3 + 3H_2O \longrightarrow Al(OH)_3 + 3H^+ + 3Cl^-$ ✓
Reaction forms mobile ions which can conduct electricity ✓

Solutions of NaCl and CH_3CH_2OH have pH = $-\log(1 \times 10^{-7})$ = 7 ✓
Solution of $AlCl_3$ has pH = $-\log(3 \times 10^{-1})$ = 0.5 ✓
For solution of CH_3COOH,
$K_a = [CH_3COO^-][H^+]/[CH_3COOH]$ ✓
Solution of CH_3COOH has a pH = 2.89 ✓
Clear, well-organised throughout, using specialist terms ✓

[Total: 14 max]

5 (a) RBr + NaOH $\longrightarrow$ ROH + NaBr ✓
   $Ag^+(aq) + Br^-(aq) \longrightarrow AgBr(s)$ ✓ [2]

   (b) Use a known mass of the haloalkane ✓ with an excess of NaOH(aq) ✓ in a pear-shaped flask ✓ fitted with a condenser ✓ heat ✓ and reflux for 40 minutes. ✓ [5 max]

   (c) Acidify with HNO_3(aq) ✓ and add an excess ✓ of $AgNO_3$(aq). ✓
   Filter the precipitate of AgBr ✓ using weighed filter paper, wash with distilled water, ✓ dry the filter paper + AgBr and weigh. ✓
   Mass of AgBr = (mass of filter paper + AgBr) – mass of filter paper ✓ [6 max]

   (d) Mass of Br in AgBr, b = 80 × mass AgBr/188. ✓
   % by mass of Br in haloalkane = b × 100/mass bromoalkane ✓ [2]

[Total: 15]

Periodic Table

The Periodic Table

Key:

relative atomic mass	1.0	atomic symbol
atomic number	**H**	name
	1	
	Hydrogen	

1	2											3	4	5	6	7	0
																	4.0 **He** 2 Helium
6.9 **Li** 3 Lithium	9.0 **Be** 4 Beryllium											10.8 **B** 5 Boron	12.0 **C** 6 Carbon	14.0 **N** 7 Nitrogen	16.0 **O** 8 Oxygen	19.0 **F** 9 Fluorine	20.2 **Ne** 10 Neon
23.0 **Na** 11 Sodium	24.3 **Mg** 12 Magnesium											27.0 **Al** 13 Aluminium	28.1 **Si** 14 Silicon	31.0 **P** 15 Phosphorus	32.1 **S** 16 Sulphur	35.5 **Cl** 17 Chlorine	39.9 **Ar** 18 Argon
39.1 **K** 19 Potassium	40.1 **Ca** 20 Calcium	45.0 **Sc** 21 Scandium	47.9 **Ti** 22 Titanium	50.9 **V** 23 Vanadium	52.0 **Cr** 24 Chromium	54.9 **Mn** 25 Manganese	55.8 **Fe** 26 Iron	58.9 **Co** 27 Cobalt	58.7 **Ni** 28 Nickel	63.5 **Cu** 29 Copper	65.4 **Zn** 30 Zinc	69.7 **Ga** 31 Gallium	72.6 **Ge** 32 Germanium	74.9 **As** 33 Arsenic	79.0 **Se** 34 Selenium	79.9 **Br** 35 Bromine	83.8 **Kr** 36 Krypton
85.5 **Rb** 37 Rubidium	87.6 **Sr** 38 Strontium	88.9 **Y** 39 Yttrium	91.2 **Zr** 40 Zirconium	92.9 **Nb** 41 Niobium	95.9 **Mo** 42 Molybdenum	– **Tc** 43 Technetium	101 **Ru** 44 Ruthenium	103 **Rh** 45 Rhodium	106 **Pd** 46 Palladium	108 **Ag** 47 Silver	112 **Cd** 48 Cadmium	115 **In** 49 Indium	119 **Sn** 50 Tin	122 **Sb** 51 Antimony	128 **Te** 52 Tellurium	127 **I** 53 Iodine	131 **Xe** 54 Xenon
133 **Cs** 55 Caesium	137 **Ba** 56 Barium	139 **La** 57 Lanthanum	178 **Hf** 72 Hafnium	181 **Ta** 73 Tantalum	184 **W** 74 Tungsten	186 **Re** 75 Rhenium	190 **Os** 76 Osmium	192 **Ir** 77 Iridium	195 **Pt** 78 Platinum	197 **Au** 79 Gold	201 **Hg** 80 Mercury	204 **Tl** 81 Thallium	207 **Pb** 82 Lead	209 **Bi** 83 Bismuth	– **Po** 84 Polonium	– **At** 85 Astatine	– **Rn** 86 Radon
– **Fr** 87 Fracium	– **Ra** 88 Radium	– **Ac** 89 Actinium	– **Rf** 104 Rutherfordium	– **Db** 105 Dubnium	– **Sg** 106 Seaborgium	– **Bh** 107 Bohrium	– **Hs** 108 Hassium	– **Mt** 109 Meitnerium	– **Unn** 110 Ununnilium	– **Uuu** 111 Unununium	– **Uub** 112 Ununbium		– **Uuq** 114 Ununquadium		– **Uuh** 116 Ununhexium		– **Uuo** 118 Ununoctium

lanthanides

140 **Ce** 58 Cerium	141 **Pr** 59 Praseodymium	144 **Nd** 60 Neodymium	– **Pm** 61 Promethium	150 **Sm** 62 Samarium	152 **Eu** 63 Europium	157 **Gd** 64 Gadolinium	159 **Tb** 65 Terbium	163 **Dy** 66 Dysprosium	165 **Ho** 67 Holmium	167 **Er** 68 Erbium	169 **Tm** 69 Thulium	173 **Yb** 70 Ytterbium	175 **Lu** 71 Lutetium

actinides

Th 90 Thorium	**Pa** 91 Protactinium	**U** 92 Uranium	**Np** 93 Neptunium	**Pu** 94 Plutonium	**Am** 95 Americium	**Cm** 96 Curium	**Bk** 97 Berkelium	**Cf** 98 Californium	**Es** 99 Einsteinium	**Fm** 100 Fermium	**Md** 101 Mendelevium	**No** 102 Nobelium	**Lr** 103 Lawrencium

Index

acid anhydrides 106
acid concentration 41, 37, 39
acid dissociation constant 37, 39, 40, 45
acid strength 36–7
acid–base 85, 123, 124
acid–base equilibria 86–7
acids 35–7, 66
activation energy 83
acyl chlorides 104–5
acylation 104–6
addition polymerisation 126, 129
addition-elimination 104–5
adjacent protons 138–9
alcohols 97, 101, 132–3, 140
aldehydes 94–8, 132, 140
aliphatic amines 119–21
alkali 35, 66
alkali concentration 41
aluminium 65, 67–9
amides 105, 140
amines 119–22, 132, 140, 141
amines as bases 120
amino acids 123–5
aqueous hydrogen/hydroxide ions 41
arene acylation 115, 141
arene alkylation 114–15, 141
arene halogenation 114, 141
arene nitration 113, 141
arene reactions 113–16, 122
arene sulphonation 116
arenes 109–12
aromatic amines 119–20
aromatic organic compounds 109–10
aspirin 106
autocatalysis 84
azo coupling 141
azodyes 121
base peak 135
base strength 36–7
bases 35–7
Benedict's solution 97
benzene 109–16, 121, 141
bidentate ligands 74
boiling points 65–6, 68, 71
bond dissociation enthalpy 51
bond fission 134
Born–Haber cycle 50, 52
Brady's reagent 96
Bronsted–Lowry acids/bases 35–6
buffer solutions 44–5
carbon chain extension 141

carbonium ion 114–15
carbonyl compounds 94–8
carboxyl group 99
carboxylates 100
carboxylic acids 99–101, 105, 132–3, 140–1
catalysis 82–4
catalysts 71, 82–4
catalytic converter 83
cell reaction 58
chemical shift 136–9
chiral atom 92
chirality 92
cis-trans isomerism 92
colorimetry 76
coloured complex ions 75–6, 85
complex ions 73–4, 77–8
concentration/time graph 22–3
condensation polymerisation 127–9
condensation reaction 96, 125
conjugate pairs 35, 44–5
Contact process 83
coordinate bond 73
coordination number 73–4, 75, 77
covalent compounds 65, 67, 68–9
d-block elements 70–2
delocalised structure – benzene 110–12
dipole 94–5, 99, 104, 123
dissociation 37
dyes 121
electrochemical cells 55–8
electrochemical series 57–8
electrophiles 113–15
electrophilic addition 112
electrophilic substitution 113, 118, 141
endothermic reactions 30
end-points 42, 43, 79
enthalpy 49–54
enthalpy change of hydration 53–4
enthalpy change of solution 53, 54
equilibria 60
equilibrium constant 33–4
equilibrium law 29
equivalence point 43
ester hydrolysis 102–3
esterification 31, 101
esters 102–3, 105, 118, 128, 132–3, 140
ethyl ethanoate equilibrium 31
exothermic reactions 30

fats 103
Fehling's solution 97
first electron affinity 50, 51
first ionisation energy 50, 51
fragment ions 134–5
Friedel-Crafts 114
functional groups 91–3, 132, 137
giant ionic lattice 49–52, 53
giant lattice 65, 68
glycerol 103
Haber process 83
half-cells 55–7
half-life 22–3
halogen carriers 114–15
halogenoalkane reactions 25–6, 122, 140
Hess's Law 49, 50, 54
heterogeneous catalysis 82–3
high resolution nmr 138–9
homogeneous catalysis 83–4
hydration enthalpy 53–4
hydrogen bonding 100, 119
hydrogen half-cell 56
hydrolysis 102–3
incomplete substitution 78
indicators 42–3
infra-red spectroscopy 132–3
initial rate 20
ionic bonding 49–52, 53
ionic charge 52, 54
ionic compounds 65, 67, 68–9
ionic product of water 40–1
ionic size 51, 54
isoelectric points 123–4
isomerism 91–2
K_a 39–40
Kekulé structure – benzene 110–12
ketones 94–8, 132
lattice enthalpy 49–52, 53
Lewis acid 74
Lewis base 74
ligand substitution 77–8, 85
ligands 73, 75
low resolution nmr 137–9
mass spectrometry 134–5
metal aqua-ions 85–7
metal chlorides 67
metal hydroxides 86–7
metal oxides 65–6
metallic lattice 71
metals 57, 70–2
methyl carbonyl group 98

Index

methyl orange 42–3
molecular ions 134–5
monomers 126, 128
multidentate ligands 74
non-metal chlorides 68, 69
non-metal oxides 65–6
nuclear magnetic resonance 136–9
nucleophilic addition 95–6, 97–8
nucleophilic addition-elimination 104–5
nucleophilic substitution 122
oils 103
optical isomerism 92, 124
organic synthesis 140–1
outer shell electrons 64, 67, 70
oxidation 55–6, 57–8, 59–60, 79–80
oxidation state 64, 67, 71, 72, 75, 80–1
oxonium ion 36
partial pressure 33
peptide link 124–5, 127
Period 3 chlorides 67–9
Period 3 oxides 64–6, 68
Periodic table 158
periodicity 64, 67
pH 38–40, 41, 45
pH changes 42–3
pH meter 43
phenolphthalein 42–3
phenols 117–18

pK_a 40
pK_{ind} 42
polar solvents 53
polarisation 52, 68
polarity 56, 58, 94–5, 99, 102, 119
polyamides 127–8
polyesters 128
polymers 126–9
polypeptides 124–5, 127
precipitation reactions 85–6
protein hydrolysis 125
proteins 124–5, 127
proton acceptor 35, 36
proton donor 35, 36
racemic mixture 92
rate constant 20–1
rate equation 20–1, 25
rate/concentration graph 24
rate-determining step 25–6
reaction mechanisms 25–6
reaction order 19–21, 23
reaction rate 19–20, 23–4
redox reactions 55–7, 59–60, 65, 67, 79–81
redox titrations 79–80
reduction 55–6, 57–8, 59–60, 79–80
relative molecular mass 134
salt bridge 55, 56
silver mirror 97
simple molecules 65–6, 67, 68

spin-spin coupling patterns 138–9
splitting pattern 139
stability constant 78
standard cell potentials 58
standard electrode potential 56–8, 59–60
standard enthalpy change of atomisation 50, 51
standard enthalpy change of formation 50, 51
stereoisomerism 91–2
strong acids 36, 38, 43, 69
strong alkali 41, 43
structural isomerism 91
tangents 23–4
temperature 30
titrations 42–3
Tollens' reagent 97
transition element complexes 73–8
transition elements 70–2, 82–4
transition metal ions 79–81, 83, 84
unidentate ligand 74
van der Waals' forces 126
water 66, 69, 77, 102–3
water ionisation 40–1
weak acids 36, 39–40, 42, 43, 44, 86, 99, 100, 117
weak alkali 43
zwitterion 123